Protocols for High Speed Networks IV

Protocols for High Speed Networks IV

Edited by

Gerald Neufield and Mabo Ito

Department of Computer Science
University of British Columbia
Vancouver
Canada

SPRINGER-SCIENCE+BUSINESS MEDIA, B.V.

First edition 1995

Originally published by Chapman and Hall in 1995
MyCopy version of the original edition 1995

DOI 10.1007/978-0-387-34885-8

A catalogue record for this book is available from the British Library

∞ Printed on permanent acid-free text paper, manufactured in accordance with ANSI/NISO Z39.48-1992 and ANSI/NISO Z39.48-1984 (Permanence of Paper).
www.springer.com/mycopy

CONTENTS

PREFACE

Welcome to the fourth IFIP workshop on protocols for high speed networks in Vancouver. This workshop follows three very successful workshops held in Zürich (1989), Palo Alto (1990) and Stockholm (1993) respectively. We received a large number of papers in response to our call for contributions. This year, forty papers were received of which sixteen were presented as full papers and four were presented as poster papers. Although we received many excellent papers the program committee decided to keep the number of full presentations low in order to accommodate more discussion in keeping with the format of a workshop.

Many people have contributed to the success of this workshop including the members of the program committee who, with the additional reviewers, helped make the selection of the papers. We are thankful to all the authors of the papers that were submitted. We also thank several organizations which have contributed financially to this workshop, specially NSERC, ASI, CICSR, UBC, MPR Teltech and Newbridge Networks.

Mabo Ito
Gerald Neufeld

Departments of Electrical Engineering and Computer Science
University of British Columbia

ACKNOWLEDGEMENTS

This workshop was sponsored by IFIP Working Group 6.1 and Working Group 6.4 in cooperation with the IEEE ComSoc. Support was also given by the New Networks Corporation, MPR Teltech Ltd, Natural Sciences and Engineering Research Council of Canada, B.C. Advanced Systems Institute (ASI), UBC Centre for Integrated Computer Systems Research (CICSR), and UBC Computer Science and Electrical Engineering Departments.

COMMITTEE MEMBERS

WORKSHOP CO-CHAIRS:

Mabo Ito
Gerald Neufeld

PROGRAM COMMITTEE:

Gregor Bochmann (Canada), Steve Deering (USA), David Feldmeir (USA), David Greaves (England), Per Gunningberg (Sweden), Marjory Johnson (USA), Bryan Lyles (USA), Hideo Miyahara (Japan), Bernhard Plattner (Switzerland), Allyn Romanow (USA), Harry Rudin (Switzerland), James Sterbenz (USA), David Tennenhouse (USA), Joe Touch (USA), Carey Williamson (Canada), Martina Zitterbart (Germany)

PART ONE

Keynote Address

1

Protocols for High Speed Networks: Life After ATM?

James P. G. Sterbenz
+1 617 466 2786
jpgs@{acm|ieee}.org jpgs@gte.com
http://info.gte.com/jpgs/jpgs.html

GTE Laboratories Incorporated
40 Sylvan Road, MS-61
Waltham, MA 02254 USA

Abstract

This paper will provide a brief history of networking, noting the transition from disjoint network infrastructure and media to the emerging integrated broadband networks. The current state of affairs and significant challenges in deploying a broadband Global Information Infrastructure based on ATM and B-ISDN will be discussed. Even if we solve all the hard technical problems and practical challenges associated with this, all we have provided is an integrated network infrastructure that is capable of transporting high bandwidth. We still have not solved the latency problem (bandwidth-×-delay product) for WAN applications, or provided for the delivery of this bandwidth to the applications through the host architecture and operating systems. The sorts of emerging applications we must support, and some of the challenges, issues, and areas of research that remain at the end system to support these new applications will be described. Finally, some of the research directions to be pursued will be briefly touched upon.

Keywords: Protocols for High Speed Networks, ATM, B-ISDN, Gigabit Networks

1. A Brief History of Networking

The history of networking and communications can be divided it into three generations, which have distinctly differing characteristics in terms of the scope of users and applications, and the integration of differing applications and media.

1.1 First Generation

The first generation lasted through roughly the 1970's and is characterised by three distinct categories: voice communication, entertainment, and data networking, each of which was carried by a different infrastructure. Voice communication was either analog circuit switched wireline telephony or analog radio transmission. Entertainment was carried by free space broadcast radio and television. Data communications was the latest entrant, and provided only a means to connect terminals to a host, either by serial link local communications (such as RS-232 or BSC) or by modem long haul connections over telephone lines for remote access.

1.2 Second Generation

In roughly the 1980's a dramatic jump in the types and scope of networking occurred, but the three categories of communication (voice, entertainment, and data) remained relatively distinct. This period took us from the experimental ARPANET to the ubiquitous Internet.

While the end user of the voice network generally continued to use analog telephone sets, the internal network switches and trunks became largely digital. Additionally, there was widespread deployment of digital PBXs (private branch exchange telephone switches). Mobile communications emerged in form of cellular telephony.

There were two significant additions to the entertainment category: the wide scale deployment of cable networks for entertainment video, and the emergence of BBS (bulletin board systems) and consumer online services (such as America Online, CompuServe, and Prodigy).

The growth in data networking during this period was significant, incorporating file transfer, remote login, and electronic mail. In the local area, shared media Ethernet and token ring networks allowed clusters of workstations and PCs to network with file and compute servers. In the wide area, store-and-forward packet routers formed the backbone of the Internet, with wide scale research and university access.

At the same time, but relatively distinct from this, connection oriented corporate networks using protocols such as BNA, DECNET, and SNA were deployed, along with the deployment of public X.25 networks (used primarily as corporate virtual private networks). Thus, even within data networking there were multiple incompatible architectures and poorly interconnected networks.

1.3 Third Generation

The first distinguishing characteristic of the third generation, which we are just beginning to enter, is an integration of the voice, entertainment, and data networking categories. Integrated media communication (data, voice, and video) runs over the same fast cell switched networks in LANs, MANs, and WANs, which will further enable applications utilising mixed media.

The second characteristic is in the scope of access: universal consumer access to the GII (global information infrastructure). This is beginning to occur with the emergence of public access internet service providers, as well as the gatewaying of the consumer online services to the Internet. The Internet is currently seeing an explosion in the number of users due to this, and some applications, such as Usenet news are becoming visibly stressed.

2. B-ISDN, ATM and Broadband Networking

The broadband network infrastructure will be constructed using B-ISDN and ATM as the core technology. We will briefly examine what B-ISDN and ATM are, what some of the desirable and undesirable characteristics are, and why they will succeed as the underlying technology for the broadband Global Information Infrastructure.

2.1 B-ISDN and ATM

B-ISDN (broadband integrated services digital network) is relatively easy to define: emerging ITU and ANSI standard compliant networks which support rates in excess of the N-ISDN (narrow band ISDN) primary rate of 64 kbps data channels, and use asynchronous transfer mode cell relay as the underlying network technology. B-ISDN tends to imply emphasis on public networks.

ATM (asynchronous transfer mode) is technically just the statistical multiplexing scheme of packets or fixed size cells, but in common use refers to the emerging switches, networks, protocols, and control mechanisms that use ATM relay of small fixed size cells for the network layer transport of data. In particular, ATM is used in the context of cell relay LANs in addition to wide area public networks.

So in general, B-ISDN and ATM are similar in use, and really refer to broadband networking based on ATM cell relay. Due to the over use and misuse of "ATM", the term "broadband networking" will be generally used after this section.

2.2 Desirable Characteristics of B-ISDN and ATM

There are a number of desirable characteristics of B-ISDN and ATM, based on fast cell switching technology. The network permits an arbitrary mesh topology of point-to-point links between switches. This allows the network to be flexible, scalable, and fault tolerant. The switches can be designed to be highly scalable, and growing the network consists of scaling the switches and corresponding links without the constraints of shared medium networks, such as rings or busses. We can use an arbitrary topology to construct giganode networks using the hierarchical organisations that will be required to reasonably address and manage them.

B-ISDN and ATM networks are connection oriented, allowing QoS (quality of service) guarantees (which is not to say either that this is easy to do in practise, or that QoS guarantees are impossible in connectionless networks). The routing overhead is performed *via* connection setup, allowing the *per* cell routing to be done easily in hardware as a simple table lookup. Finally, datagram service can be provided over a permanent connection set overlay for the purpose.

A significant advantage of B-ISDN and ATM networks is that the standards have been designed to scale (upwards) in data rates. SONET OC-3c (155 Mbps) and OC-12c (622 Mbps) are standard now, and OC-48c (2.4 Gbps) and beyond (OC-192c, OC-768c, *etc.*) will be trivial to standardise. This should be contrasted with the difficulty in arriving at agreement to a single order of magnitude increase in the Ethernet standards to 100 Mbps.

2.3 Undesirable Characteristics of ATM and B-ISDN

There are, however, a number of *undesirable* characteristics of B-ISDN and ATM, primarily due to the current standards. The biggest offender is the small cell size of 48 octets payload. This was a compromise between two small proposed cell sizes of 32 and 64 bytes, chosen by the voice community to avoid the need for echo cancellation. The fact that time scales with data rate seems to have escaped the decision making process, and we have an *almost* unworkably small cell size. In fact we may have been better off in the long run if the 16B proposal would have initially won; it would have been so small to really be unworkable and cell size could have been revisited early on.

The problem usually cited with small cell size is header overhead, which although not optimal is not the main problem. The worst effect of a small cell is the resultant short cell time (681ns at OC-12c), which require fast hardware to make decisions about the cell payload. If the cell time were somewhat longer (128+16B cell resulting in 1.8 μs for OC-12c, for example), more could be done in microcontroller software or cheaper hardware. This has pushed the availability of ATM chips, particularly for the host-network interface segmentation and reassembly, back at least two years for higher data rates. Furthermore, network interface memory design and buffering strategies would be easier to implement if the payload length were a power of two.

Another problem is the decision to require sequenced delivery and omit a sequence number from the header (necessary, given the small cell size; the AAL3/4 sequence number is intended only for error detection but not for resequencing). While it is possible to design switches and routing algorithms to enforce all cells in a stream to follow the same path, this considerably restricts the design space. Furthermore, there are a number of error scenarios which result in effective missequencing, such as lost cells and path reconfiguration. These would be easier to deal with if sequence was not required. Finally, there are a number of cases where higher bandwidth can be obtained by striping of cells across parallel links (such as when multiplexed OC-12 but not OC-12c circuits are available). This requires a mechanism for preserving cell sequence across the parallel lines, but maintaining skew within a cell time is difficult, even per hop, and thus makes this scheme hard to implement in practice.

2.4 Why B-ISDN and ATM Will Succeed

In spite of the undesirable characteristics imposed by ATM and B-ISDN standards, it seems clear that "ATM" (as loosely defined above) and B-ISDN will be the basis of the future broadband infrastructure. In spite of particular technical problems with the B-ISDN and ATM standards, the fact that there *is* a standard that had been agreed on for the physical, link, and network layers is significant. The whole point of networking is interoperability, and this outweighs technical compromises and defects. A wide range of network component manufacturers and service providers are doing, or planning to do, B-ISDN and ATM, in many cases prematurely. This is really a double-edged sword; it is this same premature embracing of "ATM" and perception of agreement that is driving us to agree on the final solution and standards. The Global Information Infrastructure means the end of proprietary network technologies to any large extent.

Having said this, we should note again that there is much to still be defined. The easy problems such as cell format, segmentation and reassembly, and physical layer protocols have been defined; while the difficult issues of multipoint virtual connection routing and traffic management are far from settled. We also shouldn't be surprised to see some revisions in the current standards. The problems with a small cell size can be overcome by a large cell definition, cell size that scales with data rate or cell groups. The connection oriented nature allows this to be a connection parameter. Similarly, a larger cell payload would allow a larger header to be used for a sequence number without further sacrificing header efficiency.

There are also RISC-like arguments that come to bear. Some of the proposed traffic management and connection routing schemes are incredibly complex (and will get worse when multipoint-to-multipoint connections are fully worked out) with the goal of obtaining optimal bandwidth usage. While bandwidth will not be free, it will become sufficiently cheaper that we need to consider what effort is worth taking to conserve it in the broadband environment, rather than basing our efforts on the cost of bandwidth in the current narrowband environment. We may see far more over engineering with simpler control mechanisms than many would now like to admit.

We also should observe that although the connection *vs.* datagram and ATM *vs.* IP arguments have frequently taken the proportions of a "holy war", there will eventually be peace. In a connection oriented "ATM" world, datagrams will have to be supported; this is currently being pursued as ABR (available bit rate) traffic. In the IP world, state will be introduced in the packet routers to allow QoS (*e.g.* [ZhDe93], [ClSh92]). Both camps are trying to solve the same problem, and the solutions will converge. The Global Information Infrastructure will be based on a B-ISDN/ATM substrate with much of the flavor of existing IP networks and applications (which is *not* to say that it will be *current* IP over *current* ATM).

3. Challenges in Deploying the Broadband Infrastructure

Even though we have general agreement on ATM cell relay as the core transport technology for the emerging broadband infrastructure, there are a number of problems that have yet to be solved and worked into the standards. These technical and practical challenges pose non-trivial barriers to the deployment of a broadband Global Information Infrastructure.

3.1 Technical Problems

There are a number of very hard problems to be solved to deploy B-ISDN and ATM networks on a large scale. Virtual connection routing is difficult due to the need to find a path with available resources to meet the users requested quality of service parameters, while optimising network resources and load balancing. The general case for dynamic multipoint-to-multipoint connections is very challenging indeed, and will require routing heuristics. There are also a number of open issues regarding signalling between switches within the network (the NNI: network node interface). Added to this are the problems of mobile addressing and the associated dynamic routing and rerouting, which will be increasingly important.

Whereas using a rate or flow specification to ensure an application's compliance to the traffic contract makes it easier to design network interfaces and to engineer switches and buffer sizes, it places the burden of determining a reasonable set of parameters on the application or user. Although this may be a simple matter for some well behaved or well understood applications such as voice or uncompressed video, it will be exceptionally difficult for other applications, such as for general case multimedia interactive applications or distributed computing.

The ABR (available bit rate) traffic class is targeted toward this, providing an IP-like datagram service over ATM networks. Unfortunately, there are a number of cases where this will be inadequate. As an example, a user requesting a file transfer using FTP may tolerate some variance in the response time it takes to get the file, and thus in the value of the average rate. On the other hand, if the file transfer is due to a click on a WWW (world wide web) browser (*e.g.* mosaic) hyperlink, then we would like to have subsecond response time (actually the first window should be available in $O(100\text{ ms})$), and thus a strict bound on the rate can be computed based on size of the first window. So even in the current IP world, communication that has been best effort will be demanded with sub-second response times by applications such as the WWW.

3.2 Unexpected Behavior

We should note that networks are extremely complex *systems of systems*, and frequently do not exhibit the behavior we would like. A significant example is the fractal or self-similar nature of network traffic that has recently been observed [LeTa94]. The assumption in broadband networking has been that many bursty sources would aggregate to relatively uniform traffic, with the corresponding benefits in reduced network resources. The observation of self-similar traffic that is bursty over several orders of magnitude in time questions that assumption. Another problem can arise in network control algorithms, an example being the systemic self-synchronisation of routing messages in IP networks [FlJa94]. This is a case of control mechanisms being difficult to predict and tune in very large scale networks. Furthermore, we will face problems as we attempt to glue implementations of existing control mechanisms together and their interaction is hard to predict (*e.g.* [MoGu95]).

3.3 Practical Challenges

There are a number of additional challenges of a more practical nature, for which the technical solutions are not easy. Network management has always been a difficult problem, but will become much more difficult in multiservice multirate giganode broadband networks.

Security, privacy, and authentication services will have to be provided. This is a standard concern in networks which trades against the desire to share information. This is not a new problem, but it is of increasing concern as the number of users having access to networks (including crackers) increases, as does the desire to conduct commerce and business across an integrated network infrastructure. Electronic fund transfers will no longer have the (perceived) physical security of private networks controlled by the banks with switches controlled by telephone carriers.

The architecture of the network and switches to allow the provision of content and the deployment of new services and application, some within the network, and some by external service providers will become an increasingly challenging problem. The current efforts in IN (intelligent networks) targeted toward the telephone network will not be sufficient for emerging integrated broadband networks, and many pieces of the current IN architecture will be subsumed by broadband signalling and multipoint connection routing. There will be a greater need for the structural and architectural aspects of IN to determine where in the network and switches function resides, and to allow the rapid deployment of new services and applications without replacing infrastructure.

These practical issues are complicated by the wide range of host and terminal equipment intelligence we should expect for a long time to come. Some of this will be legacy, but a typical home may have an large HDTV set in the living room whose primary interface is not a keyboard, a PC which also can display entertainment video, and a number of relatively dumb telephone sets.

3.4 Compatibility, Standards, and the Seamless Network

The ability to maintain backward compatibility with existing protocols and host software is critical, but can impose technical compromises on new protocols and architectures. We will have to live in a world of multiprotocol and legacy networks for a long time to come. Existing transport protocols, such as TCP and TP4 will be required to operate over new network infrastructure just to support the existing applications which use them. IP has served as a powerful unifying layer for heterogeneous sub-networks, and will also have to be supported for the foreseeable future (and indeed there will be an IP flavor to the ultimate meld of the "ATM" and "IP" worlds). Any solutions will have to balance the ability to enhance the performance and services of these existing protocols, while providing the new performance and functionality desired for emerging applications.

Similarly, standards are particularly necessary in the network world, since they provide the means to allow interoperation, but the premature adoption of poor or overly rigid standards can do more harm than good. It is a very tricky business to go from research and experimentation, to preliminary standards, to solid standards to which products can be designed.

The B-IDSN and ATM standards originally assumed a strict and relatively simple hierarchical network structure, with a public backbone network at the top. The current internetworking reality is far from this, due to the post-divestiture structure of the US telephone network, and the structure of the current Internet. The vision of a seamless LAN-MAN-WAM ATM infrastructure will not be realised for some time to come. Furthermore the need for policy based routing complicate the original assumptions.

3.5 Economic Issues

There are certainly some components of the network infrastructure that are a natural monopoly, *e.g.* we expect controls on who can dig up the street with a backhoe to provide the last mile links. On the other hand, we are moving to a more competitive world in the provision of host and terminal equipment, as well as services and network content. In the middle there is the network infrastructure, and to what degree it should be regulated.

The emerging network infrastructure should make it easier on the user, but it appears that delivery to the home will be available *via* telephone lines, CATV cable, cellular and PCS communications, satellite feeds, and possibly fiber in the power transmission lines. The question arises of how many distinct network outlets should be "in the wall". We will have to decide if the connectivity between the various local access networks is sufficient or if multiple feeds will be required with a local gateway or switch in the home. There are also a number of standard issues here relating to new technology, including providing reasonable incentives to deploy infrastructure, particularly given regulatory constraints.

Billing is considerably more difficult for broadband integrated networks than for the case of constant rate or leased line communication. It must be based on the resources required for the network, which means that in addition to the amount of data transferred and the peak rates required, it will have to be based on traffic characteristics such as burstiness. Furthermore, for broadband networks to be a success, they will have to be affordable, and the pricing will have to be non-linear, but this means that providers of bandwidth will face the reselling of fractional bandwidth. Incremental pricing of excess network capacity (*e.g.* to improve performance by prefetching or provide greater video resolution) may be a useful concept, but difficult to implement.

3.6 ATM Hype

One of the biggest problems with deploying the broadband network infrastructure is really a meta-problem. There has been so much hype with ATM, that everyone seems to want to be involved, whether or not they have anything to contribute. Although the deployment of testbeds should be encouraged, there will be overdeployment of products and networks that are not really ready for production use due to the immaturity of the standards and problems that have yet to be solved. We would have been much better just talking about how to deploy broadband technology, without creating this thing called "ATM" that everyone wants to sell and deploy, frequently prematurely. There are certainly niches where ATM *can* be deployed now, such as in some LAN environments as in backbone networks with permanent connections, but this is different from wide scale deployment in all environments. Furthermore, the users don't really care what the underlying protocols and technologies are, but rather only that the applications work and perform as desired.

4. What We Are Trying to do

Now that we have discussed the history of networking, and the current trends in B-IDSN and ATM development, we should consider what the real goals of high speed networking are:

Deliver high bandwidth at low latency to the applications that require it.

Assuming that we are able to deploy a broadband infrastructure, and have solved all of the technical challenges described, where has that gotten us? All we have accomplished is to be able to deliver high bandwidth at the network layer. Specifically we still have to deal with the speed of light latency (the bandwidth-×-delay product), and we have to deal with the transport and higher layer protocols to deliver the bandwidth to applications. This leaves us plenty to accomplish with high speed protocol research *even* if ATM/B-ISDN is a solved problem.

4.1 Application Requirements

There are two reasons to need high bandwidth: aggregation and *per* application bandwidth.

The current network infrastructure has been estimated by several sources to be O(1 Tbps). As an example, the aggregate bandwidth required during prime time in the US for interactive HDTV video on demand is O(10 Pbps) [NuPa94]. Thus it is clear that we will need a broadband infrastructure for emerging applications in aggregation. This affects the internal network design and network layer protocols more than it does the end nodes or higher layer protocols.

For the higher layers and end systems, we should consider how much bandwidth a single application can use. At the extreme end, a multi-sensory, distributed, projected three-dimensional, dynamic virtual reality session ("holodeck") would require extremely high bandwidth (It would be a useful exercise to compute the bandwidth required for this).

A more practical example is the use of the WWW (world wide web) and associated browsers (such as mosaic), which is likely to be the killer application we've been waiting for [To94]. While the network bandwidth currently supported is rather low, so is the corresponding response to files and images that are fetched by clicking on hyperlinks. It is not hard to imagine scenarios in which bursts of O(100 Mbps) would be desired by a single user, particularly when large blocks of data are fetched to render into high resolution images [Pa94].

We also need to ask what latency is tolerable. The WWW currently behaves as a request/response system, and the standard sub-second response time arguments apply. Thus, from the time a user clicks on a hyperlink to the time that the screen is full of whatever was requested should not exceed O(100 ms). Various estimates on the network topology and switch latency, coupled with the distance at most half way around the world give us a network latency of

O(10–100 ms). Thus we are potentially close to the O(100 ms) limit, without the transmission time of the data and the processing time on the local and server hosts.

So, we must at *least* deliver the right piece of information to the user with low latency at the end systems. At best, we should predict and prefetch data that will likely be requested next (*e.g.* breadth first using all the hyperlinks in the current page [To94]). We need to recognise that like memory, bandwidth is a resource, and idle bandwidth can be used to our advantage to effectively beat the speed of light.

The WWW also gives us a preview of some of the other higher layer issues that will need to be addressed. For example, as useful as URLs (uniform resource locators) are, we are already seeing dangling pointers when the resource they locate moves or goes away. Furthermore, the sheer volume of information available (relatively) instantaneously has created the need for a number of new navigation and search tools, such as web robots and spiders, which have protocol implications on their own.

5. End-to-End Issues and Principles

Now let us examine some of the issues important in the delivery of high bandwidth at low latency to applications. They assume that the broadband network infrastructure provides a high bandwidth, low latency pipeline between end hosts.

5.1 Host and Network Interface Architecture

Even though things are improving, there are still significant bottlenecks in current host systems and their network interface. There are several reasons for this, but one of the main ones is insufficient bandwidth between the network link and memory. Although researchers are beginning to realise this, it has not yet translated into many workstation architectures. Most current high performance network interfaces attach to the system bus and use DMA to transfer between the network interface and memory. In the ideal case, the network should look much more like memory than I/O [DeFa91], and data should be pipelined between memory and the network with zero store-and-forward hops [StPa93].

For integrated media, we need to consider where the streams merge and split. If data, voice, and video split at the host–network interface, transmitting them as a single stream through the network has not gained us much over using three individual streams. If the streams are to remain integrated through the host to the human interface, we need to understand how voice and video fit into the host architecture and memory hierarchy.

We should consider the distinction be between the processor–memory interconnect, peripheral attach, and the network. There certainly are similarities, and the DAN (desk area network) efforts [Fi91], [HaMcC91] attempt to exploit this. But if the internal host interconnect becomes part of the seamless ATM fabric, then every processor, memory module, and peripheral must have the intelligence of an ATM UNI (user network interface) or NNI (network node interface), or there must be a separate signalling entity. This raises issues of where and when is segmentation and reassembly done, and where raw cell (AAL0) interfaces reside.

5.2 Operating Systems

A number of the bottlenecks in high speed communications are operating system related, and they are being well documented (*e.g.* [NOSSDAV '95]). These primarily relate to avoiding unnecessary context switches, and avoiding extra data copying. In the ideal case, it is possible to have only a single context switch pair: one when the application requests the communication operation, and one when the operation is successful and the application can resume (in the case of a request for information this is when the first complete piece of information the application can use is in place on the host and available in the application user space) [StPa93].

In considering integrated communications, we note that we know how to engineer operating systems for data reasonably well. So although we know how to deal with data in the host memory hierarchy (cache, main store, backing store), what to do with audio and video objects is very much an open question. We need to understand how they fit into the memory hierarchy, and what *is* the hierarchy when we consider frame buffers and the like. Furthermore, we know how to schedule conventional processes reasonably well, but work on how to deal with the real time constraints of mixed media and stream synchronisation is just beginning.

5.3 Protocols

First, we should consider what we *don't* need in protocols for high speed networks: new protocols just for the sake of different protocols. There was a flurry of activity in new transport protocol design a few years ago, but the protocols are only one piece of the end system bottleneck. This does not mean that existing high layer protocols in their current form are adequate for high speed networks and the corresponding host and network interface architecture, but rather that when we create or modify protocols it should be for the right reasons. We should also note that the difference between a "new" protocol (e.g. TCP-like replacement) and a highly modified "old" protocol (e.g. TCP++) may be in name only. Furthermore, hacking a new protocol from an old one by simply omitting what can't be done fast doesn't get us a useful protocol.

Second, we should point out the distinction between what has to be done in support of higher data rates, and what has to be done in support of applications that tend to be enabled by these higher data rates [Pa94]. There is nothing inherently high speed about multimedia applications, particularly since we have existence proofs of such applications over current $O(10\text{Mbps})$ LANs. So a purist view of high speed protocols would be to consider what to do to an existing protocol to scale to higher data rates; if everything scales, nothing needs to be done [Pa90].

On the other hand, the advent of the broadband network infrastructure will enable the wide scale deployment of a number of new applications (including multimedia) to a degree not possible before. Thus it is reasonable to consider what must be done by the protocols to support these emerging applications on the new broadband infrastructure, so long as we understand that we are considering a number of things that are not purely "high speed" issues. Thus we wish to provide support for new integrated media (data, audio, video, and other sensory) applications over the broadband network infrastructure and the delivery of the high bandwidth directly to the application.

So what do we need in protocols for high speed networks? Generic research on high speed protocol architecture is of course important (*e.g.* [ClTe90], [Fe93], [Zi91]) as is the optimisation of protocol control mechanisms (*e.g.* [Ja88]). Furthermore, in the spirit of not inventing new protocols and interfaces just for the sake of it, the extension of existing protocols (or protocol implementations) such as TCP for high speed should be done when possible (e.g. [JaBr92], [Br92]).

Additionally, protocols which provide support for multicast and multipoint applications will be particularly important in the emerging broadband environment. B-ISDN and ATM networks provide this support natively at the network layer with multipoint connections. IP networks are beginning to provide this with the addition of multicast support [De89] and associated higher layer protocols and tools (*e.g.* [Ja], [Fr94]).

There are two performance aspects to consider here: bandwidth and latency. Since the broadband network infrastructure will support high bandwidths, the protocols must extend this end-to-end to the applications. This means:

- support for high bandwidth host and network interface architecture and operating system constructs
- sufficiently low protocol processing overhead to sustain the bandwidth and not steal excessive application CPU cycles or memory
- provision of required functionality to the application

So the protocol is really not an end in itself, but rather a "glue" that provides the functionality necessary to the application, while sitting on the host and network interface architecture (in concert with the operating system) to provide the high bandwidth pipe.

The second performance aspect is latency (really the bandwidth–×–delay product). Since we are already on the margin of sub-second latency for long-haul networks due to the speed of light (and to some degree the queueing delay in the switches), the only other thing we can do is to effectively beat the speed of light, and the only way to do this is to predict what piece of information the application or user will need so that when it is referenced it is already cached locally. By treating bandwidth as a resource, information can be prefetched, either using idle network bandwidth as available, or by using additional bandwidth that the user is willing to pay for to guarantee the enhanced response time. There are various strategies for doing this, e.g. by prefetching related memory objects [StPa90] or building a tree of possible future requests and preloading them in parallel [To89]. It is the bandwidth–×–delay product that will cause us the *most* problems in the long run, because latency it is the one thing that *cannot* scale with data rate.

Finally, given the infancy of B-ISDN and ATM standards and the open issues previously discussed, as we move to giganode gigabit networks we will continue to need research at the network layer and below on signalling, routing, network engineering, and network management and administration.

6. How to Get There

Broadly, then, we need to pursue several areas: In support of the sorts of applications enabled by high speed networks and utilise end-to-end integrated communications we need to understand multimedia operating systems and computer architectures. For WANs we need to effectively violate the speed of light, by exploring prefetching mechanisms and ways of exploiting bandwidth as a resource.

Vendors of workstations need to recognise that it is important to design high performance communication ability into the system, in particular that a high bandwidth path to memory is essential. The traditional limit on the cost of a network interface is about ten percent of the cost of the entire workstation. In an environment where networking is one of the most important thing, this fraction may well be is too small.

The network is a *system of systems*, far more complex than the computer systems we have seen in the past. Interdisciplinary approaches will be required using mathematics, traditional networking, circuits and systems engineering, computer architecture, operating systems, and applications. Furthermore the entire spectrum from theory and analysis, modelling and simulation, development and design engineering, to implementation testing and measurement will be required.

While traditional performance analysis remains important, we will have to place more emphasis on new theory and models, particularly chaos and fractals, general system theory, and game theory (*e.g.* [Sh94]).

We also will need the infrastructure to allow us to do this research. The seeding of testbeds (*e.g.* the Gigabit Testbeds in the US [CNRI94]) aim to help solve the classic "chicken and egg" problem; the chicken being the network infrastructure and the egg being the applications that run on them and enable them (or the other way around). In the future we will need to consider even more aggressive programs to encourage a sense of community among the varied disciplines necessary to develop high speed networks, and to allow reuse and interoperation of the various data, tools, and models [KlSt93]. The WWW is certainly a step in this direction, particularly in enabling the sharing of information, but doesn't provide us the mechanisms to do things like run distributed simulations and emulations, and glue together diverse tools and models.

7. Summary

We can expect the emerging broadband network infrastructure to be based on ATM cell relay technology, but there are significant technical and practical challenges to be solved in deploying a broadband Global Information Infrastructure. The easy problems have been standardised; the difficult ones, particularly multipoint signalling, connection routing, and traffic management are still open.

The ATM and IP communities are trying to solve the same problems, and the two will merge into a single community. The resultant network will have the flavor of both, with more over engineering and less predictability than currently expected in the ATM community, and more state and user QoS demands than currently expected in the IP community. Even when this is accomplished, all we have provided is an integrated network infrastructure that is capable of transporting high bandwidth. We still have not solved the latency problem (bandwidth-×-delay product) for WAN applications, or provided for the delivery of this bandwidth to the applications through the host architecture and operating systems.

So, the answer to whether high speed protocols have a life after ATM is a resounding "yes". More work needs to be done in implementing host architectures, operating systems, and network interfaces that deliver the high network bandwidth to applications that need it. The only solution to the latency problem is to beat the speed of light by predictive prefetching or preloading of information. If we are to gain much by an integrated media network, we need to extend this to multimedia host architectures and operating systems. The higher layer protocols are the "glue" that provide the necessary functionality to the application, while running on the broadband infrastructure.

Acknowledgments

Joe Touch (ISI), Greg Lauer (GTE Laboratories), Peter O'Reilly (GTE Laboratories), and Gurudatta Parulkar (Washington University in St. Louis), provided extensive helpful comments to the original presentation foils and to the text of this paper. The stimulating discussion at PfHSN'94 further contributed to the content of this paper, and I wish to thank all the participants. Discussions with Per Gunningberg (SICS) and Joe Touch (ISI) had particular influence. Finally, Gerald Neufeld and Mabo Ito provided a particularly fine venue for PfHSN'94 at the University of British Columbia, and their encouragement and understanding in the face of the turmoil associated with a relocation during this work are most appreciated.

References

Braden, R., *Extending TCP for Transactions – Concepts*, Internet RFC 1379, Nov. 1992.

Clark, David D., Scott Shenker, and Lixia Zhang, "Supporting Real-Time Applications in an Integrated Services Packet Network: Architecture and Mechanisms", *SIGCOMM'92 Conference Proceedings*, ACM, New York.

Clark, David D. and David Tennenhouse, "Architectural Considerations for a New Generation of Protocols", *SIGCOMM'90 Conference Proceedings*, 1990, ACM, New York.

Corporation for National Research Initiatives, A Brief Description of the Gigabit Testbed Initiatives, `http://www.cnri.reston.va.us:4000/public/overview.html`

Deering, S, *Host Extensions for IP Multicasting*, Internet RFC 1112, 1989.

Delp, Gary S., David J. Farber, Ron G. Minnich, Jonathan M. Smith, and Ivan Tam, "Memory as a Network Abstraction", *IEEE Network*, Vol.5 #4, July 1991, pp. 34–41.

Feldmeier, D.C., "A Framework of Architectural Concepts for High Speed Communications Systems", *IEEE Journal of Selected Areas in Communications*, Vol.11 #4, May 1993.

Frederick, Ron, *Experiences with Real-Time Software Video Compression*, available by anonymous FTP from `ftp.parcftp.xerox.com:net-research/nv-paper.ps`, 1994.

Finn, G., "An Integration of Network Communication with Workstation Architecture", *SIGCOMM'91 Conference Proceedings*, Oct. 1991, ACM, New York, pp. 18–29.

Floyd, Sally and Van Jacobson, "The Synchronization of Periodic Routing Messages", *IEEE/ACM Transactions on Networking*, Vol.2 #2, April 1994, pp. 122–136.

Hayter, M. and D. McCauley, "The Desk Area Network", *SIGOPS Operating Systems Review*, Oct. 1991, ACM, New York, pp. 14–21.

Jacobsen, Van, *et al*, `sd` (session directory), `vat` (visual audio tool), and `wb` (distributed whiteboard), available by anonymous FTP from `ftp.ee.lbl.gov:conferencing/*/`

Jacobsen, Van, "Congestion Avoidance and Control", *SIGCOMM'88 Conference Proceedings*, 1988, ACM New York, pp. 314–329.

Jacobson, V, R. Braden, and D. Borman, *TCP Extensions for High Performance*, Internet RFC 1323, 1992.

Kleinrock, Leonard, James P.G. Sterbenz, Nick Maxemchuck, Simon S. Lam, Henning Schulzrinne, Peter Steenkiste, and the HPNE group, *The National Exchange for Networked Information Systems: A White Paper*, UCLA Computer Science Technical Report CSD-930039, November 1993.

Leland, Will E., Murad S. Taqqu, Walter Willinger, and Daniel V. Wilson, "On the Self-Similar Nature of Ethernet Traffic (Extended Version)", *IEEE/ACM Transactions on Networking*, Vol.2 #1, Feb. 1994, pp. 1–15.

Moldeklev, Kjersti and Per Gunningberg, "Deadlock Situations in TCP over ATM", *Protocols for High Speed Networks IV*, Gerald Neufeld and Mabo Ito, ed., 1995, Chapman and Hall, London.

NOSSDAV '95: *5th International Workshop on Network and Operating System Support for Digital Audio and Video*, Durham, New Hampshire, April 1995.

Nussbaumer, Jean-Paul, Baiju V. Patel, Frank Schaffa, and James P.G. Sterbenz, "Networking Requirements for Interactive Video on Demand", *Gigabit Networking Workshop '94*, INFOCOM'94, Toronto, June 1994; also available as `http://info.gte.com/jpgs/paper/gbn94.ps`

Partridge, Craig, "How Slow is One Gigabit per Second?" *ACM SIGCOMM Computer Communications Review*, Vol.20, #1, Jan 1990, ACM, New York, pp. 44–53.

Partridge, Craig, ed., *Report of the ARPA/NSF Workshop on Research in Gigabit Networking*, Washington, D.C., July 1994, available by anonymous FTP from `ftp.std.com:pub/craigp/report.ps`; also available as `http://info.gte.com/jpgs/doc/arpansf94.ps`

Shenker, Scott, "Making Greed Work in Networks: A Game-Theoretic Analysis of Switch Service Disciplines", *SIGCOMM'94 Conference Proceedings*, Oct. 1994, ACM, New York, pp. 47–57.

Sterbenz, James P.G. and Gurudatta M. Parulkar, "Axon: Network Virtual Storage for High Performance Distributed Applications", *Proceedings of the Tenth IEEE International Conference on Distributed Computing Systems (ICDCS 10)*, May 1990, pp. 484–492; also available as `http://info.gte.com/jpgs/paper/icdcs90.ps`

Sterbenz, James P.G. and Gurudatta M. Parulkar, "Design of a Gigabit Host Network Interface", *Journal of High Speed Networks*, Vol.2 #1, 1993, IOS Press, Amsterdam, pp. 27–62; also available as `http://info.gte.com/jpgs/paper/jhsn93.ps`

Touch, Joseph D and David J. Farber, "MIRAGE: A Model for Ultra-high-speed Protocol Analysis and Design", *Protocols for High Speed Networks*, North-Holland, 1989, pp. 115–134.

Touch, Joseph D., "Defining High Speed Protocols: Five Challenges and an Example that Survives the Challenges", *Gigabit Networking Workshop '94*, INFOCOM'94, Toronto, June 1994.

Zhang, Lixia, Stephen Deering, Deborah Estrin, Scott Shenker, and Daniel Zappala, "RSVP: A New Resource ReSerVation Protocol", *IEEE Network*, September 1993, IEEE New York, pp. 8–18.

Zitterbart, Martina, "High Speed Transport Components", *IEEE Network*, Vol.5 #1, January 1991, IEEE, New York, pp. 54–63.

PART TWO

Quality of Service

2

On Distributed Multimedia Presentational Applications: Functional and Computational Architecture and QoS Negotiation*

B. Kerherve[b], A. Vogel[a],
G. v. Bochmann[a], R. Dssouli[a], J. Gecsei[a] and A. Hafid[a]

[a]Université de Montréal, Dpt. d'IRO,
CP 6128, Succursale Centre Ville, Montréal, H3C 3J7, Canada

[b]Université du Quebec a Montréal, Dpt. de Mathematiques et d'Informatique,
CP 8888, Succursale Centre Ville, Montreal, H3C 3P8, Canada

Abstract

Approaches to a functional and a computational architecture for distributed multimedia presentational applications are developed. These approaches are illustrated by a case study, a multimedia news-on-demand service. The concept of Quality of Service (QoS) parameters is seen to determine the new characteristics of distributed multimedia applications. Accordingly the computational architecture for distributed multimedia presentational applications is developed in a QoS driven way within the the framework of the Reference Model of Open Distributed Processing. The concept of QoS interfaces is introduced in order to handle the QoS negotiation in a general and generic way. Objects in a distributed multimedia application can negotiate their QoS parameters through these QoS interfaces. Using this approach, variants of QoS negotiation protocols are investigated.

Keyword Codes: C.2.2; H.5.1
Keywords: Network Protocols, Multimedia Information Systems

1. Introduction

The new quality of distributed multimedia applications is characterized by handling continuous media, e.g. audio and video, and by managing various medias at the same time. Distributed multimedia applications can be classified by presentational, conversational, or having both aspects. Presentational applications provide remote access to multimedia documents. Examples are news-on-demand [18] or video-on-demand [28] services. Conversational applications involve multi-directional, real-time, multimedia communications, e.g. video conference [26] or collaboration services [1]. Systems for distance education have both aspects.

In the framework of our project [34], we focus on distributed multimedia presentational applications, a multimedia news-on-demand service is selected as a case study. The

* This work was supported by a grant from the Canadian Institute for Telecommunication Research (CITR), under the Networks of Centers of Excellence Program of the Canadian Government.

purpose of the news-on-demand system is to offer an integrated, computerized multimedia news service to various customers. The contents is extracted from existing news sources such as radio, TV, wire services, print, and re-composed as (possibly personalized) multimedia objects which the clients will access. The documents contain images, text, audio and video. Potential users are government institutions, decision makers in companies, journalists, business services, etc. The system runs on a fully distributed architecture where multimedia data are stored on different sites, and users can access them from different places through the network. Functional architecture for a news-on-demand system is investigated in Section 2.

In general, multimedia applications within an open environment, i.e. heterogeneous end-systems and various types of networks, require heavily on considerations of quality of service (QoS) parameters. The connection of two end-systems by a network in the sense of the 'traditional' OSI model does not guarantee the inter-operability of the end-systems. Additionally, matching interfaces are required and within these, QoS parameters are an important factor, e.g in the case of a video-on-demand service, all the three partners, producer, network, and consumer, have to be able to process e.g. the same data rate; if the producer and the network are too fast for the consumer, the service becomes senseless. We will address this question within the Reference Model on Open Distributed Processing (RM ODP) [27]. Using the ODP concepts, a computational architecture for distributed multimedia presentational applications is developed and applied to the news-on-demand service in Section 3.

In order to clarify the responsibilities and interrelationships of the various objects and their QoS parameters, we propose in Section 4 a QoS architecture by introducing the concept of QoS interfaces. Within this architecture, mechanisms for the QoS management have been developed for our focused case study which are introduced in Section 5. Generalizations of our approach are discussed in Section 6 which also concludes the article.

2. Functional architecture

The news-on-demand system offers different functionalities to the users. In this section we first identify the different classes of users, we then present the functional architecture of the news-on-demand system and lastly we describe the different user interfaces provided by the system.

The users involved in the process of production and consumption of multimedia news can be classified into four categories: news *publishers*, news *analysts*, news *producers* and news *consumers*.

News publishers are responsible for delivering the news materials over different media and for associating descriptive and indexing information to these objects. The role of news analysts is to associate semantic information to the different monomedia objects such as key words or specific interpretation of the object. News producers build multimedia objects by linking monomedia elements and by producing control information such as presentation formats (templates) and synchronization scenarios to define the temporal order of display. The multimedia news consumer inspects the database in order to find multimedia news of interest to him. He can consult the multimedia news database in different ways: browsing, specifying an identifier of a news object or by conditional queries. These consultation modes offer a wide range of possibilities to explore the database.

The Figure 1 proposes a general functional architecture for multimedia presentational applications. This architecture is composed of three autonomous levels: the database level, which is the system's nucleus, the function level and the user interface level which can be considered as shells defined on top of the system's nucleus. The user interface shell is in charge of managing the different user interfaces which answer the specific user's needs. This level is also concerned with the user's environment, it offers services to set, modify or delete a specific user profile and takes into account the parameters of the quality of service. The function shell is responsible for the transformation of a user's specific demand to the target operations required on the multimedia database. The database level is responsible for the storage and the access to the multimedia database.

In the news-on-demand system, different user interfaces can be identified as shown within the Medialog project [6]. We define the following three different user interfaces: *monomedia news storage and analysis*, *multimedia news consultation* and *multimedia news production*. They answer specific needs of various user groups and consider the environment associated to these groups. The news-on-demand service users define their favorite environment such as the word processor, the communication environment, the parameters of quality of service as well as the tools for document annotations. This environment can be dynamically modified.

Figure 1 also shows the functional architecture for the news-on-demand system and distinguishes the different user interfaces and functions that are presented in the next three sections.

2.1. Monomedia News Storage and Analysis

The monomedia storage and analysis user interface is dedicated to the news publishers who load the database with basic monomedia objects coming from a single medium, e.g. radio, television, newspaper. This user interface offers connections to specific equipment (video camera, sound recorder, digitizers...) to capture these objects. After being captured and digitized, the objects are described and analyzed by monomedia analysts. The analysts associate registration information to the objects as well as description information describing their contents. Each medium has a predefined schema for the description of registration and description information. Generally these schemata contain keywords and a brief description.

The monomedia news storage and analysis user interface interacts with storage and analysis function, which performs three different tasks: monomedia objects *storage*, monomedia objects *description*, and monomedia objects *indexing*.

The storage and analysis function identifies the internal structures that are required for the database storage of the monomedia objects and executes the transformation of the external structure into these internal structures.

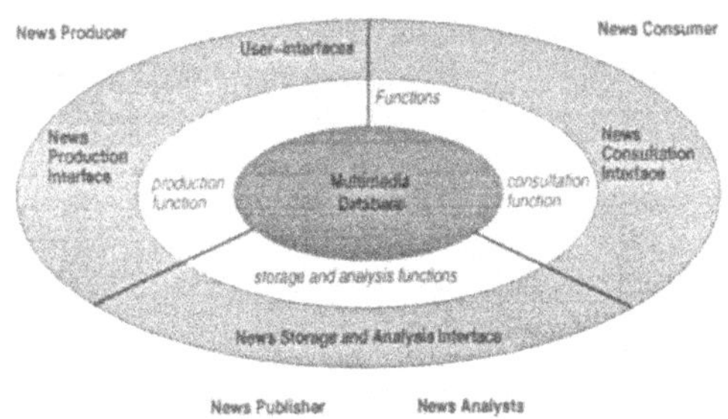

Figure 1. Functional architecture for the news-on-demand system.

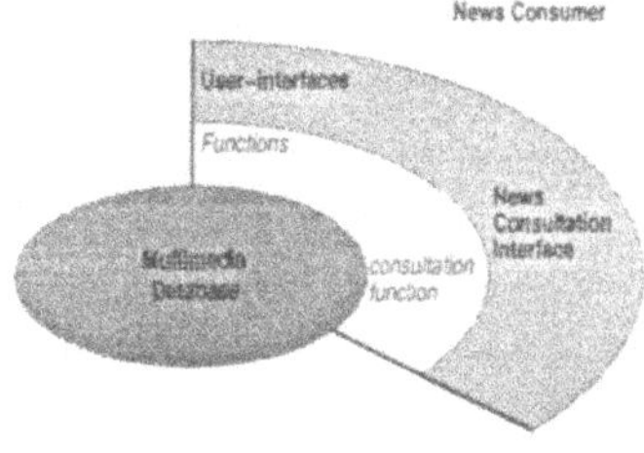

Figure 2. Considered functions of the nev on-demand system.

For example, if the HyTime standard [22] is the external structure and if an object oriented database is used, the storage and analysis function will identify the different classes which are concerned with this document and will generate the corresponding requests to be submitted to the database system. This transformation step will be different if the database system is an extended relational database system.

The storage and analysis function therefore generates the database manipulation language requests for the insertion of monomedia objects and is also responsible for the efficient storage of the monomedia objects, that means that this component of the system decides whether specific compression or indexing techniques can be used.

In order to be efficiently retrieved, the monomedia objects have to be described and classified. This description is done according to specific description schemata that are stored in the database. The storage and analysis function uses these schemata to provide the user interface level with the description information that can be used in the description and classification process.

2.2. Multimedia News Consultation

The final users of the news-on-demand service search multimedia documents in the system and can reproduce these documents. The system provides reproduction facilities and manages the copyright fees. The consultation of multimedia news documents can be separated into two different and complementary steps: the *search process* and the *display and manipulation process*.

The display and manipulation process follows the search process and offers the user the possibility to display the document and to process specific actions on the retrieved object.

In order to search multimedia documents, the users can use several retrieval techniques [2]: retrieval by keywords, browsing, guided tour, full text retrieval, similarity retrieval. Several techniques such as retrieval by keywords, browsing or guided tours are general for all the media while others, such as similarity retrieval or full text retrieval, are specific for still images or texts. In [15] we present in detail the consultation user interface which offers these different possibilities and their combination to the user. At the end of the searching process, the user can execute some operations on the document. These operations are independent and not linked to the search process. The consultation user interface must offer tools to facilitate this kind of operations. Four different tasks can be identified for the consultation function: access to the database system for retrieval functions, retrieval workspace management, internal to external model transformation, and temporal and spatial relationships enforcement.

All the information concerning the database access is transmitted to the consultation function, which generates the corresponding request formulated in the database manipulation language. At this step, the consultation function may enrich the user's requests based on his profile. The requests are then submitted to the database system and the consultation function waits for the results to send them to the consultation user interface. At this step the consultation function may keep some parts of the result in the workspace for further use. The efficient management of this workspace is an important task to be performed by the retrieval function.

While transmitting the results to the user interface level, some transformation may be required to change from the internal model supported by the database system to the external model required by the user interface level. This task is symmetrical about the one performed by the storage and analysis function. The last task performed by the retrieval function is the enforcement of spatial integration and temporal synchronization of the different components of the multimedia news document. Actually, spatial and temporal relationships between the components are an important characteristic of the multimedia news document defined through the production user interface. These relationships are stored in the database as part of the presentation component of the multimedia news document. The retrieval function is in charge of extracting these relationships from the document and to enforce them while interacting with the user interface level.

2.3. Multimedia News Production

The multimedia news production user interface is dedicated to the generation of multimedia news documents. This user interface is concerned with the creation of documents as well as with their modification. The creation of a multimedia news document needs the specification of the following three components [7,17] of the document: the *structure*, the *content* and the *presentation*.

The structure defines the objects that are part of the document and their organization, the content contains these objects (a set of monomedia objects of different types) and the presentation gives information about the layout and the synchronization of the document. The presentation part of the document specifies how the components of the multimedia news document will be presented to the final user. This specification is made through temporal and spatial relationships between objects. The temporal relationships describe the synchronization points while spatial relationships describe the organization of the components on the screen. These relationships define constraints that will be enforced by the consultation function driving the display of the document.

While defining the content of the multimedia document, the user might choose to reuse existing documents or to capture new objects. The reuse of existing documents needs a preliminary step to retrieve them, while capturing new objects such as audio or video needs the connection to specific equipment. The production user interface will therefore use the same services as the consultation user interface and the monomedia storage and analysis user interface.

Since the construction of multimedia news documents is a difficult task which can be progressively refined, the system must offer facilities permitting to modify a document. The modifications can be made on the structure, the content or the presentation part of the document.

When the user defines the components, the multimedia news production user interface captures all the information, whether with specific equipment for images, sound or video, or with specific tools such as word processors. All the elements are defined corresponding to an external multimedia document model, captured, and then transmitted to the production function for database storage. Like the storage and analysis function, the production function is also responsible for transforming the external structure to the internal structure supported by the database system and thus will share services with the former function.

In the second case, when a user builds a multimedia news document from an existing one, the production user interface connects him to a subset of the consultation user interface in order to select the pertinent documents. The production function must keep the references to those documents in a workspace in order to use them in the production process. Since the system must allow object sharing, those references will be incorporated in the new document using a composition mechanism. The production function is in charge of this document composition mechanism. It is also responsible for transforming the structure of already existing documents into other structures if they are offered at the external level.

3. Computational architecture

In this section a computational architecture for distributed multimedia presentational applications is developed and applied to the news-on-demand service. We will present an architecture within the framework of the Reference Model of Open Distributed Processing (ODP). ODP and the corresponding Reference Model was characterized by Kerry Raymond as follows [27]: "Advances in computer networking have allowed computer systems across the world to be interconnected. Despite this, heterogeneity in interaction models prevents interworking between systems. Open Distributed Processing describes systems that support heterogeneous distributed processing both within and between organizations through the use of a common interaction model. ISO and ITU-T (formerly CCITT) are developing a Basic Reference Model of Open Distributed Processing (RM ODP) to provide a coordinating framework for the standardization of ODP by creating an architecture which supports distribution, internetworking, interoperability and portability."

The main concept of the RM ODP is the viewpoint. A viewpoint is an abstraction, focused on parts of an ODP system determined by a particular interest. There are five viewpoints: *enterprise, information, computational, engineering*, and *technology*. They are defined by corresponding description languages. An introduction using a multimedia example is given in [33]. We consider here the computational viewpoint which is a functional decomposition of the system into objects that are candidates for distribution. Hence, the corresponding computational language specifies the system in terms of communicating objects. Computational objects are providing operations through computational interfaces. Interaction between computational objects is described in terms of interface binding.

For distributed multimedia presentational applications we identified two main objects: a *server* containing multimedia documents, and a *client* providing access to the server's documents. The server object is composed of several sub-objects, e.g. for different types of storage objects. When accessing a document, the client and the server will perform the roles of a producer and consumer, respectively. Both objects have several interfaces: *operational interfaces* providing operations for retrieval, access and control and *stream interfaces* supporting the continouos data transfer. From the computational viewpoint, communication aspects are only visible in terms of interface binding and quality of service parameters.

In the following we will develop a computational specification of a news-on-demand service. However, only the consultation part of the complete service, as shown in Figure 2, is specified. Figure 3 illustrates the computational viewpoint of the remaining news-on-demand service. It contains two main computational objects, a multimedia server (MM-Server) and a client (MM-Client). The server is thought to manage the different types of multimedia storages. The multimedia server is composed of sub-objects. We identified the following ones: the *database (DB) server, continuous media (CM) file servers, noncontinuous media (NCM) file servers*, and *archival storages*.

A client has initially three operational interfaces, called the search, the access and the QoS negotiation interface. The search interface provides operations (according to the search process described in Section 2.2) to retrieve information from the database, e.g. retrieve by keywords, browsing, guided tour, full text retrieval, and similarity retrieval. The access interface provides operations (according to the display and manipulation process described in Section 2.2) to access and control the access to a multimedia document, e.g. start, stop, fastforward. The QoS negotiation interface provides operations supporting the QoS negotiation and renegotiation. On demand, stream interfaces will be created to access to a multimedia document. Such a scenario is shown in Figure 3.

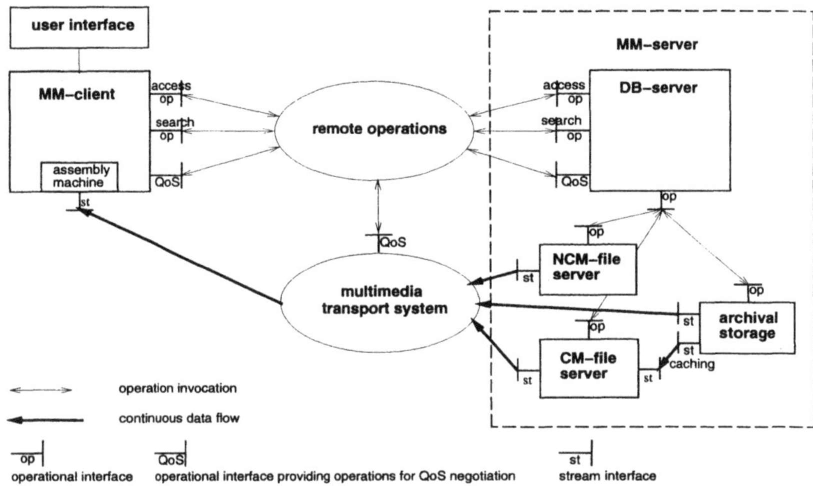

Figure 3. Computational specification of the multimedia news-on-demand service.

An incoming data stream has to be processed in order to be displayed in the appropriate way. This internal processing is indicated by the box called *assembly machine*. It includes the processing for synchronization of different streams, e.g. by the media synchronization controller [16], or processing of specialized transfer formats, e.g. MHEG-documents [25].

The server object has the operational interfaces matching the client's ones. Multimedia objects require special purpose storages because of their properties e.g. immense size (even when compression techniques are employed) and continuous character, and the enormous number of objects. Consequently, the database server is holding instead of the multimedia objects itself only references to them. According to certain administration policies, news and archived information are distinguished and intended to be stored on different objects. The objects (shown in Figure 3) called continuous media (CM) file server, noncontinuous media (NCM) file server, and archival storage are supposed to contain recently published documents in continuous and non-continuous formats and archived documents, respectively. From a technology point of view, they are implemented by conventional databases, special purpose file servers and tertiary storages, e.g. juke boxes containing CDs or tapes, respectively. From the computational viewpoint, the technology decisions are described in terms of QoS parameters.

The quality of service parameters are investigated in Section 4 in more detail. Communication aspects are reduced to the binding of interfaces within the computational viewpoint. However to illustrate the binding, we added place-holders in Figure 3. The term *remote operations* stands for the binding of operational interfaces and *multimedia transport system* for the binding of stream interfaces.

4. QoS architecture

The term QoS architecture has been limited in most previous work to considering the QoS mapping between different layers in a protocol stack. The QoS parameters concerning protocols have been investigated in the Internet community [11] as well as in the OSI community [8]. This includes research on the mappability of QoS parameters of different layers [9,5] and the mechanisms to satisfy QoS requirements, i.e. mostly resource reservation [32,23]. Also operating systems support [12,3] for multimedia applications is under research.

However, less work is done on the application level where all the various QoS parameters play together. In this article we address this problem and try to approach it within the framework of Open Distributed Processing. Similar considerations, but closer to the transmissions aspects, were done at Lancaster [5]. Our view includes also the characteristics of operating systems and particular application objects, such as databases.

For distributed multimedia presentational applications we identified three main objects which are involved in the QoS management: the MM-server, the MM-client, and the multimedia transport system, as shown in Figure 4.

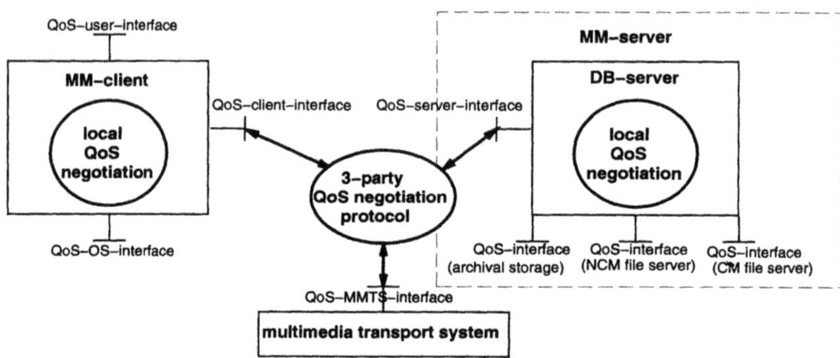

Figure 4. QoS management.

QoS interfaces are defined by tuples. The structure of the tuples follows a unique schema, a list of pairs of QoS parameters and their values. The possible values for a particular parameter are expected to be partially ordered. This assumption allows to use expressions for the value, e.g. 'frame rate greater or equal then 25 frames/sec'. Also there are two predefined values for all parameters, namely 'undefined' and 'any-value'.

An important concept which is used in the client and the server is the definition of the type of a multimedia object which we call media-type. We identified atomic and composed types. Atomic types are either mono-media types, e.g. text, images, audio or video; or multimedia objects which are multiplexed, e.g. audio and video in formats like DVI or MPEG. Composed types are constructed using atomic types as building blocks, however, additional information is required to describe how to compose them according to temporal and spatial relations. [16,31].

The following subsections consider QoS interfaces for the following objects:

- client: end-user (represented by the user interface) and operating system,
- server: database server, file server (continuous media and noncontinuous media), and archival storage
- transport system

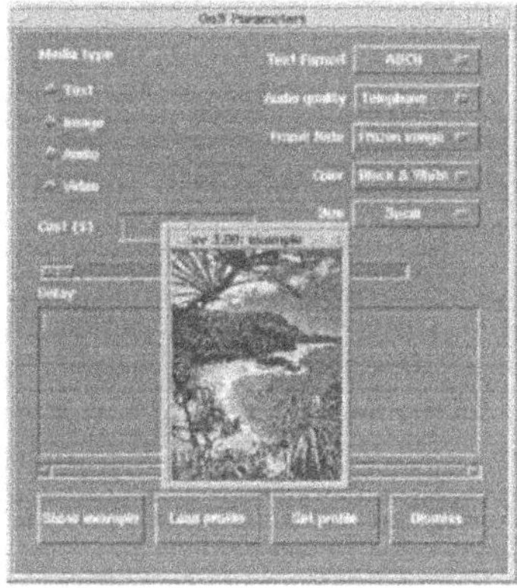

Figure 5. Quality query by example.

4.1. Client

Users-QoS interface

We suppose that a user is not willing to deal with raw numbers. We agree principally with 'quality query by example' [14]. However, our approach is to present the user a set of predefined parameters which corresponds to his/her experiences, e.g. audio quality known from the telephone or the compact disc. The user can check the offered qualities by listening to or watching examples which are provided for all QoS variants. A prototype of such an user interface, also called QoS demonstrator, is shown in Figure 5.

Table 1 shows the list of identified parameters. (We note that the possible values, mentioned in this in the foloowing tables, are only examples.)

Table 1
QoS parameters at user-interface

parameter type	possible values
media-type	{ text, image, audio, video, audio&video, composed-types }
text-format	{ ascii, postscript }
audio-quality	{ telephone, cd }
color	{ black&white, gray, color, super-color }
size	{ small, medium, large }
frame-rate	{ TV-rate,reduced-rate, frozen-images }
delay	integer (seconds)
cost	integer ($)

The actually defined tuple types are for the various medias:

<text, text-format, color, delay, cost> <image, color, size, delay, cost>
<video, color, frame-rate, size, delay, cost> < composed-type, skew, t_1, ..., t_n >

where skew is a skew or synchronization parameter and t_1, ..., t_n are QoS tuples of atomic media-types. Using these convention, an example is:
< media-type: audio&video, audio-quality: telephone, color: gray,
frame-rate: TV-rate, size: medium, delay: 1sec, cost: any >

Operating-system-QoS interface

The operating system QoS interface defines the quality of the support from the operating system. We identified two classes of QoS parameters abstracting from lower level properties such as CPU-power, memory-management and process scheduling. *Quality of devices*: including output devices, e.g. screen, loudspeaker, and storage devices used for buffering, e.g. hard disk. *Quality of software* to support the output of various medias, e.g. postscript-viewer or MPEG-player. The QoS parameters media-type, format, throughput, and guarantee class are identified. This leads to the parameters proposed in Table 2.

Table 2
QoS parameters at operating system interface

parameter type	possible values
screen	{ 1-bit, 8-bit-gray, 8-bit-color, 24-bit-color}
audio-device	{ telephone-quality, cd-quality }
disk storage	integer (bytes)
media-type	{ text, image, audio, video, audio&video, composed-types }
supported format	{ ASCII, postscript, gif, tiff, MPEG }
throughput	integer (frames/second)
delay	integer (seconds)
guarantee class	{ guaranteed, best-effort }

We take as an example the MPEG-player [29] running on a machine with a non-realtime operating system and 20 Mbyte available disk space, 8-bit-gray screen and a telephone-quality audio-device. A throughput of 10 frames/sec and a delay: 1 sec have been measured, but can not be guaranteed.
< screen: 8-bit-gray, audio-device: telephone-quality, disk storage: 20 Mbyte,
media-type: video, supported format: MPEG, throughput: 10 frames/sec,
delay: 1 sec, guarantee class: best-effort >

4.2. Server

Server-QoS interface

The server provides the access to and provides information on multimedia documents. The multimedia documents are characterized by their media-type, format, and size. Additional information is available from the storage objects which actually contain the multimedia documents. There is a throughput parameter which can be provided by a storage device to access an object. With respect to the throughput, the size of data packages, e.g. audio or video frames, is of interest. When accessing a stored object, there are also a certain delay and jitter. The delay can range from very small values, e.g. when reading from a hard disk, to quite large value, e.g. when data has to be loaded from a tertiary storage into a cache or data is preprocessed, for example converted from one format into another. A guarantee class parameter characterizes the throughput, the delay, and the jitter values with values best-effort or guaranteed. From the financial point of view, there are costs, in particular for accessing the database and for copyrights. This leads to the QoS interface as given in Table 3. According to the hierarchies shown in Figure 3, QoS parameters can be obtained dynamically from the lower layer objects using similar QoS interfaces from the tertiary storage and continuous media file server.

Table 3
QoS parameters at server-interface

parameter type	possible values
media-type	{ text, image, audio, video, audio&video, composed-types }
format	{ ascii, postscript, gif, JPEG, MPEG, DVI, ... }
color	{black&white, gray, color, super-color}
audio-quality	{ telephone, cd }
frame-rate	integer (frames/second)
size	bytes
throughput	integer (bytes/second)
delay	real (seconds)
packet size	integer (bytes) or variable
jitter	real (seconds)
guarantee class	{ guaranteed, best-effort }
cost	integer ($)

Example:
< media-type: video, format: MPEG, size: 12,345,678 byte, throughput: 0.5 Mbit/sec, delay: 15 sec, packet size: 50 kbyte, jitter: 0.001 sec, guarantee class: guaranteed, cost: $8 >

4.3. Transport System

With respect to the focused case study, a multi-media transport service is assumed. In order to define a transport service meeting our requirements, the following related work was studied: XTP (express transport protocol) [19], Tenet approach [11,23,24], RTP (real-time transport protocol) [30], OSI 95 [8], BERKOM Multimedia Transport service [4], Heidelberg Transport System [10], and ST-II [32]. Based on this study, a multi-media transport service, MMTS, is suggested. With respect to our case study, MMTS is defined by the following characteristics:

connection-oriented service type

The connection-oriented service type is motivated by the type of application. Multimedia documents require a relatively long term connection. Furthermore, resource reservation facilities are focused which also require a connection-oriented service. A fast-connection mode is not considered because the time for establishing the connection compared with the duration of the connection seems to be very short and so the problem of a fast connection is minor. Also the idea a fast connection is contradictory to the envisaged careful QoS negotiation.

unidirectional point-to-point transmission

The case study requires only a transmission of multimedia data from the server to the client. Control message exchanged between client and server in both direction may not be transmitted within the multimedia connection. Higher facilities such as remote procedure calls or traditionally transport protocol, e.g. TCP/IP, can also be used. Especially remote procedure calls fit better into the ODP architecture providing an object operational interfaces and stream (for isochronous data) interfaces.

transport service data units (TSDU) oriented

The orientation on transport service data units rather then bytes is determined by the character of multimedia data, in particular the frame-structure of digitalized audio and video. Also compressed forms maintain the frame-structure.

quality of service parameters

Resulting from the survey of related work and our own ad hoc case studies, the quality of service parameters shown in Table 3 have been identified.

The first four QoS parameters are used in the majority of comparable approaches. The guarantee parameter the parameters for throughput, delay, and jitter, namely whether they are satisfied by *best effort* or *guaranteed.* The transport service provider will charge its users. This is expressed by the cost parameter. The charge depends on the primary QoS parameters but also on other parameters as daytime, day of the week, etc. The function to calculate the cost is determined by the transport service provider. The reliability parameter determines whether the service is reliable or not. In the unreliable case, a quantitative characterization of the error-rate, e.g. as the number of lost or corrupted packages per second, is given. The considered parameters are summarized in Table 4.

Table 4.
QoS parameters at transport-service-interface.

parameter type	possible values
TSDU-maximum-size	integer (bytes)
throughput	integer (TSDUs/seconds)
delay	real (seconds)
jitter	real (seconds)
guarantee	{ best-effort, guaranteed }
cost	integer ($)
reliability	{ reliable, error-rate }

5. On negotiation protocols

The aim of a negotiation protocols is to determine all parameters in the QoS tuples of all the involved objects according to the QoS architecture, proposed in the previous section. Within the negotiation we identified the following three tasks (see also Figure 4):

- a 3-party QoS negotiation protocol between the client, server, and a multimedia transport service,
- local QoS negotiation at the server and the client, and
- renegotiation.

5.1. Three party QoS negotiation

A negotiation protocol has to include the initial negotiation (i.e. before a connection is established between client and server) and the renegotiation (i.e during a the lifetime of an already established connection). However from our point of view, i.e. above the transport layer, there is no difference in the principles of negotiation and renegotiation.

Figure 6 shows one variant of a QoS negotiation protocol, Figure 7 another. These considerations abstract from the related actions, e.g. querying. In the first case, the negotiation-agent is located at the client side, in the second variant the negotiation-agent is distributed to the client and the servers side. Other variants, e.g. changing the order of the actions in variant I and II or the negotiation-agent locating with a third party object, are possible but considered of minor importance.

The variant I protocol has the following phases:

(1) The client asks for the QoS parameters of a particular multimedia object.

(2) The server provides a set QoS tuples. Multiple QoS tuples occur when the multimedia object is available in different formats or the server provides tools to transform the format.

(3) The client negotiates locally the server's offer with the constraints from its operating system and the wishes of the user. The result of this local QoS negotiation has to be translated into a form corresponding to the MMTS QoS interface.

(4) The client requests from the MMTS a connection with the negotiated QoS parameters.

(5) The MMTS confirms or refuses. (An intelligent MMTS could report possible connections with decreased QoS parameters.)

The variant II of protocol has the following phases:

(1) and (2)
are as in variant I.

(3) The client negotiates locally the server's offer with the constraints from its operating system and the wishes of the user.

(4) The result of this local QoS negotiation is sent to the server.

(5) Translation of the requested QoS parameters.

(6) and (7)
correspond to the steps (4) and (5) of variant I, respectively.

The advantages of variant I are the complete control by the client side and reduced communications. The characteristics of network protocols with resource reservation facilities, like ST-II [32] (the resource reservation goes from the source to the sink), and the unidirectional data flow, however, are better supported by variant II.

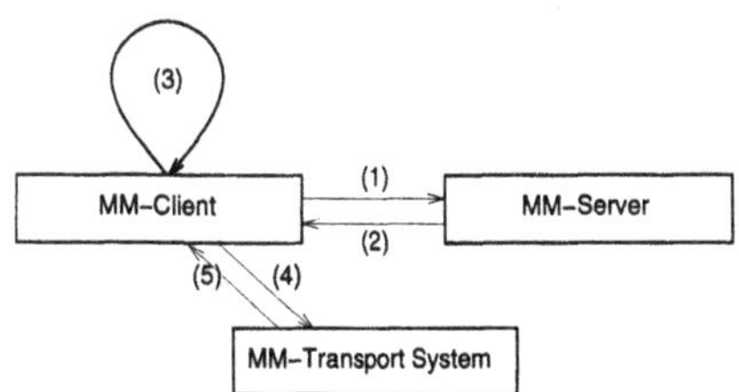

Figure 6. Variant I of QoS negotiation protocols.

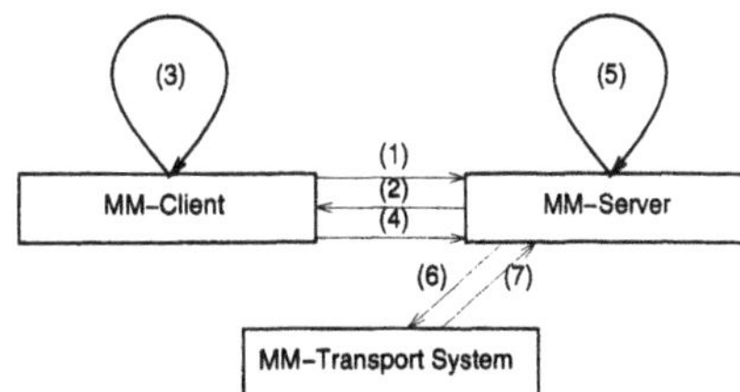

Figure 7. Variant II of QoS negotiation protocols.

5.2. Local QoS negotiation

The local negotiation includes also mechanisms for comparison and translation of QoS parameters. This is also called QoS broker [20,21]. The difficulty seems to be in the definition of algorithms for the translation. For some parameters the situation is clear, e.g. from a certain video or audio format combined with a determined frame rate a throughput requirement for the transport system can be calculated. However, especially the delay parameter which is in presentational applications very much less important than in conversational ones, depends on certain buffer mechanisms which can be used to balance the other parameters, such as jitter or errors. Experiments to determine translation functions for different transport systems are currently under study.

5.3. Renegotiation

Renegotiation do not require a new protocol. With respect to the client and the server, it is only yet another negotiation and the transport service is assumed to hide the renegotiation mechanisms.

A renegotiation can be caused by the following reasons:

- a user's demand for new QoS parameters,
- a monitors report that the transport service does not satisfy the agreed QoS parameters, or
- interrupt of one of the involved components, e.g. the transport system, because it can not satisfy the agreed QoS parameters.

6. Conclusions

We presented an approach to a functional architecture for distributed multimedia presentational applications. In our approach we introduced the concept of shells around a core, the multimedia database. We identified two shells on the user-interface level and the function level. This approach was applied to a multimedia news-on-demand service.

Furthermore, we developed a computational architecture for distributed multimedia presentational applications within the framework of the Reference Model on Open Distributed Processing. We characterized the qualities which distinguish distributed multimedia systems from (general) distributed systems in terms of QoS parameters. Accordingly, the development of the computational architecture was QoS driven. We specified a computational model of the multimedia news-on-demand service within this architecture. It should be mentioned that we developed also a formal approach, i.e. using formal description techniques, to a computational architecture for distributed multimedia systems which is presented in [33].

While other approaches consider QoS parameters specifically for communication (for the transport layer and below), we considered QoS parameters in a more general context. Our approach identifies QoS parameters in all components of the application, the producer, the consumer, and the network (transport layer). We introduced the concepts of the QoS interface through which objects can negotiate their QoS parameters. For the QoS negotiation itself, we investigated various variants of a negotiation protocol.

Currently the implementation of a prototype of the multimedia news-on-demand service is under way. The implementation environment consists of workstations connected by a local ATM switch. This includes also the implementation of an QoS demonstrator as outlined in Section 4.1. Furthermore, extended considerations of QoS negotiation and corresponding protocols for other types of distributed multimedia applications are foreseen. Currently, investigations are under way on a conversational application, the joint viewing and tele-operation service (JVTOS) [13].

Acknowledgements

We would like to thank all who have contributed to this article, in particular our co-investigators in the CITR project, the participants of the seminar on multimedia and high-speed networking held in Université de Montréal during the winter semester 1994, Klara Nahrstedt (University of Pennsylvania) and Ilka Miloucheva (Technische Universität Berlin) for fruitful discussions and comments. Special thanks to Alain Bibal, Thibault Burdin de Saint Martin and Quoc Vu for discussion, comments and contributing to the implementation of the current version of the prototype including the QoS demonstrator and a QoS negotiation protocol.

References

1. M. Altenhofen, et al., "The BERKOM Multimedia Collaboration Service" in *Proceedings of the A - CM Multimedia 93,* ed. P. Venkat Rangan, pp. 457-464, ACM Press, Anaheim (1993).
2. P. B. Berra, F. Golshani, R. Mehrotra, and O. R. L. Sheng, "Introduction Multimedia Information Systems," *IEEE Transactions on Knowledge and Data Engineering,* 5, 4 (Aug. 1993).
3. G. Blair, G. Coulson, and N. Davies, "Communications and Distributed Systems Support for Multimedia Applications," *submitted to a special issue on multimedia information systems of 'Information and Software',* Butterworth-Heinemann.
4. S. Böcking, et al., "The BERKOM MulitMedia Transport System" in *Proceedings of the 1st International Conference on ODP,* pp. 385-391, Berlin (Sept. 1993).
5. A. Campbell, G. Coulson, F. García, D. Hutchinson, and H. Leopold, "Integrated Quality of Service for Multimedia Communications" in *IEEE INFOCOM'93,* pp. 732-739, 1993.
6. C. Chapedelaine, R. Descout, and P. Billon, "Interface Design Issues in the MEDIALOG Project" in *Procedings of CASCON'93, Volume II,* ed. A. Gawman, W. M. Gentleman, E. Kidd, P. Larson, and J. Slonim, pp. 707-714, Toronto (1993).
7. S. Christodoulakis, M. Theodoridou, F. Ho, M. Papa, and A. Patria, "Multimedia Document Presentation, Information Extraction, and Document Formation in MINOS : a Model and a System," *ACM Transactions on Office Information Systems,* 4, 4 (Oct. 1986).
8. A. Danthine, "OSI 95. High Performance Protocol with Multimedia Support on HSLANs and B-ISDN" in *3th Joint European Networking Conference,* Insbruck (1992).
9. L. Delgrossi, R. G. Herrtwich, and F. O. Hoffmann, "An Implementation of ST-II for the Heidelberg Transport System" in *IEEE Globcom'92,* Orlando (1992).
10. L. Delgrossi, et al., "Media Scaling for Audiovisual Communication with the Heidelberg Transport System" in *Proceedings of the ACM Multimedia 93,* ed. P. Venkat Rangan, pp. 99-104, ACM Press, Anaheim (1993).
11. D. Ferrari, A. Banerjea, and H. Zhang, "Network Support For Multimedia," TR-92-072, The International Computer Science Institute, Berkley, CA (1992).

12. Homsy G., Govindan R., and Anderson D. P., *Implementation Issues for a Network Continuous-media I/O Server,* The International Computer Science Institute, Berkeley (1990).

13. T. Gutekunst, T. Schmidt, G. Schule, J. Schweitzer, and M. Weber, "A Distributed Multimedia Joint Viewing and Tele-Operation Service for Heterogeneous Workstation Environments" in *Proceedings of International Workshop on Distributed Multimedia Systems,* Stuttgart (1993).

14. G. Kalkbrenner, T. Pirkmayer, A. van Dornik, and P. Hofmann, "Quality of Service (QoS) in Distributed Hypermedia-Systems" in *Procceedings of RPODP'94 - Second International Workshop on Principles of Document Processing,* Darmstadt (April 1994).

15. B. Kerherve, G. v. Bochmann, R. Dssouli, J. Gecsei, A. Hafid, and A. Vogel, "Distributed Multimedia Presentational Applications Requirements," Technical Report #893, Université de Montréal (1993).

16. L. Lamont, L. Li, and N. D. Georganas, "Synchronization Architecture and Protocols for a Multimedia News Service Application," Technical Report, University of Ottawa (1994). accepted for Broadband Islands'94 Conference.

17. C. Meghini, F. Rabitti, and C. Thanos, "Conceptual Modeling of Multimedia Documents," *IEEE Computer* (Oct. 1991).

18. G. Miller, G. Baber, and M. Gilliland, "News On-Demand for Multimedia Networks" in *Proceedings of the A - CM Multimedia 93,* ed. P. Venkat Rangan, pp. 383-392, ACM Press, Anaheim (1993).

19. I. Miloucheva, "XTP and ST-II Protocol Facilities for Providing the QoS Parameters of Connection-Mode Transport Services," Research Note TUB-PRZ-W-1029, Technical University Berlin, Berlin (1992).

20. K. Nahrstedt and J. M. Smith, "Application-Driven Approach to Networked Multimedia Systems" in *Proceedings of the 18th Conference on Local Computer Networks* (1993).

21. K. Nahrstedt and J. M. Smith, "Revision of QoS at the Application/Network Interface," Technical Report MS-CIS-93-94, University of Pennsylvania (1993).

22. S. R. Newcomb, N. A. Kipp, and V. T. Newcomb, "The HyTime Hypermedia/Time-based Document Structuring Language," *Communications of the ACM,* 34 (1991).

23. C. J. Parris, G. Ventre, and H. Zhang, "Graceful Adaption of Guaranteed Performance Service Connections," TR-93-011, The International Computer Science Institute, Berkeley (1993).

24. C. J. Parris and D. Ferrari, *A Dynamic Connection Management Scheme for Guaranteed Performance Services in Packet-Switching Integrated Services,* The International Computer Science Institute, Berkeley (1993).

25. R. Price, "MHEG: An Introduction to the Future International Standard for Hypermedia Object Interchange" in *Proceedings of the First ACM International Conference on Multimedia,* ed. P. Venkat Rangan, pp. 121-128, ACM Press, Anaheim (1993).

26. P. V. Rangan, "Video conferencing, file storage, and management in multimedia computer systems," *Computer Network and ISDN Systems,* 25, pp. 901-919 (1993).

27. K. Raymond, "The Reference Model of Open Distributed Processing: a Tutorial" in *Proceedings of the 1st International Conference on ODP,* ed. J. de Meer, B. Mahr, and O. Spaniol, pp. 3-14, Berlin (1993).

28. L. A. Rowe and R. R. Larson, "A Video-on-Demand System," Project proposel, University of California at Berkeley (1993).

29. L. A. Rowe and B. C. Smith, "Performance of a Software MPEG Video Decoder" in *Proceedings of the First ACM International Conference on Multimedia,* ed. P. Venkat Rangan, pp. 75-82, ACM Press, Anaheim (1993).

30. H. Schulzrinne, "RTP: The real-time transport protocol" in *MCNC 2nd Packet Video Workshop,* 2, Research triangle park (Dec. 1992).

31. R. Steinmetz and C. Engler, "Human Perception of Media Synchronization," Technical Report 43.9310, IBM Europeen Networking Center, Heidelberg (1993).

32. C. Topolcic, "Experimental Internet Stream Protocol, Version 2 (ST-II)," Internet RFC 1190 (1990).

33. A. Vogel, "A Formal Approach to an Architecture for Open Distributed Processing," Technical Report #902, Université de Montréal (1994).

34. J. Wong, et al., "CITR Major Project 'Broadband Services'," Project Description, Canadian Institute for Telecommunication Research.

3

Implementing a QoS Controlled ATM Based Communications System in Chorus

Philippe Robin, Geoff Coulson, Andrew Campbell, Gordon Blair, Michael Papathomas and David Hutchison

Distributed Multimedia Research Group,
Department of Computing, Lancaster University,
Lancaster LA1 4YR, United Kingdom

phone: +44 (0)524 65201
e-mail: mpg@comp.lancs.ac.uk

ABSTRACT

We describe the design of an application platform able to run distributed real-time and multimedia applications alongside conventional UNIX programs. The platform is embedded in a micro-kernel/ PC environment and supported by an ATM based, QoS driven communications stack. We focus in particular on resource management aspects of the design and deal with CPU scheduling, network resource management and memory management issues. An architecture is presented which guarantees QoS levels of both communications and processing with varying degrees of commitment as specified by user level QoS parameters. The architecture uses admission tests to determine whether or not new activities can be accepted and includes modules to translate user level QoS parameters into representations usable by the scheduling, network and memory management subsystems.

Keyword Codes: C.2.4, C.2.2, H.5.1.
Keywords: Distributed Systems, Network Protocols, Multimedia Information Systems.

1. Introduction

The research reported in this paper is aimed at providing system software support for distributed real-time and multimedia applications in an environment of conventional workstations and high-speed networks. Our specific aims are as follows:-

- to support real-time and multimedia applications in a heterogeneous system consisting of PC and workstation end-systems connected by ATM, Ethernet and proprietary high-speed networks,
- to enable real-time and multimedia applications to enjoy predictable performance in both communications and processing according to user provided QoS parameters,
- to retain the ability to run standard UNIX applications alongside real-time applications.

Our approach is to use a micro-kernel operating system, specifically Chorus [1], to underpin both UNIX and real-time applications. Real-time and multimedia applications are supported by the extensions described in this paper. Alongside these, a standard UNIX SVR4 'personality' included with Chorus is used to support UNIX applications.

Our previous work in the field of distributed real-time and multimedia application support has concentrated on API issues [2], CPU scheduling issues [3], transport issues [4] and network architecture [5]. Complementary to these areas, the present paper focuses on the *resource management strategies* used in our Chorus extensions. The three major resource classes considered are *CPU cycles*, *network resources* and *physical memory*. In this paper we focus on *end-system* related communications issues rather than internet or network resource management issues (although we do cover resource allocation in the ATM network environment). Broader network and inter networking issues are discussed more fully in [5].

The paper begins by providing, in section 2, some necessary background material on Chorus. Next we present, in section 3, an overview of the architecture of our real-time support infrastructure. This consists of:-

- an *application programmer's interface* (API) at which QoS requirements can be stated,
- a *CPU scheduling framework* which minimises kernel context switches in both application and protocol processing,
- an *ATM based communications stack* which features an enhanced IP layer for inter networking,
- a framework for *QoS driven memory management*, and
- a framework for *flow** *management* which integrates the management of resources in both end-systems and the network.

We then discuss the management of CPU, communications and memory resources in this architecture. The various resource management functions are categorised as either *static* or *dynamic* as suggested in [5]. In essence, static QoS management (treated in section 4) deals with connect time issues such as QoS translation (i.e. deriving resource quantities from QoS parameters), and admission testing (i.e. determining whether new sessions can be created given their specific resource requirement and current resource availability). Dynamic resource management, which is concerned with run time issues, is not treated in detail in this paper. We offer concluding remarks in section 5.

2. Background on Chorus

Chorus is a commercial micro-kernel based operating system which supports the implementation of conventional operating system environments through the provision of 'personalities' (for example a personality is available for UNIX SVR4 as mentioned above). The micro-kernel is implemented using modern techniques such as multi-threaded address spaces and integrated message based communications. The basic Chorus abstractions are *actors*, *threads* and *ports*, all of which are named by globally unique identifiers. Actors are address spaces and containers of resources which may exist in either user or supervisor space. Threads are units of execution which run code in the context of an actor. They are scheduled according to either a pre-emptive priority based or round robin time slicing scheme. Ports are message queues used to hold incoming and outgoing messages. The inter-process communication sub-system supports both request/reply and single shot messages.

Chorus has several desirable real-time features and has been fairly widely used for embedded real-time applications. Real-time features include pre-emptive scheduling, page locking, time-outs on system calls, and efficient interrupt handling. Unfortunately, Chorus' real-time support is not fully adequate for the requirements of distributed real-time and multimedia applications, principally because there is no support for QoS specification and

---

* The term *flow* is used to refer to the end-to-end passage of data from a source application, down through the source protocol stack, across the network, up through the sink protocol stack, and eventually to the sink application.

resource reservation:-

- although it is possible to specify thread scheduling constraints relative to other threads, *absolute* statements of requirement for individual threads cannot be made,
- in the communications sub-system, the exclusive use of connectionless datagrams makes it impossible to pre-specify communications resource allocation,
- due to the use of a paged virtual memory system it is not possible to place bounds on memory access latency except by the extreme of wiring pages.

Note, however, that such limitations are not unique to Chorus: they are shared by most of the other micro-kernels in current use (e.g. [6], [7]).

3. Architecture

3.1. Application Programmer's Interface

To remedy its current deficiencies for QoS specification and real-time application support, we have extended the Chorus system call API with new low level calls and abstractions. The new abstractions, provided in both the kernel and a user level library, are the following:

- *rtports*: these are extensions of standard Chorus ports and serve as access points for real-time communications. Rtports have an associated QoS which defines timeliness constraints on communication. They also provide direct application access to buffers thus minimising copy operations.
- *devices*: these are producers, consumers and filters of real-time data which support the creation of rtports and provide the memory for their buffers. One special type of device is the *null* device which is implemented in a user level library and permits user code to produce/ consume real-time data through the use of *rthandlers*.
- *rthandlers*: these are user supplied C routines which provide the facility to embed application code in the real-time infrastructure. They are attached to rtports at run time and upcalled on real-time threads by the infrastructure when data is available/ required. They encourage an event-driven style of programming which is appropriate for real-time applications and also avoid the context switch overhead associated with a traditional *send()/ recv()* based interface.
- *QoS controlled connections*: these are communication channels with a specific QoS*. A connection is established between a source and a sink rtport according to a given QoS specification. There are two types of connection: stream connections for periodic and continuous media data, and message connections for time-constrained messages. Stream connections are *active* in the sense that they initiate the transfer of data by upcalling a source rthandler (if attached). Message connections differ in that they *passively* wait for a source thread to pass them data via an *ipcSend()* call.
- *QoS handlers*: these are upcalled by the infrastructure in a similar way to rthandlers but are used to notify the application layer when QoS commitments provided by connections have been violated.

In addition to these features, the API includes facilities for dynamically re-negotiating the

---

* QoS controlled connections are *abstractions* and are uniformly used for both remote and local communications. In the remote case, they are implemented in terms of the communications architecture described in section 3.3. In the local case, they are implemented in terms of optimised memory mapping mechanisms.

QoS of open connections and for building pipelines of 'software signal processing' modules for local continuous media processing. It also has synchronisation primitives based on eventcounters and sequencers which incorporate the notion of *deadline inheritance* [8] whereby a 'worker' thread carrying out a task on behalf of a calling thread inherits the deadline of the caller. Full details of the continuous media API are specified in [2] and [8].

3.2. Scheduling Architecture

The scheduling architecture exploits the concept of *lightweight threads* which are supported in a user level library and multiplexed on top a single Chorus kernel thread. In this context, we refer to Chorus kernel threads as *virtual processors* (VPs). The scheduling architecture is a *split level* structure [9] consisting of a single kernel scheduler (KLS) to schedule VPs, and per-actor user level schedulers (ULSs) to schedule lightweight threads on those VPs.

The advantage of lightweight threads and user level scheduling is that context switch overhead is minimal. On the other hand, the drawback of user level scheduling is that, by definition, it cannot ensure that CPU resources are fairly shared across multiple actors. This is the role of kernel level scheduling. The split level architecture combines the benefits of both user level and kernel level scheduling by maintaining the following invariants:-

i) each ULS always runs its most urgent* lightweight thread,

ii) the KLS always runs the VP supporting the *globally* most urgent lightweight thread.

The KLS/ ULS information exchange is accomplished via a combination of shared KLS/ ULS memory and kernel-to-VP upcalls [9]. The shared memory is divided into per-VP areas, each of which contains the urgency of the most urgent runnable lightweight thread known to its associated VP (along with some other information as described below). These urgency values are read by the KLS on each kernel level rescheduling operation to determine the next VP to schedule. Upcalls, implemented as *software interrupts*, are used by the KLS to inform VPs of the occurrence of real-time events in a timely fashion. Such events include timer expirations, and data arrivals from local kernel devices or from the network device. Software interrupts are always targeted at VPs but can be initiated either by kernel components (e.g. the KLS) or by library code in application actors.

The design also uses the concept of *non blocking system calls* [10] to ensure that VPs are always available to run light weight threads [8].

3.3. Communications Architecture

The standard Chorus communications stack was designed for the support of connectionless datagram services and uses retransmission strategies to enhance reliability. In contrast, our communications architecture (see figure 3) is intended to support QoS controlled connection oriented communications and configurable error control. Because of these disparate design goals, we have initially designed our stack to operate entirely separately from the existing Chorus IPC stack. However, we do intend in the future to integrate the functionality of the two stacks in a unified architecture.

3.3.1. Abstract Layering

The communications architecture enforces a strong distinction between communication for signalling purposes (i.e. connection establishment, network resource management and connection tear-down), and user data transfer purposes. The transport, AAL5 and ATM layers are common to both the signalling and the user data stack and are described below. The signalling stack specific layers comprise an upper network sub-layer for resource management

---

* The notion of 'urgency' is dependent on the scheduling policy used (e.g. it would be *deadline* for EDF scheduling and *priority* for rate monotonic scheduling). The issue of scheduling policies is deferred until section 4.3.

in IP routers and a lower network sub-layer for resource management in ATM switches. The IP layer is a subset of an existing network resource reservation protocol called RSVP [11] which we encapsulate in the IP Internet Connection Management Protocol (ICMP). The ATM signalling protocol, called ATMSig, is a subset of the ATM Forum's UNI 3.0 [12].

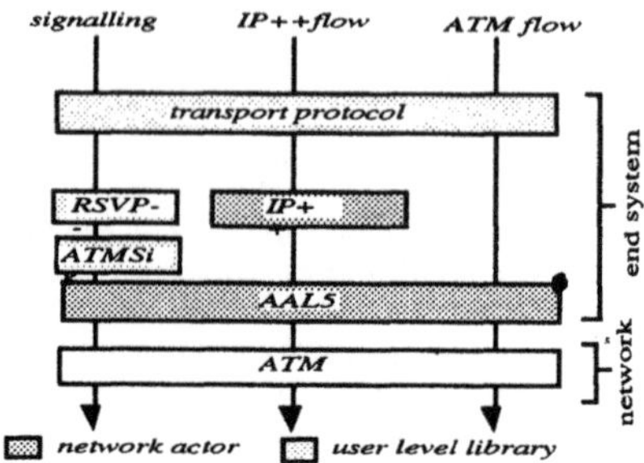

Figure 3: Communications architecture

The user data stack is positioned alongside the signalling stack. The upper architectural layer is a connection oriented transport protocol [13] which provides for QoS specification at connection time (including configurable error control), in service QoS re-negotiation, and end-to-end flow control (via a rate based mechanism). Other traditional transport layer functions such as admission control, resource reservation, performance monitoring, and dynamic QoS maintenance are supported outside the transport protocol proper by the scheduling, connection and memory management subsystems described in the remainder of this section.

The user stack's IP layer, called IP++, allows us to interwork outside the ATM network in a heterogeneous environment. It offers QoS enhanced facilities along the lines of those proposed in Deering's Simple Internet Protocol Plus (SIPP) [14]. In particular, IP++ uses a packet header field called a flow-id to identify IP packets as belonging to a particular connection or flow, and a flow-spec (see section 4.4.1) to define the QoS associated with each flow. Flow-specs are held by IP++ routers* and used to determine the resources dedicated to the router's handling of each IP++ packet on the basis of its flow-id. The state held by routers is initialised at connection set up time by the RSVP signalling protocol. Below the IP layer we use an AAL5 ATM Adaptation Layer service to perform segmentation and reassembly of IP packets into/from 53 byte ATM cells.

The lowest layer of our architecture is based on the Lancaster Campus ATM network. This delivers ATM to a mix of workstations, PCs, and multimedia devices designed at Lancaster [15]. It also interconnects a number of Ethernets and interfaces to the rest of the UK via an SMDS connection to the SuperJANET 100 Mpbs Joint Academic Network. The PCs which run the system described in this paper are connected to 4x4 ATM switches manufactured by Olivetti Research Limited (ORL) via ISA bus interface cards. The ORL ATM switches are implemented using 'soft' switching and run a small micro-kernel called *ATMos*.

3.3.2. Realisation in Chorus

In implementation, we map the abstract layered communications architecture partly onto per-actor user level libraries and partly onto a single, per machine, supervisor actor called the

---

* Flow specs are also used by the FMP (see section 3.5) to control resource reservation at the ATM level in ATM switches.

*network actor**. The transport layer of the signalling stack is implemented in the FMP actor described in section 3.5. The transport layer of user applications is implemented in the same user level library† that supports the API abstractions discussed in section 3.1. This allows the transport service interface to be provided by the library level rtport and rthandler abstractions defined in that section. The transport protocol communicates with the network actor via *asynchronous system calls* [8] for send side communications, and *software interrupts* for receive side communications.

Below the transport protocol, the rest of the communications architecture, including the ATM card device driver, is implemented in the network actor. The two signalling protocols, RSVP and ATMsig are not described here as they are considered to be outside the scope of this end-system oriented paper. In the user stack, the major complexity involved in the IP++ implementation is in supporting the routing function. This is required when the current host is neither the source nor sink of a flow but is merely routing packets from one network to another. In this case, CPU and memory resources are dedicated to flows on the basis of a flow-spec supplied by the flow management protocol (see section 4.4). Otherwise, the function of the IP++ layer is effectively null.

AAL5 is also implemented in the network actor. A software AAL5 implementation is required because our ATM interface cards only support data transfer at the granularity of ATM cells. The AAL5 implementation uses a single thread on the receive side and per-flow threads on the send side to perform segmentation and reassembly with optional checksumming. The use of per-flow threads reduces multiplexing in the stack to an absolute minimum as recommended in the literature [16]. Currently, the maximum service data unit size for the AAL5, IP++ and transport layers alike is restricted to 64Kbytes. This means that no further segmentation/ reassembly is required above the AAL5 layer+. The ATM cards generate an interrupt every time a cell is received, and every time they are ready to transmit. Communication between the interrupt service routines and the per-flow AAL5 threads is via Chorus 'mini-ports' (see section 4.4.3).

3.4. Memory Management Architecture

The purpose of the extended memory management architecture, which is built on top of the existing Chorus abstractions [17], is to ensure that applications and QoS controlled connections can access memory regions with *bounded latency*. It is of little use to offer guaranteed CPU resources to threads if they are continually subject to non predictable memory access latency due to arbitrary page faulting† . Our design encapsulates most of the QoS driven memory management functionality inside a system actor called the *QoS mapper*. The roles of the QoS mapper are:-

- supplying application actors with memory regions offering latency bounded access,
- determining whether or not requests for QoS controlled memory resources should succeed or fail,
- pre-empting QoS controlled memory from 'low urgency' threads on behalf of 'high urgency' threads when necessary, and,
- efficiently re-mapping QoS controlled memory regions from one actor to another.

---

* Note that this is distinct from the existing Chorus 'network actor' which is called the 'Network Device Manager'.

\+ It would be a relatively straightforward extension to support arbitrarily sized buffers at the API level by supporting segmentation and reassembly in the transport protocol if this proved necessary.

† Note that, in addition to buffers, it is also necessary to provide bounded latency access to code and stack regions of QoS controlled threads if QoS guarantees are to be maintained.

In addition to servicing requests from the kernel VM layer, the QoS mapper is accessed from the user level libraries implementing the connection abstraction in the intra-machine connection case. In particular, user level code invokes the QoS mapper via extended versions of the *rgnAllocate()* and *rgnFree()* Chorus system calls. These respectively allocate and free a QoS controlled region of memory at connection establishment time.

3.5. Flow Management Architecture

We have described frameworks for the management of CPU, network and memory resources but have said nothing yet of the relationship between these frameworks. It is the task of the flow management architecture, and in particular the *flow management protocol* (FMP) [5], to realise this relationship. The FMP must arrange, at connection time, for the allocation of suitable CPU, memory and network resources according to the user specified QoS of the requested connection. In end-systems, the FMP co-operates with the CPU and memory management subsystems and in the network it runs on IP++ routers and ATM switches, and coordinates itself by means of the RSVP and ATMsig protocols described in section 3.3. A central role of the FMP is to partition the responsibility for QoS support among individual resource managers. For example, for remote communications, the FMP partitions the API level *latency* QoS parameter (see section 4.1) between the network and the CPU resource managers on each end system.

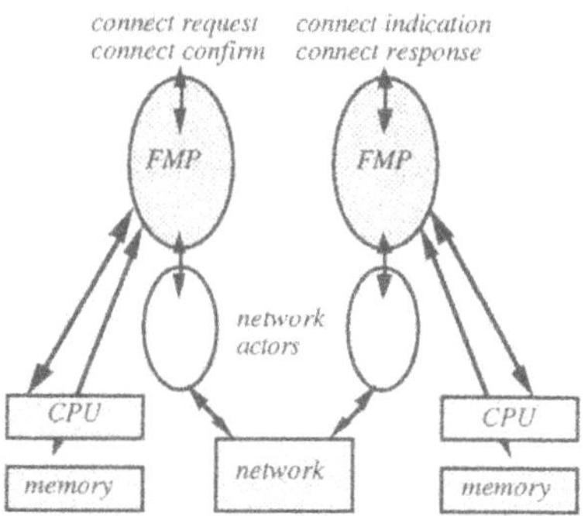

Figure 4: Flow management architecture

The FMP is also responsible for *dynamic* QoS management in flows. In this role, it can adapt to degradations in one resource by compensating in terms of another. Ideally, it will do this without either involving the application or violating overall the QoS specification. For example, an increase in jitter caused by the network can be transparently compensated for by an increased buffer allocation at the receiver - as long as the latency QoS is not thereby compromised.

The flow management architecture adopts a similar split level structure to the scheduling and communications architectures. First, when a new QoS controlled connection is requested, a *QoS translation* function (see section 4) in the user level library determines the resource requirements of the request. The output of the QoS translator is then directed to the FMP which runs in a per-machine FMP actor (see figure 4).

4. Resource Management Strategies

Prior reservation of resources to connections is necessary to obtain guaranteed real-time performance. This section describes the resource reservation framework in our system and shows how user level QoS parameters are used to derive the resource requirements of connections and make appropriate reservations. This paper concentrates on the reservation of *specific* resources (i.e. CPU, memory and network resources) rather than treating resource resevation as an integrated activity driven by the FMP.

In outline, there are two stages in the resource reservation process. *QoS translation* is the process of transforming user level QoS parameters into resource requirements and *admission testing* determines whether sufficient uncommitted resources are available to fulfil those requirements.

4.1. User QoS Parameters

The QoS parameters visible at the API level are as follows:-

```
typedef enum {best_effort, guaranteed} com;
typedef enum {isochronous, workahead} del;

typedef struct {
        com commitment;
        int buffsize;
        int priority;
        int latency;
        int error;
        int error_interval;
        int buffrate;
        int jitter;
        del delivery;
} StreamQoS;

typedef struct {
    com commitment;
    int buffsize;
    int priority
    int latency;
    int error;
} MessageQoS;

typedef union {
    MessageQoS mq;
    StreamQoS sq;
} QoSVector;
```

The two structures in the QoSVector union are for stream connections and message connections respectively. The first four parameters are common to both connection types. *Commitment* expresses a degree of certainty that the QoS levels requested will actually be honoured at run time. If commitment is *guaranteed*, resources are permanently dedicated to support the requested QoS levels. Otherwise, if commitment is *best effort*, resources are not permanently dedicated and may be preempted for use by other activities. *Buffsize* specifies the required size of the internal buffer associated with the connection's rtports. *Priority* is used for fine grained control over resource pre-emption for connections. All things being equal, a connection with a low priority will have its resources pre-empted before one with a higher priority.

Latency refers to the maximum tolerable end-to-end delay, where the interpretation of 'end-to-end' is dependent on whether or not rthandlers are attached to the rtport. If rthandlers are attached, latency subsumes the execution of the rthandlers; otherwise it refers to rtport-to-rtport latency. When rthandlers are attached a further, implicit, QoS parameter called *quantum* becomes applicable. The value of this parameter is dynamically derived by the infrastructure whenever an rthandler is attached to an rtport. It is defined as the sum of the rthandler execution time and the execution time of the protocol code executed by the same thread before/ after the rthandler is called*. To determine the quantum value, the infrastructure performs a 'dummy' upcall of the handler and measures the time taken for it to return (a boolean flag is used to let the application code in the rthandler know whether a given call is 'real' or dummy). It is the responsibility of the application programmer providing the rthandler to ensure that the dummy execution path is similar to the general case. Although the value of quantum is dynamically refined as the connection runs, an inaccurate initial value will inevitably cause QoS violations.

Error has different interpretations depending on the connection type. For stream

* Actually there is a third component to the quantum value which is the per-buffer time taken by per-connection transmit threads at the ATM level (see section 4.4.3).

connections, it is used in conjunction with *error_interval* and refers to the maximum permissible number of buffer losses and corruptions over the given interval. In the case of message connections, it simply represents the probability of buffers being corrupted or lost (error_interval is not applicable to message connections).

For stream connections, there are three additional parameters, *buffrate*, *jitter* and *delivery*, which have no counterparts in message connections. *Buffrate* refers to the required rate (in buffers per second) at which buffers should be delivered at the sink of the connection. *Jitter*, measured in milliseconds, refers to the permissible tolerance in buffer delivery time from the periodic delivery time implied by buffrate. For example, a jitter of 10ms implies that buffers may be delivered up to 5ms either side of the nominal buffer delivery time. *Delivery* also refines the meaning of buffrate. If *isochronous* delivery is specified, stream connections attempt to deliver *precisely* at the rate specified by buffrate; otherwise, if delivery is *workahead*, it is permitted to 'work ahead' (ignoring the jitter parameter) at rates temporarily faster than buffrate. One use of the workahead delivery mode is to more efficiently support applications such as real-time file transfer. Its primary use, however, is for pipelines of processing stages where isochronous delivery is not required until the last stage [2].

4.2. Resource Classes

In the following sections, we distinguish four major classes of QoS controlled connection for resource management purposes. These resource classes, named G_I, G_W, B_I, and G_W are selected on the basis of the *commitment* and *delivery* QoS parameters described in section 4.1:

Best effort { - isochronous (BI)
- workahead (BW)   Guaranteed { - isochronous (GI)
- workahead (GW)

In addition to the two best effort classes a third best effort class, B_C, is distinguished which refers to non real-time Chorus and UNIX threads out of the scope of the real-time extenstions. Additionally, all three best effort classes are often grouped together and referred to by the shorthand name B. Similarly, the guaranteed classes are collectively referred to as G.

4.3. The CPU Resource

4.3.1. QoS Translation

For admission testing and resource allocation purposes for stream connections, it is necessary to know the *period* and *quantum* of the threads associated with the connection. The period is simply the reciprocal of the *buffrate* QoS parameter and the quantum is implicitly derived at connect time as explained in section 4.1. Figure 6 illustrates the notions of period and quantum together with the related scheduling concepts of *scheduling time*, *deadline* and *jitter*.

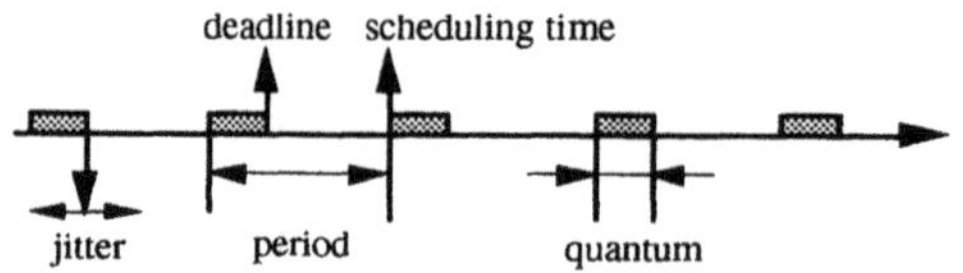

Figure 6: Periodic thread scheduling terminology

For message connections, periodic *sporadic server* threads are used at the receive side. One sporadic server per application actor is provided for each of the two applicable commitment classes (viz. G_W and B; isochronous delivery is not applicable to message connections), and each sporadic server handles all the message threads in its class. The quantum of each server is set to the *maximum* of the quanta of all the message threads in its class to ensure that adequate processing time is available for any of the server's associated threads. The period of each server is heuristically derived as follows:-

$$period = \min(recv\_latency_0, \ldots, recv\_latency_n)$$

Recv_latency$_i$ is the proportion of the total end-to-end latency allocated by the FMP to the receive end-system for message connection *i*. This method of calculating *period* is a compromise which requires less resource than a optimal period (i.e. the optimal period, 1 / *quantum*, would ensure that the server was always ready to service a message but would take all the CPU resource allocated to the class!) while offering a reasonable probability that the server will be ready when a message arrives.

4.3.2. Admission Testing

The semantics of thread scheduling for each of the three resource classes are as follows:-

- G_I: threads for these connections are scheduled to run such that the completion of a quantum is guaranteed to be completed by the logical arrival time + *j* (where *j* is the jitter QoS parameter and logical arrival time is the start of the requisite period). An extended earliest deadline first [18] (EDF) algorithm and admission test is used to ensure this behaviour.
- G_W: these are scheduled according to the preemptible EDF policy. The jitter QoS parameter is ignored and quanta may be scheduled ahead of their logical arrival time to permit workahead. Again, an admission test is performed.
- B: these are scheduled according to the preemptible earliest deadline first policy but no admission test is used.

Each of the G and B resource classes is allocated a fixed portion of the CPU resource. Note, however, that the 'firewall' that this separation implies is used only to limit the number of threads in each class - not to restrict the use of CPU cycles at run time. If there are unused resources in one class, these resources are automatically exploited by the other class at run time (see section 4.4.3).

The firewalls can be dynamically altered at run time by the programmer, but a typical configuration will allow a relatively small allocation for G threads. This is to encourage users to choose best effort threads wherever possible. Best effort threads should be perfectly adequate for many 'soft' real-time needs so long as the system loading is relatively low. The guaranteed classes should only be used when absolutely necessary in particular, guaranteed isochronous threads should only be used for connections which are delivering data to an end device intended for human perception such as a frame buffer or audio chip.

The admission tests for G_I threads are:-

$$\sum_{i=1}^{N_G} \frac{quantum_i}{period_i} \leq R_G, \quad 0 \leq R_G \leq 1 \qquad \text{i)}$$

$$\sum_{i=1}^{N_G} \frac{quantum_i}{jitter_i} \leq 1 \qquad \text{ii)}$$

The admission test for this class is a two stage process, and each of the two tests are modifications of the well known Liu/Layland test [18] (the latter guarantees that each quantum in the given set of tasks can be completed by the end of its period as long as it is runnable at the start of its period). The first test ensures that the overall resources used by all G threads are not greater than the allocated portion. N_G refers to the total number of G threads in the system and R_G refers to the portion of CPU resources dedicated to this class of threads (such that $R_G + R_B = 1$ where R_B represents the portion of the CPU resource dedicated to B threads). The second test imposes the additional constraint that each quantum must complete by the end of its user stated jitter bound rather than simply by the end of the requisite period.

For G_W threads the admission test is simply:-

$$\sum_{i=1}^{N_G} \frac{quantum_i}{period_i} \leq R_G$$

Admission tests for the sporadic servers are identical to those for G_W and B periodic threads. Each time a new message connection is created which alters the period or quantum of its server, a new admission test must be performed to ensure that the modified sporadic server can still be accommodated in the appropriate resource class.

4.3.3. Dynamic Scheduling Management

At run time, the scheduling scheme uses a combination of *priorities**, *deadlines* and *scheduling times* to capture the abstract notion of 'urgency'. The scheduler uses three distinct priority bands into which the four classes of thread are mapped. The semantics of priority are that at any given time there is no runnable thread in the system that has a priority greater than the currently running thread. Within each priority band, all threads are made runnable when their scheduling time is reached and actually run when their deadline is earlier than the deadline of all other runnable threads in the band.

The G_I class is given a single highest priority band (only critical Chorus server threads such as the pager daemon). B threads are given the next highest band and G_W threads are initially assigned to the lowest priority band. G_I threads are made runnable whenever their logical arrival time is reached (i.e. the start of the period pertaining to the current quantum). As mentioned above, G_W threads are initially assigned to the lowest priority band but they are 'promoted' to the highest band when their logical arrival time is reached. This means that they can enjoy workahead when resources allow, but not at the expense of G_I and B threads. B_W threads are also runnable before their logical arrival time but are not similarly promoted. Finally, B_I threads only become schedulable at a time indicated by the deadline minus the quantum time. This approximates isochronicity to the extent that it removes the possibility of jitter causing threads to complete *before* time although it still leaves the possibility of them completing *after* time. This overall scheme, in conjunction with the admission tests, ensures that G_I threads always meet their jitter constraints, G_W threads always *at least* meet their rate requirement, and B threads optimally share the resources left to them.

Non real-time threads in the B_C class (e.g. those from conventional UNIX applications) are assigned appropriate priorities so that they receive reasonable service according to their role. Their deadline and scheduling time are always set to *now* so that they are effective scheduled solely on the basis of their priority. As an example B_C threads fulfilling an interactive role would have relatively high priority which may be greater than that of B threads. Other B_C threads, such as compute bound applications and non time critical daemons, will have accordingly lower priorities.

---

* Note that the 'priority' in this discussion is different from the priority API level QoS parameter. In this section priority is an internal thread scheduling attribute which is not visible or directly manipulable from the API level.

4.4. The Network Resource

4.4.1. QoS Translation

The network sub-system offers guarantees on *bandwidth, delay bounds* and *packet loss.* To enable it to do this, the QoS translation function maps the API level QoS parameters onto a *flow spec* which is a representation of QoS appropriate to the IP++ and ATM levels:-

```
typedef struct {
    int          flow_id;
    int          mtu_size;
    int          rate;
    int          delay;
    int          loss;
} flow_spec_t;
```

Flow_id uniquely identifies the network level flow. It corresponds to the virtual circuit identifier at the AAL5/ATM level and the flow id in the IP++ packet header at the IP level. *Mtu_size*[*] refers to the maximum transmission unit size and *rate* refers to the rate at which these units are transmitted. These are directly derived from the buffsize and buffrate API level QoS parameters. *Delay* comprises that portion of the API level latency parameter which has been allocated, by the FMP, to the network. It subsumes both propagation and queuing delays in the network. Finally, *loss* is an upper bound probability of *mtu* loss due to buffer overflow at switches and routers. Loss is calculated from the *error* and *error_interval* API level QoS parameters and is equal to 1 - *error / error_interval.*

4.4.2. Admission Testing

In the network, only two traffic classes are recognised: *guaranteed* and *best effort* as denoted by the *commitment* API level QoS parameter. Admission testing and resource allocation are only performed for the former; best effort flows use whatever resource is left over.

For guaranteed flows, three admission tests are performed by the FMP at each switch along the chosen path: a bandwidth test, a delay bound test and a buffer availability test. If, at the current switch, the admission control tests are successful, the necessary resources are allocated. Then the FMP protocol entity in the switch appends details of the cumulative delay incurred so far, and forwards the flow spec to the next switch. Eventually, the remote end-system performs the final tests and determines whether or not the QoS specified in the flow spec can be realised.

If the required QoS is realisable, the FMP entity at the remote end-system returns a confirmation message to the initiating end-system. As it traverses the same route in reverse, the FMP *relaxes* any over-allocated resources at intermediate switches [19].

Bandwidth Test The bandwidth test consists in verifying that enough processing (switching) power is available at each traversed switch to accommodate an additional flow without impairing the guarantees given to other flows. The admission test must satisfy worst case throughput conditions; this happens when all flows send packets back to back at the peak rate. As in section 4.3.2 the admission control test is based on [18]:-

$$\sum_{i=1}^{N} t_i \cdot rate_i \leq R$$

---

* Although the discussion and admission tests in this section apply generically to both the IP++ and ATM layers, the admission tests are described here, for clarity, in an ATM context only. Mtu_size in the case of ATM cells is 53 bytes and in the case of IP++ packets is 64Kbytes. One restriction of the admission tests is that they are only applicable to switches/ routers with a *single* CPU. As we use single CPU ATM switches, this assumption is justified in our implementation environment.

Here, t_i refers to the service time of flow i in the current switch, where there are N flows and $rate_i$ is the rate of the i'th flow. R, $0 \leq R \leq 1$, represents the portion of resource dedicated to guaranteed flows.

Delay Bound Test The delay bound test determines the minimum acceptable delay bound which does not cause scheduler saturation. There are two phases in the delay bound test. First, each switch on the data path computes a local delay bound. Second, it is checked that the sum of all the local delay bounds do not exceed the flow spec's *delay* parameter.

The first phase calculation is taken from [20]:-

$$d = \sum_{i=1}^{N} t_i + T$$

Here, d is the delay incurred at the current switch. As before, t_i refers to the service time of flow i in the current switch. N represents the number of flows in a set U where U contains those flows whose local delay bound is lower than the service times of all flows supported by the current switch. T represents the largest service time of all flows in a set V where V is the complement of set U. A full proof of the theorem underlying this formula can be found in [20]

The second phase calculation is:-

$$\sum_{n=1}^{N_S} d_n \leq delay$$

This merely requires that sum of the delays at each switch is less than the delay parameter in the flow spec. N_S refers to the number of switches on the path and d_n refers to the n'th value of d obtained from the first phase calculations.

Buffer Availability Test The amount of per-switch memory allocated to a new flow must be sufficient to buffer the flow for a period which is greater than the combined queuing delay and service time of its packets. The calculation for buffer space is:-

$$buffersize = mtu\_size \lceil d \; . \; rate \; . \; loss \rceil$$

Here, *buffersize* represents the amount of memory that must be allocated at the current switch for the current flow. The combination of the queuing delay and service time is bounded by d as derived from the first phase delay formula above.

4.4.3. Cell Scheduling

The low level ATM cell scheduler runs in the context of the transmit interrupt service routine which is periodically activated by the ATM card to signal that a cell (or cells) can be copied to the card for transmission. The scheduler chooses to run one of a number of per-connection* *transmit threads* in the network actor by sending a message to a mini-port on which the transmit thread is waiting (see figure 7). The choice of thread to activate is made on the basis of priority, deadline and scheduling time. Each transmit thread is given the same priority band as its associated user level lightweight thread, and the deadline of each thread is derived from the deadline of the next cell in the thread's associated buffer. Cell deadlines themselves are derived by giving each cell in the buffer a specific temporal offset from the deadline of the entire buffer. The scheduling time of each thread becomes *now* whenever the thread has a buffer to send.

---

* Actually, only *one* thread is required for *all* threads in the G_I class because the G_I admission algorithm has ensured that the quanta of these threads do not overlap and can thus be processed sequentially (see section 4.3.2).

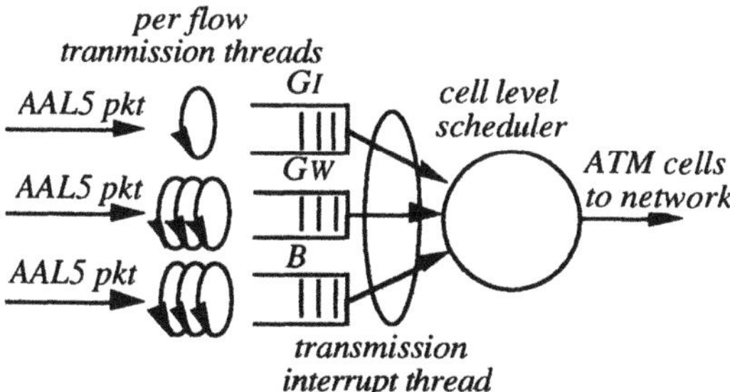

Figure 7: Cell level scheduler

The transmit threads are allocated at connection establishment time and are taken into account in the scheduling admission tests. This is done by adding a time t_{tx} to the *quantum* parameter of the connection's transmit side lightweight thread (see section 4.1). t_{tx} is calculated as *cells* x t_{cell} where *cells* is the number of ATM cells in a buffer of size *buffsize* and t_{cell} is the average time taken to transfer an ATM cell to the interface card.

4.5. The Memory Resource

4.5.1. QoS Translation

We can deduce two memory related quantities from the user supplied QoS parameters at connection establishment time: i) the number of buffers required per connection, and ii) the required access latency associated with those buffers. Buffers are implemented as Chorus memory regions.

Number of buffers To calculate the buffer requirement, the *buffsize*, *buffrate* and *jitter* QoS parameters are used. It is also necessary to take into account the network delay bound, *delay*, offered by the FMP. The network delay bound will typically permit a larger degree of jitter than the API level jitter bound and any discrepancy must be made good through the use of additional jitter smoothing buffers. Given these input parameters, the expression for the number of buffers required at the receiver is:-

$$buffers = buffrate\ (delay + quantum + \frac{jitter}{2})$$

In this formula, the expression in the brackets represents the maximum time for which any single buffer must be held. *Delay* is the delay bound specified in the network level flow spec while *quantum*, *jitter* and *buffrate* are API level QoS parameters. *Jitter* is divided by two because the jitter parameter expresses both lateness and earliness and it is only the lateness component that need be taken into consideration.

Only one buffer is required at the sender due to the structure of the send-side communications architecture: each buffer is assumed to be 'on the wire' before the start of the next period.

Region access latency There are basically two qualities of memory access available in the standard Chorus system. These relate to the access latency of swappable pages and the access latency of locked pages. The latency bound of the former is a function of i) the delay due to the RPC communication between the VM layer and the mapper, and ii) the delay associated with the external swap device*. The latency bound of the latter is much smaller and is a function of the system bus and clock speed.

We assign either swappable or locked regions to connections on the basis of their resource class as follows:-

- G_I: buffer regions allocated to these connections are locked and non-preemptible.
- G_W: buffer regions for these connections are locked but are potentially preemptible by memory requests from G_I connections if memory resources run low.
- B: buffer regions for these connections are assigned from standard swappable virtual memory. These regions may be explicitly locked by the API library code but are subject to pre-emption from by both G_I and G_W connections. The decision as to whether the library code should lock buffers or not is determined by the *priority* API level QoS parameter.

The QoS mapper can deduce the class of each memory request on the basis of the *commitment*, *delivery* and *priority* QoS parameters which are initially passed to the *rgnAllocate()* system call and retained to validate future operations on regions.

4.5.2. Admission testing

In its admission testing role, the QoS mapper maintains tables of all the physical memory resources in the system. In a similar way to the KLS, it also maintains firewalls and high and low water marks between resource quantities dedicated to the different connection classes. The B section is used by all standard and non real-time applications as well as best effort connections.

If no physical memory is available to fulfil a request from a G_I connection, the QoS mapper can *preempt* a locked memory region from an existing B or G_W connection. Similarly, G_W connections can preempt locked regions from B connections. The QoS mapper chooses for preemption the buffer associated with the lowest priority connection in the lowest class available. The effect of preemption is simply to transform locked memory into standard swappable memory. This, of course, may result in a failure of the preempted connection's QoS commitment. However, a software interrupt is delivered to the ULS of a thread whose memory has been preempted so that if QoS commitments are violated, the connection concerned can deduce the likely reason.

5. Conclusions

We have described the design of a QoS driven communications stack in a micro-kernel operating system environment. The discussion has focused on resource management aspects of the design and in particular we have dealt with CPU scheduling, network resource management and memory management issues. The architecture minimises kernel level context switches and exploits early demultiplexing so that incoming data, even at the cell level, can always be treated according to the QoS of its associated API level connection. It also eliminates data copying on both send and receive (except for unavoidable copies to/from the ATM interface card). On send, the user's buffer is mapped to the lower layers which process it *in situ*, and, on receive, the lower layers allocate a buffer and map it to the transport layer which subsequently passes it to the application by passing the address of the buffer as an argument to an rthandler.

At the present time we are experimenting with an infrastructure consisting of three 486 PC's

---

* We intend in the future to look at the possibility of bounding the access latency to swappable pages (e.g. through specialised page replacement policies and disc layout strategies), but our present design simply considers the access latency of swappable pages to be unbounded.

running Chorus and connected to an Olivetti Research Labs ATM switch via ISA bus ORL ATM interface cards. The PCs also contain VideoLogic audio/ video/ JPEG compression boards as real-time media sources/ sinks. The current state of the implementation is that the API, split level scheduling infrastructure, transport protocol and ATM card drivers are in place. In the next implementation phase we will refine the QoS driven memory management scheme and add heterogeneous networking with IP++ support.

We would also like to experiment with an ATM interface card with on-board AAL5 support. This would limit the flexibility of our current design and would not allow us to experiment with ATM cell-level scheduling, but we could better evaluate the performance potential of the system if SAR functions did not have to be carried out in software. Apart from the severe performance hit, an architectural limitation of our current cell-level card it that it obstructs the ideal strategy of a single, non-multiplexed, per-connection thread operating all the way up/ down the stack. This is because SAR must be carried out asynchronously with higher level protocol processing and thus more than one thread is required. A related drawback is that the receive side AAL5 kernel thread in the network actor is impossible to schedule correctly due to the need to copy cells off the card as soon as possible. With a card featuring on-board AAL5 and DMA for data movement these drawbacks would be eliminated.

There remain a number of important issues which we have yet to tackle. One point that remains to be addressed is the need to synchronise real-time data delivery on separate application related connections (e.g. for lip sync over audio and video connections). Along with our collaborators at CNET, Paris, we are currently investigating the use of real-time controllers written in the Esterel real-time language for this purpose [21]. Another issue, which is being addressed in a related project at Lancaster, is the requirement for QoS controlled multicast connections. We already know how we can support multicast at the API level, but our ideas on engineering multicast support in the micro-kernel environment are still immature. A final issue is the incompleteness of the dynamic QoS management design. In particular, we would like to extend our design to include access latency bounds on swappable memory regions and also to accommodate comprehensive QoS monitoring and automated reconfiguration of resources in the event of QoS degradations.

Acknowledgement

The research reported in this paper was funded partly by CNET, France Telecom as part of the SUMO project, and partly under UK Science and Educational Research Council grant number GR/J16541. We would also like to thank our colleagues at CNET, particularly Jean-Bernard Stefani, Francois Horn and Laurent Hazard, for their close co-operation in this work.

The support of the Swiss FNRS for Michael Papathomas through grant no. 8220-037225 is gratefully acknowledged.

References

1. Bricker, A., Gien, M., Guillemont, M., Lipkis, J., Orr, D., and M. Rozier, "Architectural Issues in Microkernel-based Operating Systems: the CHORUS Experience", *Computer Communications*, Vol 14, No 6, pp 347-357, July 1991.

2. Coulson, G., and G.S. Blair. "Micro.-kernel Support for Continuous Media in Distributed Systems", To appear in Computer Networks and ISDN Systems, Special Issue on Multimedia, 1994; also available as Internal Report MPG-93-04, Computing Department, Lancaster University, Bailrigg, Lancaster, U.K. . February 1993.

3. Coulson, G., Blair, G.S., Robin, P. and Shepherd, D., "Extending the Chorus Micro-kernel to Support Continuous Media Applications", *Proc. Fourth International Workshop on Network and Operating System Support for Digital Audio and Video*, Lancaster University,

Lancaster LA1 4YR, UK, October 93.

4. Campbell, A., Coulson, G. and Hutchison, D., "A Multimedia Enhanced Transport Service in a Quality of Service Architecture", *Proc. Fourth International Workshop on Network and Operating System Support for Digital Audio and Video*, Lancaster University, Lancaster LA1 4YR, UK, October 93.

5. Campbell, A., Coulson, G. and Hutchison, D., "A Quality of Service Architecture", ACM Computer Communications Review, April 1994.

6. Accetta, M., Baron, R., Golub, D., Rashid, R., Tevanian, A., and M. Young, "Mach: A New Kernel Foundation for UNIX Development", *Technical Report* Department of Computer Science, Carnegie Mellon University, August 1986.

7. Tanenbaum, A.S., van Renesse, R., van Staveren, H. and S.J. Mullender, "A Retrospective and Evaluation of the Amoeba Distributed Operating System", *Technical Report*, Vrije Universiteit, CWI, Amsterdam, 1988.

8. Coulson, G., G.S. Blair, P. Robin, and D. Shepherd, "Supporting Continuous Media Applications in a Micro-Kernel Environment." in Architecture and Protocols for High-Speed Networks. Editor: Otto Spaniol. Kluwer Academic Publishers, 1994.

9. Govindan, R., and D.P. Anderson, "Scheduling and IPC Mechanisms for Continuous Media", Thirteenth ACM Symposium on Operating Systems Principles, Asilomar Conf. Center, Pacific Grove, California, USA, SIGOPS, Vol 25, pp 68-80, 1991.

10. Marsh, B.D., Scott, M.L., LeBlanc, T.J. and Markatos, E.P., "First class user-level threads", Proc. Symposium on Operating Systems Principles (SOSP), Asilomar Conference Center, ACM, pp 110-121, October 1991.

11. Zhang, L., Deering, S., Estrin, D., Shenker, S and D. Zappala, "RSVP: A New Resource ReSerVation Protocol", *IEEE Network*, September 1993.

12. ATM User Network Interface Specification Version 2.4, August 5th, 1993.

13. Campbell, A., Coulson G., García F., and D. Hutchison, "A Continuous Media Transport and Orchestration Service", *Proc. ACM SIGCOMM '92,* Baltimore, Maryland, USA, August 1992.

14. Deering, S., "Simple Internet Protocol Plus (SIPP) Specification", Internet Draft, <draft-ietf-sipp-spec-00.txt>, February 1994.

15. Scott, A.C., Shepherd W.D. and A. Lunn, "The LANC - Bringing Local ATM to the Workstation", *4th IEE Telecommunications Conference*, Manchester, UK, 1993, also available as Internal Report ref. MPG-92-33, Computing Department, Lancaster University, Lancaster LA1 4YR, UK, August 1992.

16. Tennenhouse, D.L., "Layered Multiplexing Considered Harmful", *Protocols for High-Speed Networks*, Elsevier Science Publishers B.V. (North-Holland), 1990.

17. Abrossimov, V., Rozier M. and Shapiro M., "Generic Virtual Memory Management for Operating System Kernels", SOSP'89, Litchfield Park, Arizona, December 1989.

18. Liu, C.L. and Layland, J.W., "Scheduling Algorithms for Multiprogramming in a Hard Real-time Environment", Journal of the Association for Computing Machinery, Vol. 20, No. 1, pp 46-61, February 1973.

19. Anderson, D.P., Herrtwich, R.G. and C. Schaefer. "SRP: A Resource Reservation Protocol for Guaranteed Performance Communication in the Internet", *Internal Report*, University of California at Berkeley, 1991.

20. Ferrari, D. and D. Verma, "A Scheme for Real-Time Channel Establishment in Wide Area Networks", *IEEE J. Selected Areas in Comm.*, Vol 8 No 3, April 1990.

21 Hazard, L., Horn, F., and J.B. Stefani, "Notes on Architectural Support for Distributed Multimedia Applications", *CNET/RC.W01.LHFH.001*, Centre National d'Etudes des Telecommunications, Paris, France, March 91.

4

Statistical Sharing and Traffic Shaping: Any Contradiction?

Yee-Hsiang Chang

Hewlett-Packard Laboratories
1501 Page Mill Road, Palo Alto, CA 94304-1126, USA

Statistical sharing to achieve a high network utilization is a major motivation behind the packet-switched network. This idea takes advantage of the bursty nature of traffic sources to achieve a better network utilization. In recent years, various control mechanisms have been proposed for future high-speed networks to support the network traffic management. Some of the control schemes advocate traffic shaping to reduce the burstiness of the traffic, and others insist on maintaining the bursty nature for better statistical sharing. This paper looks into the true meaning of statistical sharing, and tries to shed light on the design of better control mechanisms by using a simple queueing model to show the relationship between the congestion and the network utilization.

1. MEANING OF STATISTICAL SHARING

The theoretical foundation for statistical sharing is based on the law of large numbers [1-2], which is described by Kleinrock as, *"the collective demand of a large population of random users is very well approximated by the sum of the average demands required by that population"*. That is, the stable state in the network utilization is achieved when the number of users is large, in which case each individual traffic balance its burstiness traffic with others. This stable state is the key to provide good sharing, which is known as *statistical sharing*. On the other hand, if the population is small and each traffic is bursty, an unstable condition is produced and results in bad sharing. One example is in Figure 1, the multiplexing of traffic A and B reduces the overall variance. If there are more traffic multiplexing together, the variance goes down further. However, when two bursts from traffic A and B arrive at the same time to a point that the network can not sustain, delays and losses are likely to happen. This produces a network congestion, in which case the network is overloaded and loses its ability to provide user-requested services.

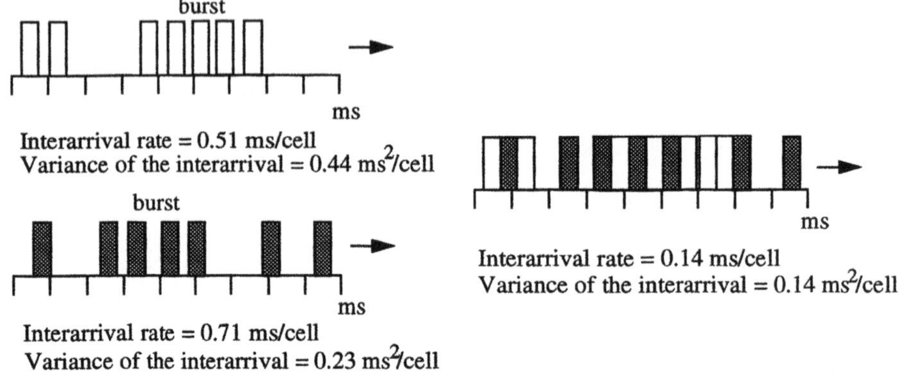

Figure 1. Multiplexing of Traffic Reduces the Overall Variance.

The goal of the network design is to have good sharing (high network utilization) and also maintain guarantees to services. Good sharing can be achieved also by a deterministic way. The deterministic traffic has the potential to obtain even better sharing than the bursty one, since the deterministic traffic is stable in nature. For example, good sharing can happen in the traditional telephone networks with constant-bit-rate circuits (Figure 2)†. When all the channels are fully occupied and multiplexed together, the network utilization is 100%.

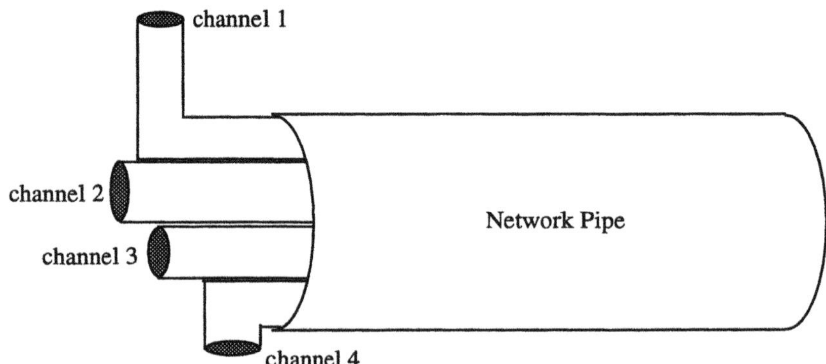

Figure 2. Good Sharing Under the Constant-Bit-Rate Network as Each Channel Fully Utilizes its Capacity.

So, why the trend today moves toward statistical sharing instead of maintaining the traditional circuits? This is because many real-life applications tend to be bursty, which waste bandwidth with the traditional circuit-switched networks. Can we do traffic shaping within the

† Note that Figure 2 uses a conceptual way to show the sharing among different channels; the actual multiplexing (e.g., time division multiplexing) is different.

tolerance of each application to achieve better sharing? The answer is yes because the traffic is more stable especially when fewer bursty users share the same link. We argue that the final solution for the sharing is a combination of both statistical (due to application's nature) and deterministic ways (due to traffic shaping).

In general, statistical multiplexing reduces the long-term average randomness, but increases the potential to have a severe short-term randomness. As the example in Figure 1, although multiplexing reduces the overall variance, the bursts from both streams generate a bigger burst, which potentially causes buffer overruns and generates delays in the down-stream nodes. Traffic shaping helps to smooth out the short-term randomness.

One misconception among some literatures states that maintain the burstiness of the traffic is the key for statistical sharing.[†] This is obviously not true. The key for statistical sharing to work is to reach the stable state with large population. Maintain the burstiness of the traffic does not generate a stable condition. On the other hand, more deterministic the traffic is stabilizes the overall traffic even with a small population, and achieve better sharing.

2. TRAFFIC CHARACTERISTICS AND TO WHAT DEGREE WE CAN CHANGE IT

There is a limit for traffic shaping due to applications' traffic characteristics. What is this limit? To answer, we need to classify communication requirements for different applications. There are three different types of real-time communications. The first one is called *hard real-time or hard guaranteed.* For this type of the application, a maximum time limit is set and all the communications are required to finish within this limit. One example is the communications for real-time control signals on embedded systems (such as the space shuttle). The second one is called *soft real-time or statistical guaranteed.* For this type of the application, the time limit is the same as the first case but can be achieved statistically (not 100% guarantee). One example of this type is the communication for meta computing. Meta computing requires fast communications to work on a wider area. However, the statistical fluctuation of the message transmission is not fatal to the application. The third type is *play-back applications* defined by Clark ed. al. [5]. This type has even less real-time requirements from the network than the two previous cases - it relies on the end systems to adjust. The best examples of play-back applications are voice and video communications. At the transmission source, the voice or video signal is first packetized, and then transmitted over the network. The receiver buffers the incoming messages to remove jitter, and play the voice or video back at the designated play-back points. The receiver can adjust the play-back point within a range according to the network condition. The play-back applications will be the vast majority in the future [5]. In this paper, we mainly look at the traffic shaping issue on this type of the real-time communications.

The play-back point is adjusted between two limits. At one end is the time that the application has a zero performance gain if the message arrives earlier than this point. For example, the audio communication requires the round-trip delay within 400 ms [6]. If the network provides a lesser time, there is no performance gain to users. This time limit is marked as T_l in Figure 3. At the other end is the time that the application is intolerable to the delay if the message arrives later than this limit (see Figure 3). For example, if an audio delay is more than 5 seconds†, the interactive communication is completely unacceptable. We use T_h in Figure 3 to represent this time.

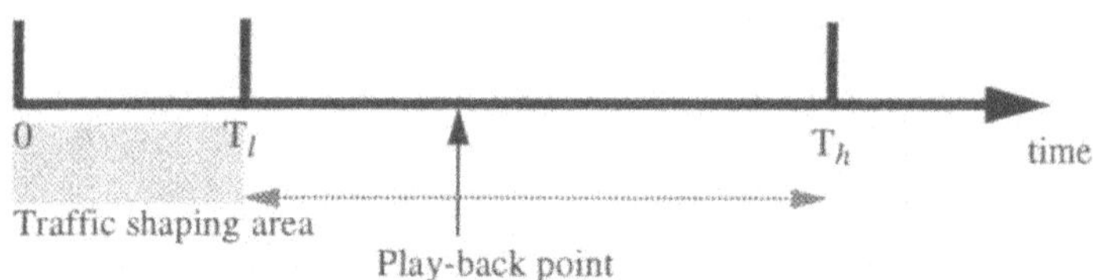

T_l The time limit that the application does not achieve any better performance if the time used is less than this number.
T_h The time limit that the performance does not be tolerable if the time used is more than this number.

Figure 3. Timing Requirements and the Adjustable Area for a Play-Back Application.

Between T_l and T_h is the range that the play-back point can adjust. If the network is temporarily congested, the play-back point moves toward T_h to recover late messages. When the network is less congested, the play-back point moves back to T_l to achieve a faster response. Traffic shaping should introduce no further delay than its T_l for a message. In general, traffic shaping adjusts the delay for each message (within its delay budget) in order to achieve less delay and better sharing for the overall traffic.

3. THE STYLES OF TRAFFIC SHAPING AND THE PERFORMANCE MODEL

Traffic shaping can be exercised in two places for a packet network. One place is to reduce or increase the peak packet size, and the other one is to un-smooth or smooth the packet/cell inter-arrival time (Figure 4). For ATM with fixed-size cells, traffic shaping is for the latter case (Figure 4b).

---

† More specifically, the paper from [3] makes the comments about Leaky-Bucket from [4] as "A simple model, described in [4], works in the following way: each switch at the network entrance puts packets from each data flow into a corresponding bucket which has a fixed size. The bucket opens periodically to emit packet for transmission. When the bucket is full, incoming packets are discarded. ... The first version of Leaky-Bucket reduces statistical multiplexing because packets are transmitted at a constant rate rather than whenever the channel is available. ... The VirtualClock algorithm avoids those drawbacks by merely ordering packet service without reducing statistical sharing." The point is that a constant bit rate (CBR) channel, which is the ultimate shaped traffic, does not hurt sharing or statistical sharing with other channels, but help sharing instead. The point from [3] about the message should be sent "whenever the channel is available" is tough to achieve because the local resource availability does not guarantee the remote network resource availability. Also note that Leaky-Bucket from [4] does not generate a CBR channel, but a shaped traffic.

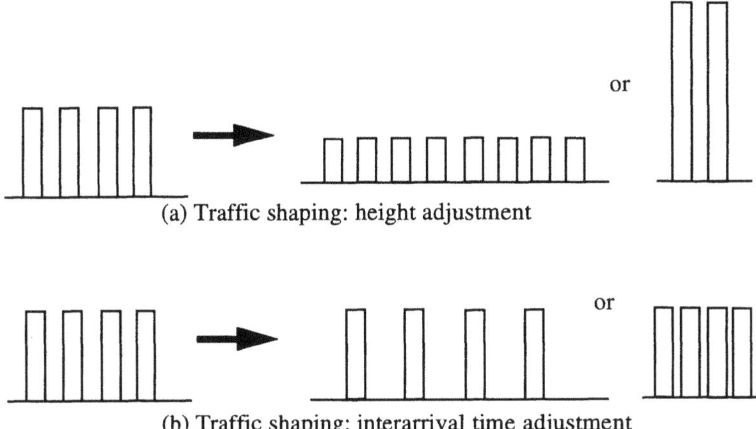

(a) Traffic shaping: height adjustment

(b) Traffic shaping: interarrival time adjustment

Figure 4. Different Ways of Traffic Shaping

If we look at the interarrival time of packets, it is a renewal process, which is best modeled by a general arrival. Following, a simple queueing model is employed to prove the significance of traffic shaping, and demonstrate a way to maintain a high network utilization while eliminating the congestion at the same time.

3.1 The performance model (G/D/1)

Here, a simple model is used to show the performance of a network with traffic shaping. Figure 5 presents a network configuration with several switches and three connections across the network. These connections go through one of the links together, where is the potential bottleneck. If the traffic control and management of the network allow more connections across this very link at the same time, the network achieves a higher utilization. This link is modeled by a simple queue (Figure 6). Its characteristics represents the whole network under congestion. The link can actually be any link in the network under a heavy load.

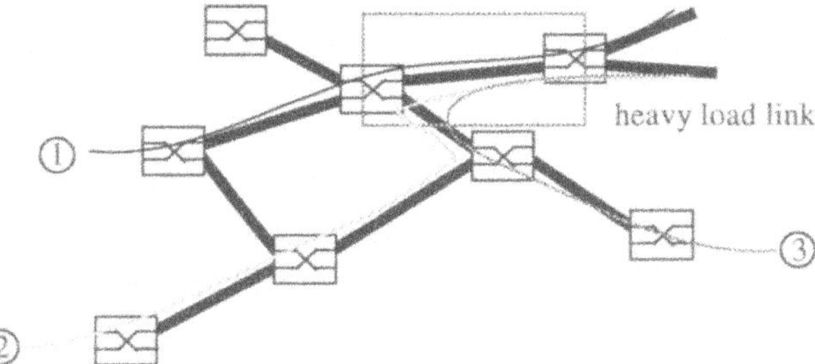

Figure 5. Network Model that Uses a Heavily Load Link to Represent the Whole Network.

† There is no empirical data for this. 5 seconds is just an example here.

The network is modeled with a deterministic service rate and a general arrival rate (G/D/1) (Figure 6). The deterministic service rate is assumed because the future switches will be fast and take a constant time to process each fixed-size cell. The arrival process consists of the traffic from various incoming sources and can be occasionally faster than the output link speed, which results in a queueing effect. We also assume no peak rate allocation. If the network employs the peak rate allocation, the total aggregated speed is lower than the link speed at any instance, and no queue is formed. The simple queueing model is used to derive a generalized close-form solution that can demonstrate the basic relationship of various parameters. Note that this model is a steady-state solution. More complicated transient analysis is left for further study.

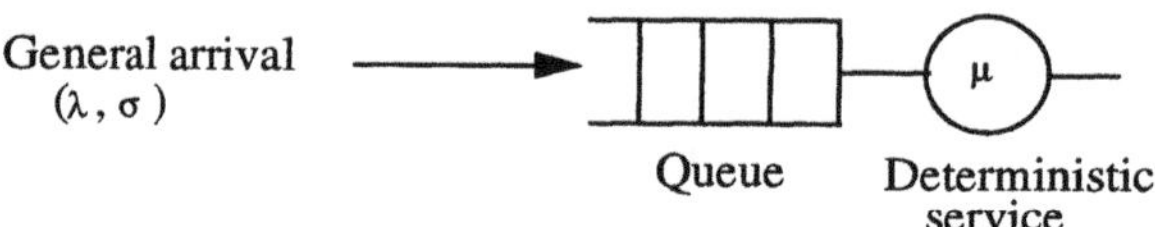

λ Average arrival rate.
σ Standard deviation of the arrival process.
μ Average service rate.

Figure 6. Network Model with General Arrival and Deterministic Service (G/D/1).

Other assumptions are used. First, the effect of admission control is neglected. We assume a perfect admission control to calculate the performance upperbound. Second, the policing function is assumed enforced at the boundary of the network, which no unexpected data can flow in. Third, the scheduling discipline is FIFO and no priority. FIFO is employed because it is the simplest scheduling discipline to implement and exists many ATM switches today. Also, FIFO provides the best sharing among the scheduling disciplines [5]. No priority is used because the goal is to see the performance under only one class. In other words, we are looking at the performance of traffic at the highest priority.

The network utilization is obtained by observing the number of packets or cells in the queue. If there is always one packet or cell being serviced by the network all the time, the network utilization is 100%, which means a constant flow of packets or cells moving at the link speed. So, the

Network utilization = $1 - P_0$.

P_0 is the probability of zero packet or cell in the network service point. P_0 can be obtained by the following derivation because of the steady state condition in the queueing network.

E(number of arrivals in time T) = E(number of departure in time T)

$E(x)$ is the expected value of random variable x.

T is a long period of time.

Since T is a long time, the number of the arrival packets/cells (λT) during this period should be equal to the number of departure packets/cells ($\mu T(1-P_0)$ to maintain the steady state. λ and μ represent the average arrival and service rate for the queueing system.

$$\lambda T = \mu T(1 - P_0)$$

$$P_0 = 1 - \frac{\lambda}{\mu} = 1 - \rho$$

$$\text{Network utilization} = 1 - P_0 = \rho \quad (1)$$

Note that the network utilization is determined only by the average arrival and service rate. By solving the G/D/1 queue, we get the average queue length ($E(Q)$) [7-8], which is

$$E(Q) = \sigma^2\lambda^2/2(1-\rho) \quad (2)$$

The waiting time (W) in queue is

$$W = E(Q)/\lambda = \sigma^2\lambda/2(1-\rho) \quad (3)$$

The queue length is an indication of the network congestion, which normally means the server lags behind for packet or cell processing and results in a queueing delay. The waiting time in the queue contributes to part of the total delay (this time plus the propagation delay and the processing delay constitute the overall delay for a message.) The propagation delay is fixed once the distance is determined. The part that can be controlled is the queueing delay. Today, this number is not trivial. The propagation delay across the US continent (3000 miles) via Internet is about 22 ms. The measured minimum one-way delay is around 50 ms (using *ping* program), and the average delay is around 80 ms.† The queueing delay represents the difference between the average delay and the minimum delay, which constitutes 60% over the minimum delay. Furthermore, 27% of the packets are lost due to the congestion.

We can see in (2) and (3) there are three parameters that affect the network queue length and waiting time. They are σ, λ, and μ. σ is the standard deviation of the arrival process. The value of σ indicates the burstiness of the arrival process. If the value is big, the inter-arrival time varies very much. If the value is small, most of the interarrival time is similar. It is clear that reducing the burstiness of the traffic reduces the queue length and avoid congestion. Reducing the λ and increasing the μ also reduce the queue length. However, the latter case also reduces the network utilization.

So, to keep a good network utilization, the design should rather reduce the variance (or bursty) of the traffic than reducing the arrival rate or increasing the service rate in a long-term sense. This suggests that traffic shaping is desirable.

† This number is measured between California and North Carolina at 1pm Pacific Time Zone June 13, 1994.

4. THE TRAFFIC MANAGEMENT DESIGN AND THE PERFORMANCE UPPERBOUND

With the basic model in mind, let us see the fundamental principals for the network control mechanism. Here, we only consider the sharing among all real-time traffic (no priority used). One way to achieve a good network utilization is to multiplex non-real-time traffic with the real-time traffic to fill the unused network capacity. This is a good idea if there always exists enough non-real-time traffic. By only dealing with the real-time traffic to achieve a high network utilization, the design also does well with or without multiplexing non-real-time traffic.

The design goal of the control mechanism: *To achieve a high network utilization and congestion-free at the same time.*

The congestion is directly related to the queue length. The control mechanism should be designed to reach the balance point between the highest possible network utilization and low network congestion. This balance point is not fixed and has very much to do with the traffic pattern of the sources. There is a dynamics among the following parameters. As mentioned, we assume to have the FIFO as the scheduling mechanism and a perfect admission control to simplify the model.

- network utilization,
- scheduling,
- congestion (queueing delay, queue length),
- traffic pattern (λ, μ, and σ),
- admission control, and
- control mechanisms (control burstiness, or rate).

Since the parameters λ and μ contradict the requirements of a shorter queue length and a higher network utilization, it does not make sense to control them at the first place. We argue that the control mechanism should be in two levels. At the first level, the network should exercise traffic shaping all the time to reduce the possible burstiness. This is a traffic management function, which can result in a shorter queue length/delay time and a higher network utilization in the long term. If the first level traffic management can not eliminate the congestion condition and the congestion becomes a more serious situation, the second level control mechanism should be employed. At the second level, by sacrificing the network utilization, the network or the end systems should reduce λ or increase μ to reduce the queue length. This is generally a flow control function [8-9].

The G/D/1 example is a simple model to show the relationship among several major parameters. A more complicated model is desirable to demonstrate more detail information about how to reach this balance point. Following, a performance upperbound analysis is done using the simple G/D/1 model. A simple shaping scheme is introduced for the purpose of demonstrating the shaping effect and finding the performance upperbound, and not for the purpose of proposing such a scheme. We also employ the JPEG data as a generic video traffic for the same purpose.

4.1 Upperbound analysis

If the nature of a real-time class limits the maximum waiting time (for example, the voice can not tolerate too much delay, T_h), and the first level control mechanism (traffic shaping) can only function up to T_l, we can obtain a performance upperbound for the network utilization, which represents the best value that the network utilization can reach.

For example, using the JPEG compressed Star War movie† as a generic video traffic pattern for input. The movie stream has the following statistical values:

Average rate = 5.34 Mbps (13895.46 cell/sec)
Peak rate = 15.06 Mbps

The internal speed is assumed to be 200 Mbps.†† The internal speed in the sender is very likely to be faster than the rate passing through the network, because the former rate depends on the cpu, memory, and the internal bus speed, which is normally a order of the magnitude faster than the network speed. The internal switch speed is also several times faster than the switch port rate to avoid blocking within the switch.

As JPEG frames generated in the source, they are packetized (or segmented) into cells. The interarrival time among cells is based on the internal speed in the sender (Figure 7). In the case of a 200 Mbps internal speed, the variance of the JPEG stream is

Variance = 2.890781 ms^2/cell

This value is also used to represent the switch internal speed in this paper.

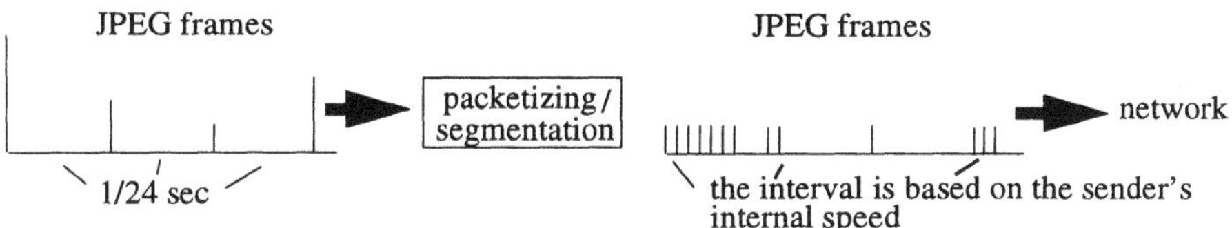

Figure 7. The Internal Rate is Faster than the Delivery Rate

With above numbers, the network utilization upperbound is calculated in two conditions: with and without shaping. For the case without shaping, the performance upperbound for the utilization against the delay (using the equation (3)) is shown in Figure 8. If an application has a 200ms delay budget (e.g., voice communications), and the propagation delay is assumed insignificant in this case, the maximum network utilization is around 0.86 after subtracting the reassembly delay (one interval - 41.67ms†††) at the receiver. This one interval reassembly delay is unavoidable. However, depending on the variation of the traffic (σ), this reassembly delay might need to increase to recover the late cells or just drop the frame.

---

† The JPEG compressed starwar movie data is from Bellcore.

†† The internal speed is the data rate inside the source and the data rate inside the switch.

††† In our case with 200 Mbps internal speed, the cells from the same frame remain inside the 1/24 sec range after packetizing.

For the case with shaping, there are two solid curves in Figure 8 that present the result. In both cases, the traffic shaper has the knowledge that a frame arrives in every 1/24 sec (41.67ms), and uses 41.67ms or 83.34ms as a window to shape the traffic. For the example of using 41.67ms as the shaping window, the traffic shaper buffers the data every 41.67ms, then sends the cells out in a constant rate in the next frame interval (Figure 9). Note that the numerical result shows that the traffic shaper does not need to align on the frame boundary (Table 1) if the traffic is random enough†, which makes the shaping function easier to implement.

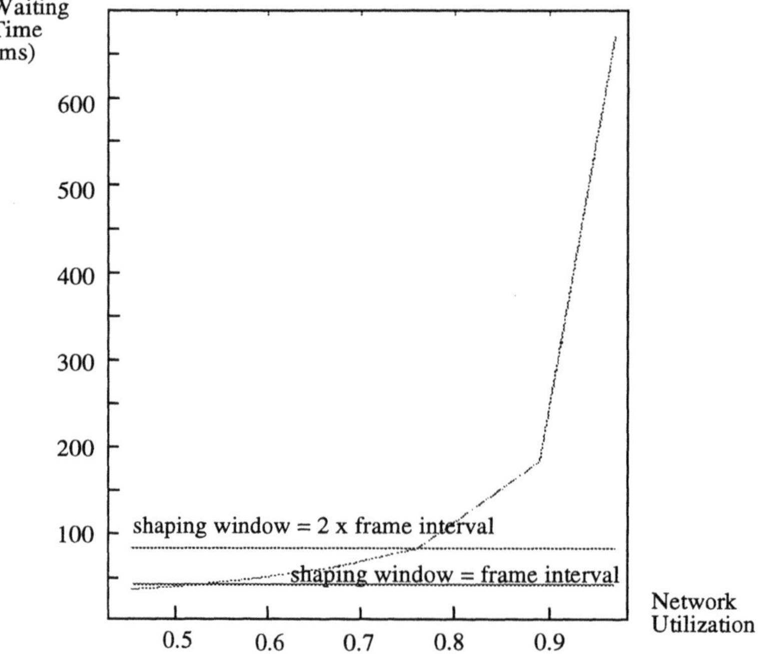

Figure 8. Queueing Delay vs. Network Utilization with and without Shaping.

† If the shaping window size is fixed, the average and the variance of the interarrival time are the following.:

$$average = \sum P_i x_i = \frac{x_1}{w} \cdot \frac{x_1}{x} + \frac{x_2}{w} \cdot \frac{x_2}{x} + \ldots = \frac{x_1^2 + x_2^2 + x_3^2 + \ldots}{wx}$$

$$variance = \frac{x_1^3 + x_2^3 + x_3^3 + \ldots}{w^2 x}$$

where
w: the frame interval
x: the sum of x_1, x_2, ...
$x_1, x_2, \ldots$: the cell number within the shaping window

Shifting the window changes the distribution of cells in each interval. However, as long as the distribution is random enough, and the number of the interval is large, the overall values are similar.

The disadvantage of traffic shaping is that a constant delay is generated, 41.67ms or 83.34ms in the examples; however, due to the reduction of the variance, the waiting time reaches a fixed value even under the heavy load.

Table 1
Window Shifting Effects

	Align on frame boundary	Shift 1/4 window size	Shift 1/2 window size	Shift 3/4 window size
Average of cell interarrival (ms)	0.073295	0.073294	0.073294	0.073294
Variance of cell interarrival (ms^2)	0.000288	0.000283	0.000283	0.000283

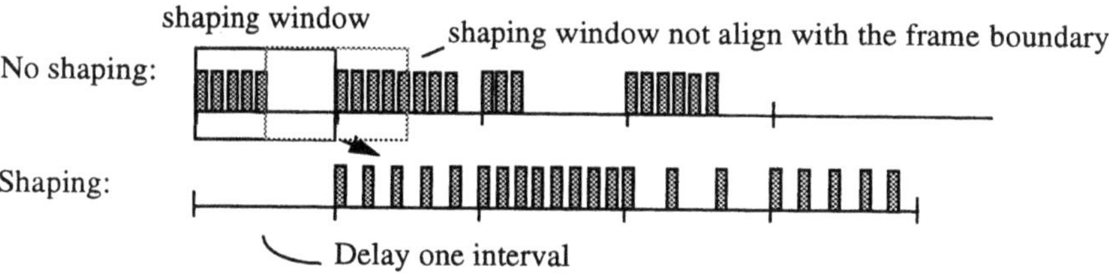

Figure 9. The Traffic Shaping Scheme and the Associated Delay.

The shaping window can be reduced or increased. The biggest case covers all the data and generates a constant-bit-rate stream (CBR). The smallest case results in no shaping at all. When the window is small, the buffering delay is small, but the variance is big. On the other hand, when the window is big, the buffering delay is big, but the variance is smaller.

5. WHERE ARE THE POSSIBLE PLACES TO EXERCISE TRAFFIC SHAPING

The ability to do traffic shaping lies on the knowledge of the application traffic pattern and required guarantees. This fits well with the basic concept of the ATM traffic management that requires a negotiation of the traffic contract. In the current standard, a signaling message carries this information across networks for admission. Having this knowledge throughout networks, the traffic shaping function can be exercised in various places.

The most convenient place for traffic shaping is to associate this function with the network policing mechanism. The policing mechanism makes sure the application sends data according to promised. For each packet, a delay budget can be specified by the application to the policing mechanism to inform the shaping range.

The traffic shaping function can also be put into the intermediate nodes [10-11]. This idea has better chance to perform more effective traffic shaping due to using all the nodes. The most complicated case is the per-VC traffic shaping in every node. However, the hardware complexity and processing overhead must be taken into consideration.

6. CONCLUSION

In this paper, the meaning of sharing is addressed. We conclude that by using traffic shaping to introduce more deterministic traffic behavior is desirable to stabilize the network condition and achieve good sharing. We also propose that the future network should use a combination of both statistical (due to application's nature) and deterministic sharing (due to traffic shaping). Then, we look at several types of real-time applications, especially the play-back one, to see how much shaping is allowed to do. An analytical model is used to show the relationship between the network congestion and the network utilization. From the model, it is shown that more deterministic the traffic is, less congestion the network is, and the high network utilization is achieved. From this analytical model, we then do the performance upperbound analysis, discuss the design of control mechanisms by proposing a two-level control scheme, and where can we exercise these control mechanisms in high-speed networks.

REFERENCES

1. Kleinrock, L., Queueing Systems Vol II: Computer Application, John Wiley & Sons, New York, 1976.
2. Wolff, R.W., Stochastic Modeling and the Theory of Queues, Prentice-Hall, 1989.
3. Zhang, L., "VirtualClock: A New Traffic Control Algorithm for Packet-Switched Networks," ACM TOCS, 1990.
4. Turner, J., "New Directions in Communications (or Which Way to the Information Age?)," IEEE Communications Magazine, October 1986.
5. Clark, D.D., Shenker, S., and Zhang, L., "Supporting Real-Time Applications in an Integrated Services Packet Network: Architecture and Mechanism," SIGCOMM 1992.
6. Emling, J.W. and Mitchell, D., "The Effects of Time Delay and Echos on Telephone Conversations," Bell System Technical Journal, November 1963.
7. Kleinrock, L., Queueing Systems Vol I, John Wiley & Sons, New York, 1975.
8. Mukherjee, A., Landweber, L.H., and Faber, T., "Dynamic Time Windows and Generalized Virtual Clock: Combined Closed-Loop/Open-Loop Congestion Control," INFOCOM '92, 1992.
9. Makrucki, B.A., "On the Performance of Submitting Excess Traffic on ATM Networks," IEEE GLOBCOM '91, December 1991.
10. Boyer, P.E., Guillemin, F.M., Servel, M.J., and Coudreuse, J.-P., "Spacing Cells Protects and Enhanced Utilization of ATM Network Links," IEEE Network, September 1992.
11. Verma, D., Zhang, H., and Ferrari, D., "Delay Jitter Control for Real-Time Communication in a Packet Switching Network," in Proceedings of TriComm '91, 1991.

PART THREE

Architecture

5

A high performance Streams-based architecture for communication subsystems

Vincent Roca [a, b] and Christophe Diot [c]

[a] BULL S.A.
1, rue de Provence; BP 208; 38432 Echirolles cedex; France
e.mail: V.Roca@frec.bull.fr; fax: (33) 76.39.76.00

[b] LGI - IMAG
46, avenue Felix Viallet; 38031 Grenoble cedex; France

[c] INRIA
2004 route des Lucioles; BP 93; 06902 Sophia Antipolis; France
e.mail: christophe.diot@sophia.inria.fr; fax: (33) 93.65.77.65

Abstract
During the last few years several ideas have emerged in the field of communication stack architectures. Most of them question the use of the OSI layered model as a guideline to protocol implementation. In this paper we apply some of these ideas to a Streams-based TCP/IP stack. We define a lightweight architecture that aims at reaching high performances while taking advantage of Streams benefits. In particular, we introduce the notion of communication channels that are direct data paths that link applications to network drivers. We show how communication channels simplify the main data paths and how they improve both Streams flow control and parallelization. We also propose to move TSDU segmentation from TCP to the XTI library. It brings further simplifications and enables the combination of costly data manipulations. Early performance results and comparisons with a BSD TCP/IP stack are presented and analyzed.

Keyword Codes: C.2.2
Keywords: Network protocols

1 INTRODUCTION

As networks proceed to higher speed, there is some concern that the Network to Presentation layers will present bottlenecks. Several directions of research are followed in order to make up the performance gap between the lower hardware layers and the upper software layers [Feldmeier 93b]. Some of them question the standard OSI model for protocol implementation.

This will be discussed in section 3.1.

Such considerations as portability, modularity, homogeneity and flexibility can lead to choose Streams as the environment of the communication subsystem. This is the choice that has been done for Unix System V. Yet this environment also adds important overheads. Our goal in this paper is to define design principles likely to boost the performances of Streams-based communication stacks while preserving as much as possible Streams assets. These principles will have to be beneficial both to standard mono-processor and to symmetric multi-processor (SMP) Unix systems. In order to support our proposals, we describe the changes we have introduced in a Streams-based TCP/IP stack. This choice of TCP/IP is not restrictive and the architecture presented here can be used for other protocol stacks.

This paper is organized as follows: because the Streams environment is central to our work we present its main concepts in the second section. In the third section we review experiments and ideas found in the literature that question layered implementations. We also analyze in depth a standard Streams-based stack and show its limitations. In the fourth section we present an architecture to solve these problems and discuss performance measurements. Then we conclude.

2 THE STREAMS ENVIRONMENT

2.1 The basics of Streams

We present here the main features of Streams. More details can be found in [Streams 90]. The notion of stream, or bidirectional (read and write, also called input and output) channel is central to Streams. A stream enables the user of a service to dialog with the driver that supplies that service. Communication is done through messages that are sent in either direction on that stream. In addition to drivers, Streams defines modules that are optional processing units. Modules can be inserted (pushed) then removed (popped) at any time on each stream. They are used to perform additional processing on the messages carried on that stream, without having to modify the underlying driver. Within a driver or module, messages can be either processed immediately by the interface functions that receive them (the put() routines), or queued in the queues associated with the stream (read and write queues) and processed asynchronously by the service() routines.

Figure 1 represents a basic Streams configuration where two applications dialog with a driver through two different streams. On one of them, an additional module has been pushed. In this example, the Streams framework is entirely embedded in the Kernel space of the Unix system. If this is usually the case, this is not mandatory: this environment can be ported to the User space of Unix, or to various different operating systems and intelligent communication boards. The Stream-Head component of Figure 1 is responsible of the Unix/Streams interface and in particular of the User space/Kernel space data copy.

Streams offers the possibility to link several drivers on top of each other. Such drivers are called multiplexed drivers because they multiplex several uppers streams onto several lower streams (see Figure 2). Streams can be used for many I/O systems, but in the case of communication stacks, a standardized message format is associated to each interface between adjacent layers of the OSI model. These normalized interfaces are called TPI (Transport Protocol Interface), NPI (Network Protocol Interface), and DLPI (Data Link Protocol Interface).

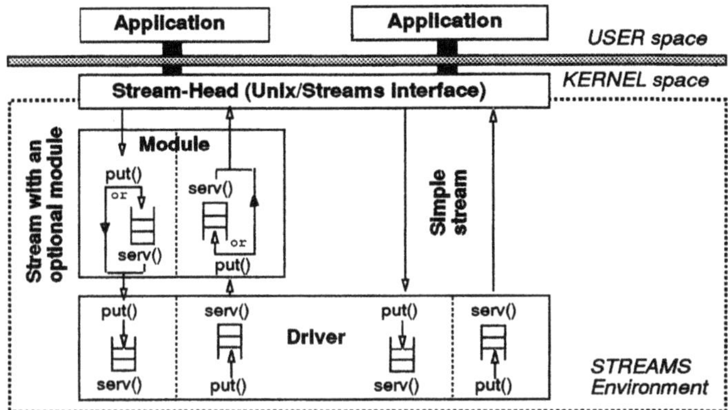

Figure 1: A basic Streams configuration.

Other implementation environments exist: the BSD style environment associated to the Socket access method is the most famous one but, unlike Streams, it does not impose a strict structure on protocol implementation. The x-Kernel [Hutchinson 91] is an experimental environment similar to Streams in the sense it provides a highly structured message-oriented framework for protocol implementations.

2.2 The parallelization of Streams

The Streams environment has been extended to facilitate the development of Streams components in SMP systems. This extension has been widely commented in [Campbell 91], [Garg 90], [Heavens 92], [Kleiman 92], [Mentat 92a] and [SunOS 93]. These extensions all define several levels of parallelism:

. ***Global level:*** only a single thread is allowed in the driver code. Drivers coming from non parallel systems will run with minimal changes.
. ***Queue pair level:*** only a single thread is allowed for a given queue pair (read and write queues of a stream). This is used for components that only share data between the input and output flows.
. ***Queue level:*** a separate thread is allowed in each queue. This is used by components that maintain no common data.

The simplest solution to implement these levels consists in using locks within the Streams framework [Kleiman 92]. A thread that wants to perform some work on a given queue must first acquire the associated lock. A better mechanism that maximizes CPU usage consists in using synchronization elements: when a thread cannot perform some work for a queue -

detected thanks to the associated synchronization element - the request is registered and the thread goes elsewhere. This work will be handled later by the thread that currently owns the synchronization element. An intelligent use of these parallelism levels should ideally remove the need of additional locks. In fact, standard Streams-based TCP/IP stacks make heavy use of private synchronization primitives to the detriment of efficiency (see section 3.2.1).

Two concepts are common in parallel Streams:

. ***Horizontal parallelism:*** this is the queue or queue pair parallelism. Several contexts of a given driver can be simultaneously active.
. ***Vertical parallelism:*** it is similar to the usual notion of pipeline parallelism. It requires that messages are systematically processed in service routines.

3 STATE OF THE ART

3.1 Layered architectures and performance

The use of the OSI model as a guide to protocol implementation has been more and more questioned during the past few years for several reasons:

. A layered protocol architecture as defined in the OSI model often duplicates similar functionalities. [Feldmeier 93a] identifies: error control, multiplexing/demultiplexing, flow control and buffering.
. Hiding the features of one layer to the other layers can make the tuning of the data path difficult. [Crowcroft 92] describes an anomalous behavior of RPC that comes from a bad communication between the Socket and TCP layers. The layered implementation principle is greatly responsible of this design error.
. The organization in several independent layers adds complexity that may not be justified [XTP 92] [Furniss 92].
. The ordering of operations imposed by layered architectures can prevent efficiency [Clark 90].

To remedy these problems, several solutions have been proposed:

. Adjacent layers may be gathered: This is the approach followed by the XTP protocol [XTP 92] that unifies the Transport and Network layers in a Transfer layer. Extending the transport connection to the network layer removes many redundancies and facilitates connection management. Another example is the OSI Skinny Stack [Furniss 92], a "streamlined" implementation of a subset of the Session and Presentation layers. It has initially been designed for an X Window manager over OSI networks. Upper layers are merged in a single layer, the invariant parts of protocol headers are pre-coded and the analysis of the incoming packet is simplified.
. [Feldmeier 90] and [Tennenhouse 89] argue that multiplexing should be done at the network layer. A first benefit is that the congestion control of the network layer, if it exists, can also provide flow control functionality for upper layers. Second, because the various streams are distinguished, the QOS specified by the application can be taken into account

by the network driver. Eventually the context state retrieval is minimized when multiplexing is performed in a single place.

. [Mentat 92b] describes a lightweight architecture for Streams-based stacks where transport protocols are implemented as modules pushed onto streams. This feature greatly reduces the overhead due to layering (see sections 3.2.1 and 4.1).

. Starting from the fact that memory access overhead may become predominant over computation on RISC architectures, [Clark 90] proposes the ILP principle (Integrated Layer Processing) that aims at reducing data manipulations (encryption, encoding, user/ kernel copy, checksum...). Instead of doing them in separate loops in different protocols, each data word is read once and all the required manipulations are performed while data is held in the CPU registers.

3.2 A case study: a standard Streams-based TCP/IP stack

We have first analyzed a standard Streams-based TCP/IP stack where each protocol is implemented as a multiplexed driver (see Figure 2).

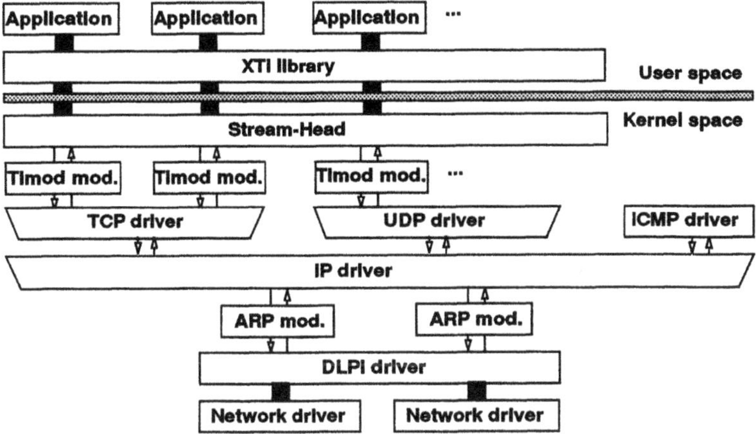

Figure 2: A standard Streams-based TCP/IP stack.

XTI is the transport library defined by the X/Open organization [XTI 90]; the Timod module works in close collaboration with XTI; the DLPI driver performs the interface between upper Streams-based components and the lower network drivers.

3.2.1 Limitations of this architecture

Importance of access method and interface overheads

In [Roca 93] we have shown that the access method overhead (XTI, Stream-head) and the driver's upper and lower interfaces overhead[1] are extremely costly. In a Streams-based XTP stack working in a loopback configuration, only 25 to 45% of the total processing time is spent for core protocol processing.

An analysis of buffer management

A detailed analysis of buffer management on the output path shows that:

. Old versions of the access method (XTI) and of the Stream-Head impose a limitation to the application TSDU size. Large TSDUs are first segmented by XTI to 4 kilobytes and each segment is sent independently to the transport provider.

. Because TSDUs are copied in kernel buffers by the Stream-Head without any regard to the packet boundaries (not yet known), it is not possible to reserve room for the protocol headers. A separate buffer must be allocated and linked to the data segment.

. As mentioned in section 2, the format of Streams messages are normalized at each interface. A TSDU received by TCP consists in a buffer containing a transport data request (or T_DATA_REQ) TPI primitive followed by the data buffers. A packet created by TCP must be preceded by a buffer containing a network datagram request (or N_UNITDATA_REQ) NPI primitive. Because the TSDU boundaries differ from the packet boundaries, there is no possibility to reuse the T_DATA_REQ buffer to initialize the N_UNITDATA_REQ primitive.

. As packets may overlap several TSDUs or be only part of a TSDU, every operation in TCP outgoing list is based on expensive offset calculations. Overlapping also increases the number of buffers allocated to hold (through pointers and offsets to avoid data copy) duplicated data.

Data buffering in Streams-based transport protocols

A BSD transport protocol implementation has direct access to the sending and receive data lists (sockets). This is not the case with a Streams-based stack where there are two receive data lists: one of them consists in the read queue of the Stream-Head. A Streams-based TCP driver cannot know what amount of data, waiting to be given to the receiving application, is present in this read queue. The second list is an internal TCP list used to store received data when the Stream-Head read queue is full. The receive window is estimated by examining this local TCP input queue, now often empty. This feature can increase the number of acknowledgments: window updates may now be sent for each packet received (every two packets if we take TCP optimizations into account) instead of every application buffer filled.

Another consequence is that data waiting to be sent to the application can be greater than the receive window! This is the case when both the Stream-Head read queue and the internal TCP list are full.

By default, and unlike the Socket strategy, the Stream-Head does not try to optimize the application receive buffer filling. A second consequence of giving received data immediately to the Stream-Head is that application buffers often contain only one packet worth of data.

---

1. By upper interface overhead we mean the time spent to identify the message type and queue/remove this message. By lower interface overhead, we mean the time spent to allocate and initialize a message block that identifies the message type, and then send it. Some of these operations may be by-passed in some cases; message blocks may be reused from one driver to the next one and messages may be processed immediately without queuing them.

Streams flow control

Streams flow control is based on the examination of the next queue of the stream. If the queue is saturated, the upstream driver/module is informed of it and stops sending messages. Because a multiplexed driver multiplexes several upper streams on several lower streams, upper and lower streams are not directly linked. Therefore Streams flow control cannot see across a driver. If the DLPI driver is saturated, applications won't be blocked until all the intermediate queues are saturated (see Figure 3[1]). In the meanwhile, a lot of memory and CPU time re sources will be devoted to non-critical tasks to the detriment of DLPI. Similar remarks can be done on the input path.

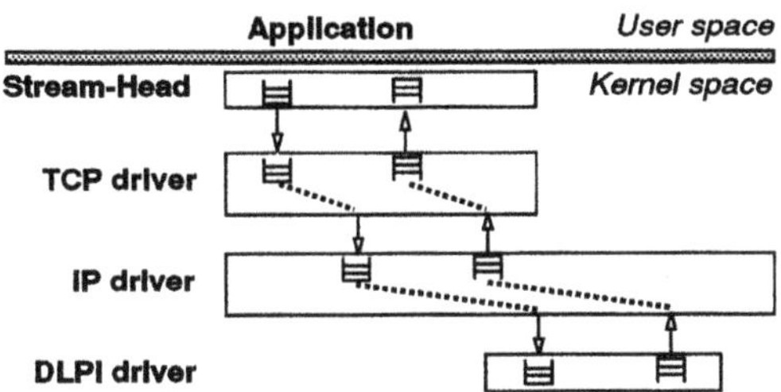

Figure 3: Streams flow control in a standard Streams based stack.

Parallelization of the stack

The parallelization of multiplexed drivers with few access points (DLPI, connectionless protocols such as IP) creates problems because queued message processing are serialized (Figure 4 - left). In order to preserve parallelism, the standard solution consists in multiplying the access points to those drivers, namely to open several streams (Figure 4 - right). The problem can be solved but at the expense of additional complexity.

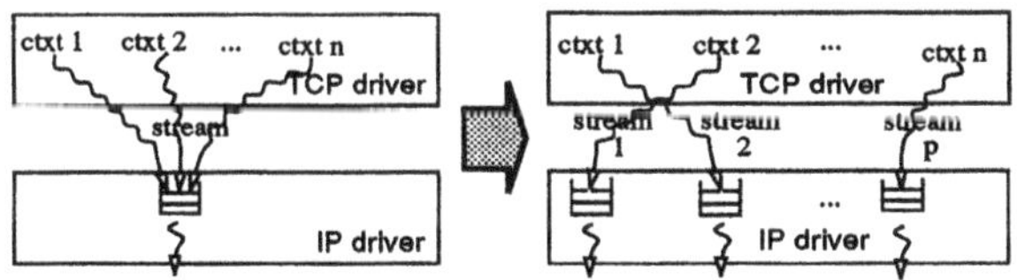

Figure 4: Parallelization of the IP driver.

Another problem is the important use of locks within the stack. [Heavens 92] describes the various locks used by the TCP driver: each control block structure (or TCB) is protected by a "mutex" lock, the chain of TCB is protected by a readers/writer lock, and the hash table used to demultiplex incoming packets is protected by a write lock. This solution does not take advantage of the synchronization facilities offered by Streams (see section 2.2).

---

1. Figure 3 is a simplified vision of reality. The actual situation is yet close to it.

4 AN IMPROVED ARCHITECTURE FOR STREAMS-BASED COMMUNICATION STACKS

We have shown in the previous chapter why standard Streams-based communication stacks are inefficient. We now present two design principles that reuse some of the ideas of section 3.1, and that may shatter the myth that "Streams-based stacks are slow":

. the communication channel approach, and
. an evolution to the XTI library.

4.1 The Communication Channel approach

In this approach all the protocols are implemented as Streams modules instead of Streams multiplexed drivers. When an application opens a transport endpoint a stream is created and the adequate protocol modules are automatically pushed onto this stream. We call "Communication Channel" the association of a stream and its protocol modules, because we create a direct path (i.e. without any multiplexing/demultiplexing operation) between an application and the data link component. Note that there are as many communication channels as there are transport level endpoints. Figure 5 illustrates this approach in case of a TCP/IP stack. It must be compared to Figure 2.

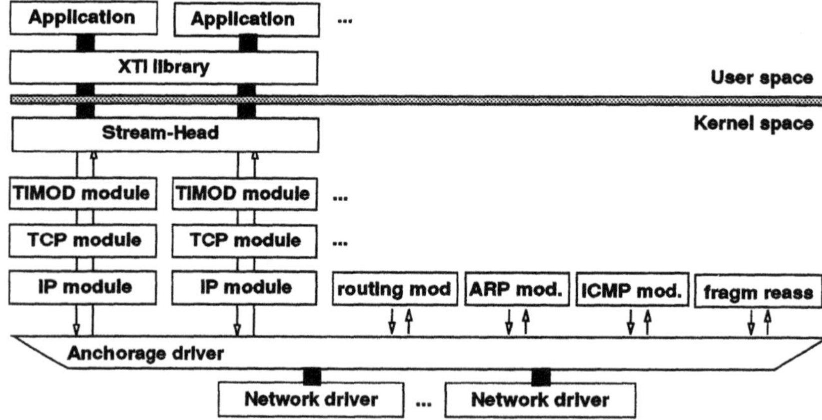

Figure 5: Architecture of a TCP/IP stack using the Communication Channel approach.

The only driver now used is the bottom Anchorage driver. Its goals are:

. to be an anchorage point for the communication channels,
. to serve as common interface to the lower network interfaces,
. to determine the destination channel for each incoming packet.

This third point has to be developed. In a standard TCP/IP stack, packets received from an

Ethernet network are demultiplexed in three steps: first the Ethertype field of the Ethernet header enables a switch onto IP or ARP. Then IP packets are demultiplexed according to protocol field of the IP header and given to the corresponding transport protocol. Finally, the transport protocol assigns the packet to the connection concerned. This mechanism is incompatible with our approach where a communication channel is created per transport connection; inserting an incoming packet on a given channel means it is destined to the associated connection. The search for the right channel requires that the initial three stage demultiplexing is gathered and moved to the Anchorage driver (see section 4.3.1).

The general problem of demultiplexing incoming packets could be greatly simplified when using new networks like ATM or AN1 [Thekkath 93]. They have fields in their link-level headers that may be directly associated to the destination channel.

Tasks that are not related to a given transport connection (i.e. forwarding of packets coming from one LAN and destined to another LAN, ARP and ICMP message processing, IP fragment reassembly) cannot be associated to a communication channel. They are handled by dedicated modules pushed onto special streams, out of the main data path. Multiprotocol configurations are also possible. Incoming packets are demultiplexed by the Anchorage driver in the same way and handed to the appropriate channel.

This architecture is an extension to [Mentat 92b]. In that case, IP is still a multiplexing/ demultiplexing component. An advantage is to enable the use of standard DLPI drivers and to avoid the problems we have with packets not related to a local transport connection (see above). Yet this is also a limitation in the sense that it does not bring as far as possible the concept of communication channel i.e. of direct communication path. In particular the DLPI and IP components are still linked by a single stream which may prove to be a bottleneck in a SMP implementation.

4.1.1 The benefits of this approach

We can identify three kinds of benefits to this approach:

Simplified data path

The main data path is greatly simplified by this architecture: a module is much simpler than a driver since multiplexing is no more required. Protocols needed by an application are selected once and for all.

Another asset is the reduced use of Streams queues: there is now at most one queue pair per module, whereas two queue pairs (upper and lower) can be used in case of a driver. Anyway, the use of queues and service procedures is optional. When the outgoing and incoming flows are not flow-blocked, messages are never queued within a channel. It saves several queueing/ service procedure scheduling/dequeueing operations. It also minimizes process switching: all outgoing or incoming packet's processing are performed by the same kernel thread. This is similar to the x-Kernel [Hutchinson 91] strategy that attaches processes with messages rather than protocols.

Then, as each path is distinguished, they can be optimized separately. For instance, the routing component can work on raw buffers containing the incoming packets that need routing. It saves the need of formatting and decoding messages in accordance with the DLPI interface. At the same time, communication channels can still work on well formatted DLPI messages.

Another point is that ARP (used to perform Internet to Ethernet address translation) has been removed from the main data path. On the outgoing side, the Anchorage driver calls a functions that performs the address translation if necessary (it depends on the network nature). On the incoming side, the Anchorage driver automatically identifies ARP packets and sends them to the ARP module. On the contrary, in standard stacks, each incoming packet needs to cross ARP (see Figure 2).

Then, the demultiplexing of incoming packets destined to well established connections uses a hashing algorithm [Kenney 92]. Other packets still use the original linear lookup algorithm of TCP (see section 4.1.2).

Finally we implemented a mechanism to solve the problem of independence between transport protocols and their input data lists located in the Stream-Head (see section 3.2.1). By default TCP now retains data until the Stream-Head tells it, with an M_READ message, that the application wants to read data, and until enough data has been received to fill the application buffer. Then TCP sends the required amount of data to the Stream-Head which in turn sends it to the application. Of course if the PUSH flag is set in the TCP header, data is immediately sent to the Stream-Head.

Extended Streams flow control

Because the protocol modules are pushed onto a single stream, the Streams flow control is now extended to the whole communication stack. Before doing any processing, we first check if a component within our channel is saturated. If yes we immediately take appropriate measures, i.e. we stop any processing on the current message or we throw this message away. For instance, an incoming UDP datagram can be freed in the Anchorage driver, as soon as the application has been identified, if this latter is saturated. On the contrary, in a standard stack this datagram would cross DLPI and IP, be demultiplexed by UDP and then freed. Having a Streams flow control extended to the whole stack insures that saturation situations are quickly solved. A direct consequence is that more connections can be simultaneously handled.

Better parallelization

The multiplication of access points to the IP and Anchorage components is a natural consequence of this approach. The horizontal parallelism available at transport level is now extended to the whole communication stack. This is true for the outgoing side as well as the incoming side where processing is parallelized as soon as packets have been affected to the right communication channel. The second benefit is that the use of Streams synchronization mechanisms is now possible in transport protocols. Because the demultiplexing of incoming packets has been moved elsewhere, transport modules can take advantage of the queue pair synchronization level (see section 2) which makes output and input processing for each transport context mutually exclusive.

4.1.2 The technical problems arisen

Demultiplexing of incoming packets in the Anchorage driver
As the Anchorage driver is now in charge of demultiplexing incoming packets, this latter:

. must know the protocol header formats of the transport, network and physical layers.
. must know the transport connections in order to compare the identifiers of the packets (Ethertype, local and foreign addresses, transport protocol identifier, local and foreign port numbers) with that of the connections.

The transport protocol informs the Anchorage driver of the local and foreign addresses/port numbers used on a connection as soon as possible, i.e. when the communication channel is established and the foreign address known:

. In case of a TCP active open this is done during the connection request.
. In case of a passive open TCP needs to wait until the connection is accepted. Before the connection is accepted, there is no channel associated to this embryonic connection. Incoming messages are then oriented to a common default TCP/IP channel and are demultiplexed by the default TCP module. These packets are demultiplexed two times: by the hashing algorithm of the Anchorage driver, then by the linear lookup algorithm of the default TCP module. This is penalizing during the setup stage but it favors well established communications.

A similar problem may occur during a graceful connection close: the communication channel can be released whereas there remains unsent or unacknowledged data. In that case the default channel is used.

The case of connectionless protocols like UDP is simpler. Because the foreign address can change at any time, the UDP context search only relies on the destination port number. This piece of information is known as soon as the application has bound itself and is immediately communicated to the Anchorage driver.

Reassembly of IP fragmented packets
The demultiplexing of incoming packets requires the analysis of all the protocol headers. In case IP has fragmented the packet, the transport header may not be present and if the fragment has reached destination, its reassembly is required before it can be affected to the right channel. This is handled by a special module (see Figure 5). But it is well known that IP fragmentation is costly and should be prohibited [Kent 87]. [RFC 1191] describes a mechanism to find the Maximum Transmission Unit (MTU) along an arbitrary Internet path.

Parallelization of the routing tasks
This approach naturally parallelizes communication channels, but other management channels are not parallelized. This is not a problem for ARP or ICMP that have little traffic, but it may be serious if this system is used as a router. The simplest solution consists in multiplying the access points to the routing module in the same way as IP is parallelized in standard stacks (see section 3.2.1).

4.1.3 Performance analysis

Test methodology

In this section we present performance measurements of our Streams-based stack when doing bulk data transfers over a single TCP connection and compare it with a BSD TCP/IP stack. We work in loopback mode with both the sending and receiving applications on the same machine which is a RISC monoprocessor system. Because the relative priority of the sending and receiving processes has an important impact on throughput when working in loopback mode, we set all process priorities to the same fixed and favored level. Both stacks use 16 kilobytes windows, no TCP or IP checksum, the receiving application buffer size is set to 5888 bytes, and the MTU of the loopback driver is set to 1536 bytes.

Configuration setup time

In our stack the creation of a transport access point requires a stream to be opened and modules to be pushed onto this stream. With a traditional Streams-based or BSD stack, the opening of a transport access point is local to the application and to the protocol. Experiments yielded a ratio 24 between these two solutions.

Table 1 : Transport access point creation time.

BSD TCP/IP stack (socket system call)	0.16 ms
Our Streams-based TCP/IP stack (t_open system call)	3.91 ms

TCP/IP throughput

Figure 6 represents the performances of our Streams-based compared to that of the BSD stack and table 2 the behavior above 8 kilobytes. In spite of the additional overhead created by the message-based communication of Streams (see next section), performances are very similar.

The BSD curve shows a sharp throughput increase after 935 and 5031 (i.e. 4096 + 935) byte TSDUs. This value of 935 is the boundary between "small" buffer requests satisfied by allocating up to four 256 byte mbufs and "large" buffer requests satisfied by allocating a single 4096 byte cluster. In order to optimize buffering, the socket layer tries to compress mbufs: if the last mbuf of a socket is not filled and if the new mbuf is small enough to fit there, then data is copied and the new mbuf freed.

Streams has a better memory management since it maintains a pool of buffers of several sizes. The smallest suitable buffer is allocated for each request. Memory is used optimally and performances are higher.

Note that these tests only highlight the data path simplification asset of our solution. As TCP already regulates data transfers, the extended Streams flow control does not intervene here.

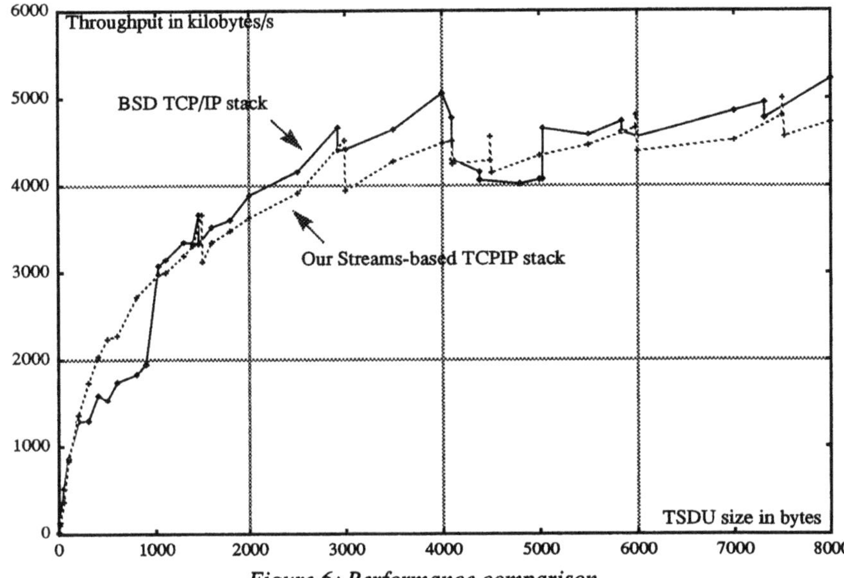

Figure 6: Performance comparison.

Table 2: Performance comparison above 8 kilobytes.

TSDU size (in bytes)	BSD TCP/IP stack (in kilobytes/s)	Our TCP/IP stack (in kilobytes/s)
12 000	5 045	4 988
16 000	5 274	5 424
24 000	5 243	5 372
32 000	5 170	5 452

4.2 Doing segmentation at User level

Segmentation of TSDUs is usually performed by transport protocols. We have shown in section 3.2.1 that it creates several complications. We propose here to move this functionality to the XTI library.

4.2.1 The benefits of this approach

Performing TSDUs segmentation in the XTI library, before crossing the User/Kernel boundary has two kinds of advantages.

First, the main data path of TCP is simplified:

. We can take advantage of the reorganization of buffering imposed by the User/Kernel data copy to reserve room for the future protocol headers.
. The T_DATA_REQ buffers associated to the segments can be reused to initialize the N_UNITDATA_REQ.
. Data manipulations in TCP outgoing list (duplication and retransmission) are now based on complete buffers (or on list of buffers if the MSS is greater than the maximum buffer size). We have the relation:

one segment <=> one buffer [1]

. because the segment boundaries are known, it is possible to combine the physical data copy with the checksum calculation as proposed in [Clark 90].

The second kind of advantage is the reduced number of User/Kernel interface crossings when dealing with small TSDUs. Indeed TCP coalescing feature is also moved to XTI; XTI now waits until enough data is received before sending anything to TCP. With applications like FTP that use one byte TSDUs when working in ASCII mode, this is particularly benefic. The case of isolated TSDUs is handled by the PUSH mechanism: if a small TSDU must be sent immediately, [RFC 793] specifies that the application needs to set the PUSH flag. Our XTI recognizes this flag[1] and sends data to TCP at once.

4.2.2 The technical problems arisen

We have supposed so far that TSDUs' segments of size MSS (Maximum Segment Size) will never be resegmented by TCP. This is usually the case. Yet, it may be required to send fewer bytes than initially expected (when its sending window is zero, TCP sends a 1 byte packet to probe the peer window). In that case, we allocate a new buffer, copy data to send in that buffer and insert it in the outgoing data list. Relation [1] is thus respected.

Because data can be stored in XTI and the control returned to the application, a strict buffer management policy must be adopted to prevent data corruption. There are several possible solutions: either the application systematically works on new buffers, or XTI copies not transmitted data into an internal buffer. The drawback of the first solution is to require the modification of applications buffer management which is contrary to our goal. If the last solution is more satisfactory, it also leads to a third copy of some data (the other two copies are at the User/Kernel and Kernel/device boundaries). Experiments (see section 4.2.3) have shown it is not a problem.

This evolution of XTI compels TCP to inform XTI of the MSS negotiated during connection opening. This is done on the first transmission request. XTI informs TCP it will perform TSDU segmentation and in the acknowledgment TCP returns the MSS in use.

As mentioned above, efficiency requires that applications tell XTI when data transfer is

1. [XTI 90] specifies that the PUSH flag found in the Socket library cannot be used through the XTI interface. Instead of adding this facility we slightly modified the semantics of the T_MORE flag that delimitates the TSDU boundaries to make its absence implicitly equivalent to PUSH. This is possible because TCP is stream oriented and does not use the notion of TSDUs.

finished. Because we can't only rely on this mechanism, we have added a timer based mechanism in XTI to force the sending of data after a certain idle time.

Finally, an evolution of the putmsg() Streams system call is needed to tell the Stream-Head to create several Streams messages, one per segment, within a single system call. This is some kind of writev() system call that, in addition, recognizes message boundaries.

4.2.3 Performance analysis

TCP/IP throughput

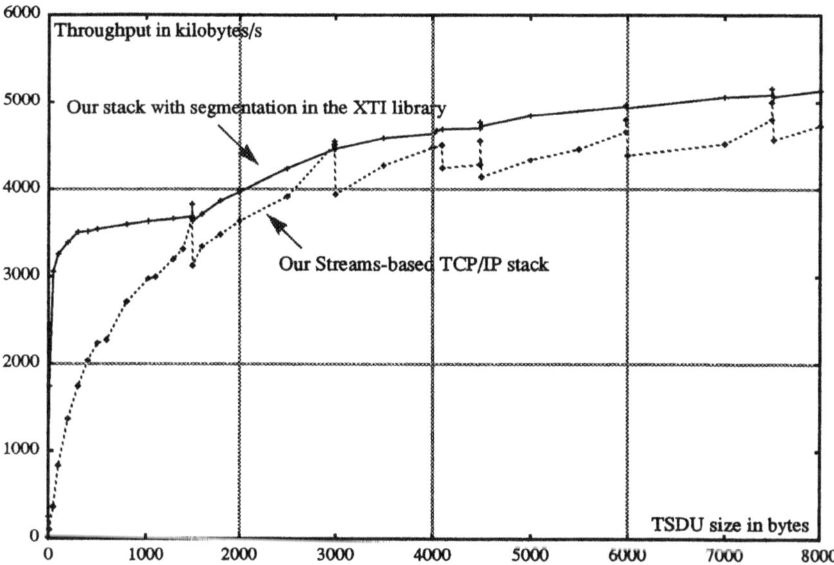

Figure 7: Traditional vs. XTI-based segmentation of TSDUs.

Performance results are presented in Figure 7. We see that doing segmentation in the XTI library makes the throughput curve radically different: instead of a steady increase of throughput with performance degradations after multiples of MSS, the curve reaches immediately a first knee, and then looks rather smooth with peaks for TSDU sizes equal to a multiple of MSS. We never experience sudden performance degradations after particular values of the TSDU size. There are two reasons for it:

- because XTI retains data until it can fill a packet, using small TSDUs only increases the number of calls to the XTI sending primitive. This is less expensive than doing a system call, hence the first knee.
- because we chose to return the control of TSDU buffers to the sending application, it is required to copy data that can not be sent immediately in an internal buffer of XTI. The

only TSDU sizes that do not yield additional data copies are the multiple of MSS, hence the peaks. The size of these peaks gives an idea of the overhead induced by these additional data copies: relatively small.

Note that protocol checksums are still disabled. The possibility of gathering the data manipulation loops (copy and checksum) is neither reflected in Figure 7 nor in Table 2.

Detailed measures

Table 3 shows the processing time distribution of our TCP module for data transmission requests. We compare the two TSDU segmentation policies. With XTI segmentation, room has been reserved for protocol headers which saves a buffer allocation/liberation. It amounts to 8 μs (allocation, third row) plus 9 μs (liberation, not shown in Table 2). The simplification of data duplication adds another 13 μs saving. This saving increases when the segment overlaps several buffers in TCP outgoing list: two or more new buffers must be allocated to hold the additional segment parts. This often occurs if segmentation is performed in TCP, never when segmentation is performed in XTI. A total of 29 μs are saved (in comparison, the sending of a 1024 bytes TSDU requires 260 μs from the application to the loopback driver when disabling checksums). This estimation takes into account neither the reduction of the number of system calls nor the additional work required in the XTI library and Stream-head.

Table 3: TCP performance analysis when doing TSDU segmentation in TCP vs. XTI.

		Segmentation in TCP		Segmentation in XTI	
Upper TPI interface	reception and analysis of the T_DATA_REQ message	5 μs	6%	5 μs	9%
TCP processing	insertion of the TSDU in TCP output list	6 μs	73%	7 μs	63%
	control and initialization of TCP header	25 μs		17 μs	
	duplication of data to send	25 μs		12 μs	
Lower NPI interface	creation of an N_UNITDATA_REQ	13 μs	21%	13 μs	28%
	sending of the message to IP	3 μs		3 μs	
		total: 77 μs		total: 57 μs	

The additional data copy required in the XTI library when data cannot be sent to TCP immediately amounts to 22 μs for a 1024 byte TSDU. This figure compares very favorably with the 147 μs required to send a small TSDU to TCP when this latter has not enough data to generate a packet.

5 CONCLUSIONS AND WORK IN PROGRESS

In this paper we have presented a lightweight architecture for Streams-based communication stacks. We applied various ideas that have emerged during the last few years and that intend to solve some of the problems created by standard layered architectures. We have shown that a condition to reach a good performance level is to avoid designing a multiplexed driver for each protocol, as the OSI model and Streams may urge us to do. Using Streams modules instead of drivers:

. simplifies the main data path and increases performances,
. improves the Streams flow control within the stack; memory and CPU resources are affected to important tasks. A direct consequence is that more connections can be supported.
. and improves the parallelization of the stack; we take full advantage of Streams synchronization facilities, save the need of additional locks in transport protocols and increase the parallelism available at the network layer.

These improvements required to reorganize some of the protocol tasks and in particular the demultiplexing of incoming packets. We also modified the XTI library in order to make it perform the TSDU segmentation, a task usually done by TCP. This solution:

. enables some more simplifications on the output path,
. reduces the number of crossings of the User/Kernel boundary, and
. enables the combination of the User/Kernel data copy with checksum calculation.

One could think to apply these enhancements to a BSD TCP/IP stack. The concept of communication channel cannot be easily applied to a BSD stack which lacks the notions of stream and of pushable processing components. On the contrary, moving the TSDU segmentation in the Socket library is possible and does not yield other problems than those mentioned in section 4.2.2.

Portability of applications has become an important issue. In this regard if the modifications we made to our TCP/IP stack are important, they are also totally hidden to applications. Applications still open, use and close transport endpoints *exactly* as they used to do.

For the present we have shown that our Streams-based stack performs as well and sometimes better than a BSD stack. This is encouraging when considering that the message oriented aspect of Streams creates large overheads. On the contrary the BSD stack is an "integrated" stack; its function-call based communication between protocols induces a small overhead. The performance gains obtained so far are not decisive enough. Future work will unable us to go further in this quest for performances. It includes the integration of data manipulation functions, experiments highlighting the impact of our extended Streams flow control, and performance studies on multiprocessor systems. We will also compare our work with other implementation techniques, in particular with user level protocol libraries.

ACKNOWLEDGEMENTS

The authors thank the people at Bull S.A. who helped us and provided us a working environment. Special thanks to Michel Habert from Bull and Christian Huitema from INRIA for the interest they showed in that project and for their advice.

REFERENCES

[Boykin 90] J. Boykin, A. Langerman, "Mach/4.3 BSD: a conservative approach to parallelization", USENIX, Vol 3, No 1, Winter 1990.

[Campbell 91] M. Campbell, R. Barton, J. Browning, D. Cervenka & ali, "The parallelization of UNIX System V Release 4.0", USENIX, Winter'91, Dallas, 1991.

[Clark 89] D. Clark, V. Jacobson, J. Romkey, H. Salwen, "An analysis of TCP processing overhead", IEEE Communication Magazine, June 1989.

[Clark 90] D. Clark, D. Tennenhouse, "Architectural considerations for a new generation of protocols", ACM Sigcomm '90, Philadelphia, September, 1990.

[Crowcroft 92] J. Crowcroft, I. Wakeman, Z. Wang, "Layering considered harmful", IEEE Network, Vol 6 No 1, January 1992.

[Feldmeier 90] D. Feldmeier, "Multiplexing issues in communication system design", ACM SIGCOMM'90, September 1990.

[Feldmeier 93a] D. Feldmeier, "A framework of architectural concepts for high-speed communication systems", IEEE Journal on Selected Areas of Communication, May 1993.

[Feldmeier 93b] D. Feldmeier, "A survey of high performance protocol implementation techniques", Research Report, Bellcore, February 1993.

[Furniss 92] P. Furniss, "OSI Skinny stack", draft paper, February 1992.

[Garg 90] A. Garg, "Parallel Streams: a multiprocessor implementation", USENIX, Vol 3, No 1, Winter 1990.

[Heavens 92] I. Heavens, "Experiences in fine grain parallelization of Streams based communication drivers", Technical Open Forum, Utrecht, Netherlands, November 1992.

[Hutchinson 91] N. Hutchinson, L. Peterson, "The x-Kernel: an architecture for implementing network protocols", IEEE Transactions on Software Engineering, Vol 17, No 1, January 1991.

[Kay 93] J. Kay, J. Pasquale, "The importance of non-data touching processing overheads in TCP/IP", ACM Sigcomm'93, New-York, September 1993.

[Kenney 92] P. McKenney, K. Dove, "Efficient demultiplexing of incoming TCP packets", Computing Systems, Vol 5 No 2, Spring 1992.

[Kent 87] C. Kent, J. Mogul, "Fragmentation considered harmful", ACM SIGCOMM'87, August 1987.

[Kleiman 92] S. Kleiman, J. Voll, J. Eykholt, A. Shivalingiah & ali, "Symmetric multiprocessing in Solaris 2.0", COMPCON, San Francisco, Spring 1992.

[Mentat 92a] "Mentat Portable Streams", Mentat Inc., commercial document, 1992.

[Mentat 92b] "Mentat TCP/IP for Streams", Mentat Inc., commercial document, 1992.

[RFC 793] "Transmission Control Protocol", Request For Comments, September 1981.

[RFC 1191] J. Mogul, S. Deering, "Path MTU discovery", Request For Comments, November 1990.

[Roca 93] V. Roca, C. Diot, "XTP versus TCP/IP in a Unix/Streams environment", Proceedings of the 4th Workshop on the Future Trends of Distributed Computing Systems, Lisbon, September 1993.

[Streams 90] "Streams Programmer's Guide", Unix System V Release 4, 1990.

[SunOS 93] "Multi-threaded Streams", SunOS 5.2, Streams Programmer's Guide, May 1993.

[Thekkath 93] C. Thekkath, T. Nguyen, E. Moy, E. Lazowska, "Implementing network protocols at user level", ACM Sigcomm '93, New York, September 1993.

[Tennenhouse 89] D. Tennenhouse, "Layered multiplexing considered harmful", Proceedings of the IFIP Workshop on Protocols for High-Speed Networks, Rudin ed., North Holland Publishers, May 1989.

[XTI 90] "Revised XTI (X/Open Transport Interface): Developers' specification", X/Open Company, Ltd., 1990.

6

Protocols for Loosely Synchronous Networks*

Danilo Florissi and Yechiam Yemini

Distributed Computing and Communications (DCC) Lab,
Computer Science Building, Columbia University, New York NY 10027, USA
Email: df@cs.columbia.edu, yy@cs.columbia.edu

Abstract

This paper overviews a novel transfer mode for B-ISDN: Loosely-synchronous Transfer Mode (LTM). LTM operates by signaling periphery nodes when destinations become available. No frame structure is imposed by LTM, thus avoiding adaptation layers. Additionally, LTM can deliver a spectrum of guaranteed quality of services. New Synchronous Protocol Stacks (SPSs) build on LTM by synchronizing their activities to LTM signals. Such signals can be delivered directly to applications that may synchronize its operations to transmissions, thus minimizing buffering due to synchronization mismatches. SPSs can use current transport protocols unchanged and, potentially, enhance them with the real-time capabilities made possible through LTM.

Keyword Code: C.2.1; C.2.2; C.2.3
Keywords: Network Architecture and Design; Network Protocols; Network Operations

1. INTRODUCTION

Emerging Broadband Integrated Service Digital Networks (B-ISDNs) will have to integrate traffic requiring a broad range of guaranteed Quality of Services (QoS). The network transfer mode must be able to provide guarantees on delay, jitter, and loss to address the needs of data, voice, or video applications. Additionally, certain applications may require synchronization of remote activities and transfers. For example, synchronization is required among remote real-time computations or applications that use the network as a massively parallel computing resource. Current transfer technologies, the Synchronous Transfer Mode (STM) [4,9] and the Asynchronous Transfer Mode (ATM) [2,4], are limited in providing full coverage of these requirements. For example, ATM networks do not support guaranteed synchronization and offer a limited form of QoS guarantees.

This paper introduces a new transfer mode for B-ISDNs, Loosely-synchronous Transfer

* This research has been supported in part by ARPA contract F19628-93-C-0170, NSF contract NCR-91-06127, and Brazilian Research Council (CNPq) grant 204544/89.0.

Mode (LTM). An LTM network enables transmissions by a source to a given destination during certain periodic time intervals or *bands*, much like STM networks. During bands, a source can transmit packets of arbitrary protocol structure and size (within set bounds). Unlike STM networks, the unit of transfer is not fixed, the size of bands is typically much larger than the transmission time of a unit of transfer and the periodicity of bands can be flexibly controlled. Once transmitted, an LTM packet can experience contention with other packets, as in ATM networks, as it moves towards the destination. Unlike ATM networks, the level of contention and with it the expected delay, jitter, and loss probability can be strictly controlled. Unlike ATM networks too, an LTM network does not require packets to be of a fixed size and structure, eliminating the need for adaptation layers at interfaces.

The main questions that LTM seeks to address in novel ways are: (1) how to synchronize source and destination with the network; (2) how to transfer multiprotocol frames without fragmentation and reassembly; (3) how to control and guarantee QoS; and (4) how to accomplish efficient bandwidth sharing.

The main goal of this paper is to describe the organization and functions of the interface stack of LTM networks. The primary purpose of this *synchronous stack* is to support isochronous application-application flow of packets. Two goals guide the design of this stack: (1) preserve as best as possible existing internet stacks; and (2) extend these stacks with a network-driven source and destination synchronization.

The first goal is accomplished by not creating a specialized packet structure for LTM networks (as in an ATM cell or in an STM frame). Instead, we treat the LTM as a media access layer and handle layers above through standard packet encapsulation techniques. It is important to note that the synchronous stack does not perturb packet structures or operations of current protocol entities. Indeed, an existing stack can be easily located above the LTM MAC through appropriate conversion of Service Access Points (SAPs). The synchronous stack extends the functionality of these existing stacks by providing an orthogonal service of synchronizing motions of packets through the stack and the network.

The second goal is accomplished by providing novel bottom-up synchronization signals from the network through the stack. Source and destination applications can synchronize their activities to the periodicity and size of network bands. For example, a video application can generate frames to synchronize with bands over which they are transmitted. Furthermore, through appropriate top-down signalling to the network, applications can exercise control over bands periodicity and size.

This paper is organized as follows. Section 2 presents an overview of Isochronets (a switching architecture that implements LTM) and LTM. Section 3 overviews SPS and how multiple traffic classes can be supported. Section 4 compares LTM with STM and ATM. Finally, Section 5 concludes.

2. ISOCHRONETS BACKGROUND

The goal of this section is to describe one particular existing LTM network: Isochronets [5,10]. It is important to emphasize that other LTM implementations exits, one example being the Highball network [7].

2.1. Architecture and Principles of Operations

Isochronets seek to provide flexible control of contention to accomplish desired QoS by routing network traffic along routing trees leading to respective destination nodes. Bandwidth is time-divided among, and synchronized along routing trees. The basic construct for bandwidth allocation is a time-band (*green-band*) assigned to a routing tree. Figure 1 depicts a network topology with the routing tree (marked with directed thick links) leading to the dark node. The graph on top plots traffic motion from source to destination through the gray nodes. The Location-axis shows the location of a given frame at the time marked in the Time-axis. During the green-band (shaded area in the graph), a frame transmitted by a source will propagate down the routing tree to the destination root (a typical frame motion is depicted using a line within the shaded area). If no other traffic contends for the tree, the frame will move uninterrupted, as depicted by the straight line.

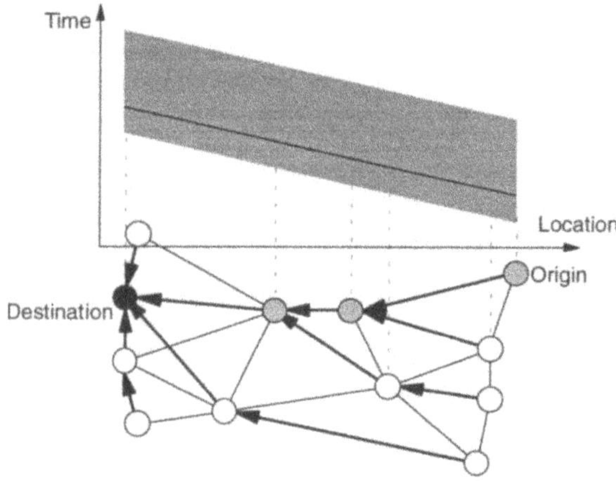

Figure 1. Green-band.

The green-band is maintained by switching nodes through timers synchronized to reflect latency along tree links. Synchronization is per band size, which is large compared to frame transmission time. It can thus be accomplished through relatively simple mechanisms. Routing along a green-band is accomplished by configuration of switch resources to schedule frames on incoming tree links to the respective outgoing tree link. A source sends frames by scheduling transmissions to the green bands of its destination.

Bands are allocated periodically as portions of a *cycle*. They need not occupy the same width throughout the network. Indeed, one can view a green band as a resource that is distributed by a node to its up-stream sons (as long as the bands allocated to sons are scheduled within the band of the parent). In particular, if the bands allocated to two sons do not overlap, their traffic does not contend. By controlling band overlaps, switches can fine-tune the level of contention and statistical QoS seen by traffic.

One may view these mechanisms to schedule traffic motions by way of band allocations as a media-access technique. The entire network is viewed as a routing medium consisting of routing trees. Bandwidth is time- and space-divided among these routes. Sources need access respective trees during their band times, seeing the network as a time-divided medium, much like Time Division Multiple Access (TDMA) [9]. This technique, accordingly, is called *Route Division Multiple Access (RDMA).*

A *contention band* is a band that may be shared by multiple sources simultaneously. Its name is derived from the fact that multiple sources may decide to use the band at the same time and thus *contend* for intermediate tree links. When contention occurs, the collision resolution mode used is designated in terms of the signs "–", "+", and "++". In *RDMA–*, only one of the colliding frames proceeds, while the others are discarded. In *RDMA+*, one colliding frame proceeds while the others are buffered, but only up to the band duration. *RDMA++* operates similarly to RDMA+, but also stores frames beyond band termination, rescheduling them during the next band.

Isochronets use *priority bands* and *multicast bands* in addition to contention bands. Priority bands are allocated to sources requiring absolute QoS guarantees, similar to a circuit service. Traffic from a priority-source is given the right of way, by switches on its path, during its priority band. Unlike circuit-switched networks, however, priority sources do not own their bands. Contention traffic may access a priority band and use it whenever the priority source does not. During a multicast band, the routing tree is reversed and the root can multicast to any subset of nodes.

Bands are thus shared resources that may be passed from intermediate nodes to subtrees. Nodes may decide to pass only portions of their bands to their sons. Also, the portions may be of different sizes. Thus, the final band allocation scheme may be designed taking advantage of the rich structure enabled by the band allocation possibilities.

A few observations regarding Isochronets are in order. Multiple simultaneous routing trees can schedule transmissions in parallel (have simultaneous green bands), depending on the network topology. Figure 2 shows two non-interfering routing trees.

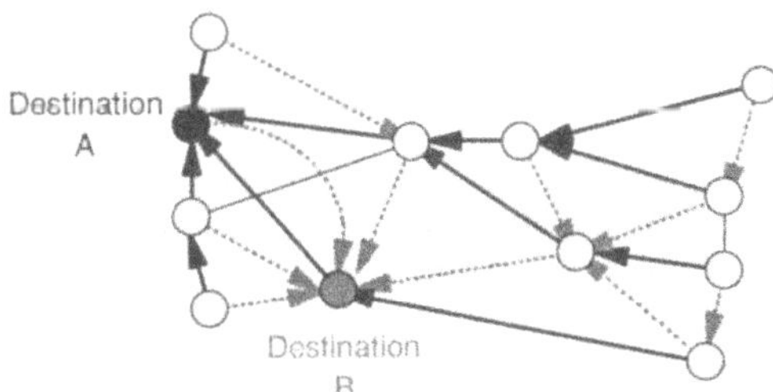

Figure 2. Multiple non-interfering trees.

No header processing is necessary in Isochronet nodes. Frames on incoming links can be mapped into the corresponding outgoing link based solely on the current routing tree structure which, in turn, may be derived from the current time. Thus switching can be accomplished without any processing that is dependent on frame contents. This means that Isochronets may operate at any link speed.

Since no frame processing is performed at intermediate nodes, all stack layers above the media-access layer are delegated to interfaces at the network periphery. That is, Isochronets may transport any frame structure without adaptation because frames do not need to be parsed to derive routing information. A typical stack organization for Isochronets is depicted in Figure 3. Interconnection of Isochronets can be accomplished by way of media-layer bridges using extensions of current well-understood technologies.

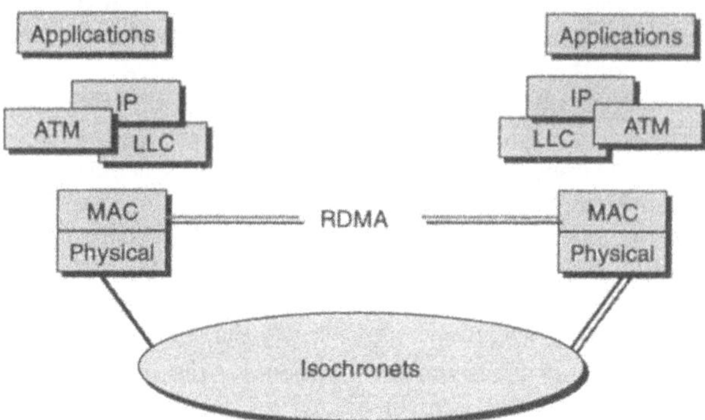

Figure 3. Multiple protocol stacks in Isochronets.

The following is a typical Isochronet operation scenario. A set of end nodes is connected to a backbone Isochronet network. The backbone periodically enables destinations in a cyclic manner until all destinations are covered. The end nodes interact with the Isochronet backbone switches by accessing the bands to deliver frames and by requesting services. Services are requests for band allocation with QoS demands in the form of band type, size, and periodicity.

An interesting question to be solved in RDMA is how to use contention or priority bands when they enable multiple backbone destinations. That is, the end node is attached to a RDMA backbone switch that signals contention or priority bands to multiple destinations. For example, a contention band may enable more than one non-interfering tree, as depicted in Figure 2. If the end node has only one link to the switch, its use must be multiplexed among all enabled destinations. One possibility is to partition the band at the periphery node among all destinations and signal each destination individually. This is equivalent to time-dividing the link between the periphery node and the attached switch among destinations. Another possibility is to use a local frame addressing scheme between the periphery node and the backbone. Each frame would contain the intended destination address so that the attached switch at the backbone can decide how to forward the frame. Notice that such addressing scheme is local between the periphery node and the attached switch, and is not used inside the RDMA backbone.

2.2. Isochronets Support LTM

This section defines the main properties of LTM networks. The nomenclature used follows the one in Isochronets.

LTM networks issue *synchronization signals* that can be used to schedule frame motion among source-destination applications. The period of time between signals is called a *band*. Bands embody two global network status: (1) a connection to destinations in the network, and (2) a certain QoS associated to the band. Bands are repeated periodically in a *cycle*.

Depending on the QoS offered, bands can be of three kinds: (1) contention, (2) priority, and (3) multicast. During a contention band, access is shared through some fair competition and the only guarantee provided is that the network will seek to optimize bandwidth use. During a priority band, the network will provide a circuit service to the destination. That is, frames from the sources will not be affected by any contending traffic. During a multicast band, the network provides contention free multicast to a group of destinations.

LTM can adapt to traffic characteristics and mimic advantageous characteristics of both STM and ATM. Similarly to STM, QoS can be guaranteed. That is, through the allocation of priority to destinations in the network, LTM can deliver the requested QoS. For example, end-end delay can be bound by the time waiting for the priority band (which in turn is bound by the cycle duration) plus the transmission and propagation delay in the network. Additionally, since sources get priority and not exclusive use of resources, bandwidth utilization is improved in LTM when compared to STM.

Furthermore, nodes at network periphery do not need to synchronize their clocks globally, as in STM. Necessary synchronization information is given by the network and nodes need to synchronize only locally with the network. Also, the necessary accuracy is much lower when compared with STM because nodes need to know only what is the current band (which usually lasts for a long time). For example, an 8 bit slot in a STM frame at 2.4Gb/s transmission rate lasts 3.3ns. Typical bands last between a few hundreds of nanoseconds up to a few scores microseconds.

Similarly to ATM, diverse traffic classes may be serviced by LTM. The transfer of a given traffic class is bound to periods in which the network is offering the most appropriate characteristics for the service requirements. For example, video traffic must be sent during periods when priority to the correct source/destination is enabled. Data traffic may be sent during any period in which the correct destination is enabled.

LTM may nevertheless achieve accurate traffic synchronization and potentially avoid buffering in the network. In ATM, such buffering is necessary to compensate synchronization mismatches due to network resource multiplexing. Since in LTM global network information is known, sources may tune traffic generation in order to minimize contention buffering in the network.

In the context of this work, LTM is to be used as the transfer technique in the backbone network. Signaling information is supplied by the attached switches to the periphery nodes that can use it to implement their protocol stacks.

3. THE SYNCHRONOUS PROTOCOL STACK

The SPS is the stack at peripheral nodes attached to a backbone network that uses LTM. In addition to the normal data flow between stack layers, SPS implements a bottom-up flow of synchronization signals from the underlying LTM network. These synchronization signals can be used at any layer, including the application, to implement synchronization functions. The general structure of SPS is depicted in Figure 4. This section briefly describes each SPS component.

The Physical Layer (PL) does not need to be bound to any special technology (i.e., electronic or photonic implementations may be employed) since LTM does not rely on any particular frame structure. The data link and network layers are collapsed into the LTM Media Access

(LTM-MAC) layer that uses LTM as the transfer mechanism. The Transport Layer (TL) is responsible for the allocation and control of network resources, such as bands with necessary QoS. The Application Layer (AL) interacts with the TL requesting necessary classes of services.

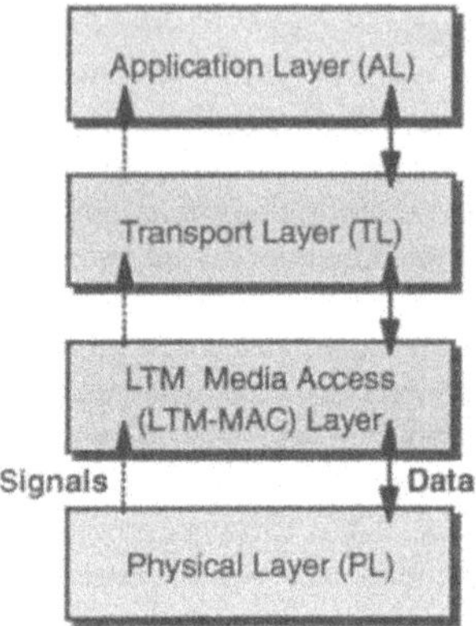

Figure 4. Synchronous Protocol Stack (SPS) structure.

Any traditional TL protocol currently used over STM or ATM may be used over the LTM-MAC directly with no changes other then adapting to the LTM-MAC SAPs. In addition, the signaling information from LTM-MAC can be used to enhance the functionality provided by any traditional TL for real-time service provision. These issues are detailed in Section 3.2.4.

A unique feature of SPS is that synchronization signals (and not only data) may be reflected all the way to the AL. The network may thus inform its current status directly to applications which may then schedule its operations to transmissions. For example, video traffic may be scheduled to generate frames when the proper band begins in each cycle.

In the general case, SPS may be a portion of the overall protocol stack, as depicted in Figure 5. Stack layers need to be able to operate in real-time to handle signaling from the MAC-LTM. The protocol stack is divided in two portions: a lower real-time protocol stack that can handle signals from LTM, and an upper non-real-time protocol stack that does not have real-time service provision. The *SPS boundary* is the interface between both stacks. The interface is responsible for buffering requests to overcome operational lack of synchronization between both stacks. Synchronization of operations with transmissions can be guaranteed only below the SPS boundary. The figure shows data being sent from protocol stack A to B, using the signaling from the LTM network. Notice that signaling is not passed above the SPS limit.

The extreme cases of this scenario are two. In the first, the SPS boundary is above the application layer, and thus LTM signals can be relayed up to applications. This can be the case when machine hardwares and operating systems provide real-time support, that is, when it is possible to predict upper bounds on execution times. In the second, the SPS boundary is at the interface with the LTM network and no signaling from the LTM network is forwarded to upper protocol layers. This could be the case, for example, when traditional protocols are to be implemented on conventional machines without real-time support.

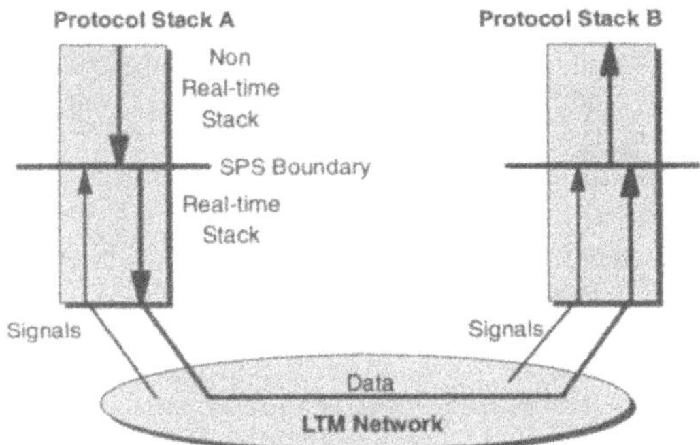

Figure 5. Handling signals in the protocol stack.

3.1. The LTM-MAC Service Access Points

This section summarizes the LTM-MAC SAPs. Notice that the LTM layer may be implemented using any technology as long as the LTM-MAC SAPs are kept unchanged and all protocol layers above the LTM-MAC can operate independently of the specific mechanisms used to implement the LTM.

The LTM-MAC SAPs are summarized in Figure 6. The first service is *OUT_BAND.signal(band, size)*. It is a signal from LTM to mark the beginning of an outgoing band. The *band* parameter has the form *<band_id, type>* where *band_id* is an identifier for the band and *type* is one of the QoS associated with the band (contention, priority, or multicast). For example, *<5, c>* means that the band identifier is *5* and its type is contention. The *size* parameter is the length of the corresponding band in nanoseconds.

Similarly, *IN_BAND.signal(band, size, protocol)* signals the beginning of an incoming band. Additionally, since the MAC may multiplex multiple transport entities above it, the *protocol* parameter identifies the protocol that should service the band.

OUT_BAND.signal(band,size)	Signals the beginning of an outgoing band.
IN_BAND.signal(band,size,protocol)	Signals the beginning of an incoming band.
ESTABLISH_BAND.request(size, periodicity,type,destination,band_id)	Allocates a band of given size, periodicity, and type to a given destination.
ESTABLISH_BAND.response(reason)	Answers band allocation requests.
RELEASE_BAND.request(band_id)	Releases a band.
RELEASE_BAND.response()	Answers band release requests.
DATA.request(frame)	Sends a frame through the current band.
DATA.indication(frame)	Signals reception of a frame.

Figure 6. LTM-MAC SAPs.

The next services are used to establish a band. *ESTABLISH_BAND.request(size, periodicity, type, destination, band_id)* requests the establishment of a band of a given *size*, *periodicity*, and *type* to the respective destinations. The periodicity is the amount of time between occurrences of the same band. If it is 0, the band is allocated only once in a cycle. The band identifier for the connection is returned in the *band_id* field. The *destination* field denotes not only the destination machine address, but also the destination protocol used in the IN_BAND.signal SAP at the destination.

The manner in which the LTM-MAC is going to achieve band allocation is dependent on its internal operations. Contention bands are allocated as a portion of the LTM supplied contention band to the given destination. To allocate priority or multicast bands, the signaling of the backbone network must be used to negotiate the allocation and inform all nodes involved. Priority bands are allocated as a portion of the respective LTM supplied contention band. *ESTABLISH_BAND.response(reason)* indicates if the band was established or not (and in the latter the *reason* for failure). Band requests may fail because not enough resources are available to allocate the requested QoS.

SAP *RELEASE_BAND.request(band_id)* is used to release previously allocated bands. *RELEASE_BAND.response()* is used by LTM to signal when the request is finished.

Finally, the last services are used to send and receive frames through LTM. *DATA.request(frame)* sends the user supplied *frame* through the network. *DATA.indication(frame)* is used by LTM to signal the arrival of a *frame*.

3.2. Transport Protocols

This section discusses how current transport protocols (IP, ATM, etc.) can use LTM-MAC SAPs directly to implement their functionality. Additionally, the signalling features enabled by LTM are used to show how such protocols can be extended to implement asynchronous, synchronous, and isochronous services.

3.2.1. Mapping Destination Addresses onto LTM-MAC Bands

Transport protocol addresses need to be mapped into appropriate band identifiers at the network periphery to enable transmissions through the LTM-MAC. Translation tables are used to this effect. The fields in a translation table are destination address and band identifier. For each reachable destination address, the corresponding band identifier (that is, the identifier for the band that routes to that destination) is given.

The next question to be addressed is how such translation tables are set initially. That is, transport layer entities need to be able to find band identifiers connected to desired destination addresses. A variation of the ARP protocol of the Internet Protocol [3] stack is used to implement this function. The protocol works as follows. A special frame containing the address of the requesting TL entity is transmitted during a band to request the TL address of the associated destinations. The peer TL entities recognize the special frame and reply with their TL addresses, using the band to the requesting entity. If the band identifier of the requesting TL entity is not already known by a replying TL entity, its reply is sent during all bands in the cycle to cover all possible requesting entities. Notice that by sending such request frames during all bands in the cycle, all destinations are covered and the table in the requesting TL entity is accordingly initialized.

The address translation mechanism described may be implemented more efficiently (in terms of bandwidth use, that is, avoiding broadcasts of requests) by using a special name server accessible through a special band identifier. The server keeps the current mapping and answers requests for address resolution.

3.2.2. LTM Supports Asynchronous Traffic

Asynchronous traffic requires no time constraints or loss guarantees on frame delivery. Examples of applications that generate such traffic are electronic mail delivery, file transfers, etc. Since these applications do not have hard timing constraints, frame loss may be overcome by retransmission. Asynchronous traffic can be directly supported on top of contention bands.

For example, to implement a file transfer application, a contention band can be used to transfer each portion of the file. When errors occur, they are detected at the destination TL entity that requests retransmission using the reverse band to the source.

The Internet Protocol. The traditional Internet Protocol (IP) [3] is a an example of asynchronous communication protocol. To implement IP, a contention band is established to each destination through the ESTABLISH_BAND.request SAP. Packets received from upper layer entities are buffered in the IP layer according to their band identifiers (which is computed from frame destination addresses). When the beginning of an outgoing band is signalled to the IP layer, it forwards the respective buffered frames. Similarly, when the beginning of an incoming band to an IP entity is signalled, the entity receives the frames and forward them to upper layer entities.

3.2.3. LTM Supports Isochronous Traffic

In *isochronous traffic*, frames must be played-back (that is, used) with minimal jitter between them (that is, frame access should happen at constant intervals) and some loss may be tolerated.

Such services can be implemented on SPS by making the TL compile requests into two parameters available through ESTABLISH_BAND.request: priority band size and periodicity (that is, how many cycles apart should the band be allocated). The mapping is performed according to the QoS requested, that is, depending on the requested jitter and bandwidth. The priority band periodicity is determined by the jitter requirements and buffering capacity at the destination. The priority band size is then computed from the requested bandwidth, link capacity, and granted periodicity.

For example in a video transmission, if the cycle period is 125μs, allocation can be implemented as follows. A priority band can be allocated for each frame every 264 cycles. Alternatively, a smaller priority band can be allocated for each frame with higher frequency, depending on the amount of buffers available at the destination. As long buffering space for 1 frame is available, the allocation can be done such that every 264 cycles 1 complete frame is delivered. A typical video transmission in this scenario is depicted in Figure 7. When an application receives a signal from the LTM-MAC, it is awaken and it transmits a video frame (or portion of a video frame). After that, the next video frame (or portion) is generated by the application which then goes to sleep waiting for the next signal. The signals thus pace the application to generate isochronous frames.

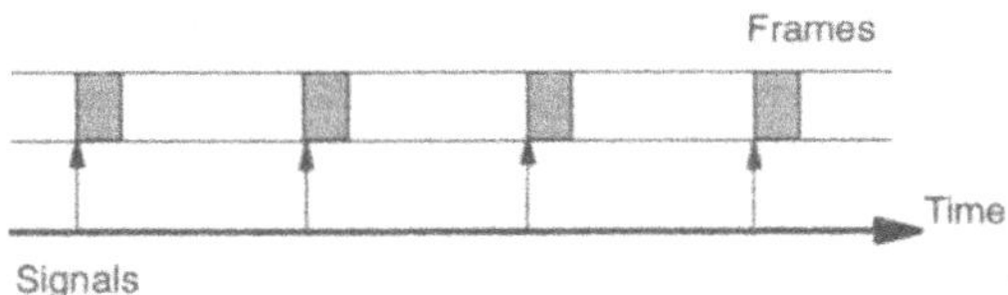

Figure 7. Isochronous transmissions.

Another possibility is to profit from the fact that some loss may be tolerated in this kind of communication. The TL may then allocate two kinds of contention bands: one for asynchronous traffic and another for isochronous traffic. The contention band for isochronous services can be used according to distributed protocols that allocate resources by maximizing multiplexing constrained to the tolerable loss allowed by applications, as it is done in the context of ATM networks [4]. That is, portions of the contention band for isochronous traffic are allocated not to guarantee lossless communication, but to deliver low probability of loss. In this manner, the portion of the band to be allocated is smaller than would be necessary for no loss. When sending frames, the TL always gives priority to isochronous traffic over asynchronous traffic.

After the band allocation phase is completed, the TL receives signaling information from LTM-MAC when corresponding bands begin. It then schedules signaling to applications when the corresponding priority bands are due. When signaled, applications may send data to TL which uses LTM-MAC to transmit them. Potentially, traffic generation may be scheduled to begin only when signaling is received from TL, thus minimizing buffering.

The ATM Adaptation Layer. The ATM Adaptation Layer (AAL) [2,4] protocols can be implemented using two alternative ESTABLISH_BAND.request options: priority or contention band. The ATM virtual path and virtual channel identifiers are translated into band identifiers. ATM services not requiring QoS are implemented on top of contention bands, in similarity to the IP protocol stack. QoS demanding services must be implemented on priority bands. The following overviews how each AAL protocols can be implemented.

The AAL 1 is intended to service constant bit rate applications such as uncompressed video transmissions. ALL 1 services can be directly implemented using a priority band with periodicity equal to the necessary sampling rate. If the sampling rate is too small and each sample contains more information than what can be allocated in one band, the sampling rate may be increased and the band size decreased. For example, to accommodate 100Mbits/s video transmissions, a band of size 3.3Mbits can be allocated every 33ms or a band of size 100Kbits can be allocated every 1ms.

The AAL 2 is intended for variable bit services. Such services can be accomplished in several ways, depending on the error rate to be allowed in the communication. One possibility is to allocate two contention bands, one for normal contention traffic, and another to service variable bit rate services, as explained previously. Another possibility is to guarantee error-free delivery by allocating a priority band.

The AAL 3/4 and 5 are intended for data communications sensitive to loss, but not to delay. This is the ideal application for a contention band, as explained for IP.

Notice that all frame structures of the various AAL protocols can be sent directly to LTM-MAC, without adaptation. This is because LTM does not rely on any particular frame structure to perform its operations.

3.2.4. LTM Supports Synchronous Traffic

In *synchronous traffic*, it is necessary to guarantee maximum end-end delay (that is, the delay to the destination may fluctuate, but must be bound by a pre-negotiated value) and error-free communication. This kind of traffic is supported by allocating a priority band.

For example, a virtual high-speed multiprocessor machine can be implemented using a set of machines interconnected by a network such as Isochronets. This application requires sporadic exchange of small amounts of data for interprocess communication. The transfer, nevertheless, needs to be reliable (error-free) and done in a timely fashion due to the high-speed of the processors. Priority bands can be pre-allocated for this sporadic communication in every cycle. The bandwidth size is computed from the maximum bandwidth required between processors.

Most observations from Section 3.2.3 in the context of scheduling isochronous traffic generation according to the synchronization signals from LTM are applicable for synchronous traffic as well.

Synchronous IP. An important feature in SPS is that the signals that are input from the LTM-MAC can be used to extend existing TL protocols towards providing synchronous services. For example, the IP suite can be extended with new SAPs to the application layer to support synchronous transport. Such SAPs would be implemented using priority bands at the LTM layer.

3.2.5. Compiling Higher Level QoS Parameters

Higher level QoS parameters such as end-end delay, loss, and jitter need to be compiled into the elements made available by the LTM-MAC layer, that is, type of band, band size, and band periodicity. Such compilation is performed by TL protocols, depending on the high-level QoS parameters they offer. In this section presents an example of how the translations can be performed.

In the example Protocol 1, two parameters (delay and bandwidth) are used by the application layer to request transport layer services: maximum end-end delay and bandwidth needed (Step 1). The variable P in Step 2 is used to always allocate only a portion of the LTM supplied band, to avoid compromising the whole band with one request, if possible. The allocation begins backwards in Steps 3 and 4 searching for idle portions in the cycles from the deadline (T+D) up to the current time (T). In Step 5, the allocation is tested. If it was successful, that is, the first allocated priority band begins at least at time T+O (where O is the overhead necessary before the first frame can be sent) the application is informed about the allocation. If not, new allocations are tried with a new vale for P. If, after all values for P have been tried, no feasible allocation exists, the failure is communicated to the application.

1. Let the requested delay be D, and the requested bandwidth be B.
2. Assign 50% to P.
3. Mark location T+D (where T is the current time) in the time line.
4. Search each cycle backwards beginning from T+D and fill at most P percent of the idle portion of the band assigned for the source and destination pair. The search is performed by requesting the LTM-MAC to allocate a portion of the requested size of the band to the destination. The search begins with the cycle in which the band ends before and closest to T+D. It ends with the one in which the band begins after and closest to T.
5. Let E be the instant in the time line when the first found portion of the band begins. If $E \geq T+O$ (where O is the overhead until the first transmission can happen), the allocation is feasible. Stop and inform the application. If $E < T+O$, the allocation is not feasible go to Step 6.
6. If P is less than 100%, add 10% to P and go to step 3. If P is 100%, stop. The requested service cannot be delivered. Stop and inform the application.

Protocol 1. Example end-end delay and bandwidth QoS compilation.

A few observations are important in the example described. Firstly, optimizations can be performed, but were not adopted for simplicity. For example, O could be estimated to avoid the situation in which the allocation succeeds, but the feasibility test fails. Secondly, this is only one possibility for mapping end-end delay and bandwidth requests into priority bands. Each transport layer protocol may have its one translation algorithm, most suitable for the services it intends to provide.

Notice that the implementation of requests for QoS in terms of jitter and bandwidth can be accomplished using Protocol 1 by substituting the required maximum jitter for the maximum delay. Also, care must be taken to request a periodic allocation, instead of a single allocation (where the period is input as the delay in the protocol).

4. RELATED WORK

This section compares LTM with STM and ATM as solutions for B-ISBN.

Plain STM generates a periodic fixed-size frame. The frame is divided in fixed-sized slots (usually of size 1 byte or a multiple of 1 byte) that can be used by sources to transmit information. Once allocated, bandwidth is guaranteed for the connection, thus delivering good QoS in terms of guaranteed end-end delay, and no jitter or loss. It is necessary, nevertheless, to keep a virtual global clock in order to synchronize all nodes in the network to the global frame and slots within the frame.

The main problems of adopting STM for B-ISDN is the lack of flexibility in the slot size and in supporting on-demand service allocation. Applications such as voice communication require small slots (usually 8 bits per frame), while video communication would best profit from large slots. If the slot size is defined too big, network bandwidth may be wasted while if it is too small, it may be difficult to allocate broadband services. STM lacks provision for asynchronous traffic as well (e.g., on-demand packet switching). When such traffic needs to access the network, slots must be allocated in the whole path from source to destination with unacceptable end-end delays.

Flexibility in bandwidth allocation is the main force pushing ATM as a solution for B-ISDN. In ATM, information is partitioned into fixed-size cells that are sent asynchronously to the destination. Destinations are recognized by using identifiers in the cells (as opposed to being identified by the location in a frame as is the case in STM) and, as a consequence, no global clock synchronization is required. Nonetheless, virtual connection (channel or path) establishment is necessary to allocate identifiers. Bandwidth can be flexibly allocated based on source demands by scattering incoming traffic into cells.

The main problems of ATM are limited support for asynchronous or synchronous communication and the trade-off between guaranteed QoS and efficient network utilization. The main drawback of asynchronous communications over ATM is that they need to be preceded by the virtual connection establishment phase, which involves end-end delays. The connection establishment phase in many applications may take longer than the transfer phase, which makes asynchronous communications inefficient both in end-end delays and in resource utilization. Some work [4,6] has been done to overcome this problem by allocating permanent virtual channels for the purpose of sending asynchronous traffic, but these solutions may require complex management of identifiers for all source/destination possibilities and may poorly use network resources.

The necessary QoS parameters for synchronous or isochronous communications are negotiated during connection establishment. Nevertheless, if all network resources are to be allocated to guarantee QoS, ATM will poorly use network resources, similarly to what happens in STM. For example, video coding algorithms usually generate variable bit rate outputs. To guarantee no loss during a video section, ATM would need to allocate resources for peak bit rate. But, the peak to mean bit rate ratio is usually high, which means that network utilization may become poor. Due to this problem, resource allocation in ATM networks is usually performed based on a lower QoS than the one requested. The idea is that multiplexing several connections and granting lower QoS to each may deliver high QoS for the multiplexed ensemble while accomplishing high network resource utilization. Unfortunately, it is not clear what is the actual QoS delivered to a particular connection under this regimen. Usually such QoS can only be characterized using a probability distribution, with a low (but existent) chance of severe service degradation for some connections.

LTM merges the flexibility in bandwidth allocation of ATM with the support for guaranteed QoS communications found in STM. As opposed to ATM or STM, where network resources have to adapt to traffic characteristics, in LTM traffic can adapt to network operations. That is, the network is in charge of informing sources about its current status so that sources may adapt its traffic generation accordingly. The unit of transfer in LTM is not pre-set to a fixed structure or size. QoS (delay, jitter, or loss) may be guaranteed or offer a controlled form of probabilistic guarantees. In addition, LTM offers synchronization signals that can be used by sources to schedule operations, thus minimizing buffering in the network.

By giving updated status information, LTM enables scheduling of traffic generation and thus may potentially minimize network buffering due to synchronization errors. Synchronous services may be achieved by local control exchange between the network and periphery nodes, without incurring end-end delays as is necessary in traditional protocol stacks.

5. CONCLUSIONS

This paper introduced a novel transfer mode: Loosely-synchronous Transfer Mode (LTM). LTM operates by signaling periphery nodes when destinations become available and encompasses advantages of both Synchronous Transfer Mode (STM) and Asynchronous Transfer Mode (ATM). Similarly to STM, synchronous communications with guaranteed QoS can be supported directly on LTM. Bandwidth allocation flexibility, one great advantage of ATM, can be found in LTM as well. Nevertheless, many of the problems introduced by STM and ATM are overcome by LTM: (1) no frame structure is necessary for communication; (2) traffic adapts to network status (instead of the other way around); (3) buffering in the network may be significantly lowered by correlating traffic generation to network status; (4) strict QoS is supported directly; and (5) synchronization signals are provided to the protocol stacks at periphery nodes.

Isochronets are candidate hardware infrastructures to implement LTM. Isochronets divide network bandwidth among routing trees and allocate periodic time intervals (bands) during which the trees are enabled. Routing is achieved by sending frames during bands in which the target destination is the root of an enabled tree. No frame-dependent processing is necessary to route frames in Isochronets, thus making their operations independent of any specific protocol stack and their implementation independent of any specific technology (such as electronic or optical).

The Synchronous Protocol Stack (SPS) is a novel protocol stack that uses LTM as its Media Access (MAC) mechanism. Because synchronization signals flow in SPS from LTM upwards to the application, SPS may incorporate protocols to support asynchronous, synchronous, and isochronous communications. Traditional transport layer protocols may be directly implemented in SPS. Additionally, such protocols can be extended to offer real-time and multicast services.

REFERENCES

1. Balraj, T. S., and Yemini, Y., "PROMPT—a destination oriented protocol for high-speed networks", in *Protocols for High-Speed Networks, II,* ed. M. J. Johnson, North Holland, 1990.
2. Boudec, J. Y. L., "Asynchronous Transfer Mode: a tutorial", *Computer Networks and ISDN Systems*, vol. 24, no. 4, May 1992.
3. Comer, D. E., *Internetworking with TCP/IP*, Volume I, Second Edition, Prentice Hall, 1991.
4. De Prycker, M., *Asynchronous Transfer Mode: solution for Broadband ISDN*, Second Edition, Ellis Horwood, 1993.
5. Florissi, D., "Isochronets: a high-speed network switching architecture (thesis proposal)", Technical Report CUCS-020-93, Computer Science Department, Columbia University, 1993.
6. Gerla, M., Tai, T.-Y., and Gallassi, G., "LAN/MAN interconnection to ATM: a simulation study", in *Proceedings of INFOCOM*, IEEE, 1992.
7. Mills, D.L., Boncelet, C.G., Elias, J.G., Schragger, P.A., and Jackson, A.W., "Highball: a high speed, reserved access, wide area network," Tech. Rep. 90-9-1, Electronic Engineering Dept., University of Delaware, 1990.
8. O'Malley, S. W., and Peterson, L. L., "A highly layered architecture for high-speed networks", in *Protocols for High-Speed Networks, II,* ed. M. J. Johnson, North Holland, 1990.
9. Tanenbaum, A. S., *Computer Networks*, Second Edition, Prentice Hall, 1988.
10. Yemini, Y. and Florissi, D., "Isochronets: a high-speed network switching architecture", in *Proceedings of INFOCOM*, IEEE, San Francisco, CA, USA, April 1993.
11. Zimmer, W., "FINE: a high-speed transport protocol family and its advanced service interface", in *Protocols for High-Speed Networks, III,* eds. B. Pehrson, P. Gunningberg, and S. Pink, North Holland, 1992.

CoRA - A Heuristic for Protocol Configuration and Resource Allocation

Thomas Plagemann, Andreas Gotti and Bernhard Plattner
ETH Zurich, Computer Engineering and Networks Laboratory (TIK)
Gloriastrasse 35, CH-8092 Zurich, Switzerland

Abstract

Da CaPo (**Dyna**mic **C**onfigur**a**tion of **Pro**tocols) provides an environment for the dynamic configuration of protocols. Implementations of single protocol mechanisms (called modules) represent the building blocks. This paper describes the heuristic CoRA (**Co**nfiguration and **R**esource **A**llocation) developed in the context of Da CaPo. CoRA configures protocols at runtime with respect to application requirements, properties of offered network services and available resources in the end systems. The goal of the configuration is to support a wide range of Quality of Service (QoS) requirements with protocols that are optimally adapted to what is needed (i.e., to increase protocol performance by decreasing protocol complexity). Generally, the problem of protocol configuration is quite complex because the set of all possible configurations might be rather large. The classification of building blocks and a measure for the resource usage of building blocks are combined in a structured search approach enabling CoRA to find suitable configurations under real-time constraints.

Keyword Codes: C.2.2, D.2.1, I.2.8
Keywords: Network Protocols, Requirements/Specifications, Problem Solving, Control Methods, and Search

1. Introduction

Recent developments in data communications are dominated by advances in two (mutually spurring) areas: high-speed networking and distributed multimedia applications. It is well known that the communications bottleneck in modern high-speed networks is located in the end system. Multimedia applications increased the set of different requirements (in terms of throughput, end-to-end delay, delay jitter, synchronization, etc.). These needs may not all be directly met by the networks; end system protocols have to enrich network services to provide the quality of service (QoS) required by applications. Obviously, fixed end system protocols are not able to support the wide range of different application requirements on top of current networks (ranging from modem lines up to gigabit networks) without including overhead (i.e., unnecessary functionality) for multiple combinations of the cross-product application requirements and networks.

The aim of the Da CaPo (**Dyna**mic **C**onfigur**a**tion of **Pro**tocols) project is to improve the described situation by configuring end system protocols. Protocols will be optimally adapted to application requirements, to offered network services and to available resources in the end systems. Configuration serves to support a wide range of QoS requirements and to increase protocol performance by decreasing protocol complexity. The properties of network services and end system resources as well as the application requirements are described in a common syntax. We developed a heuristic called CoRA (**Co**nfiguration and **R**esource **A**llocation) to configure suitable protocols at runtime. Mainly, a classification of building blocks and their resource usage are considered in CoRA to reduce the complexity of the configuration task and so decrease configuration time.

Application and QoS driven protocol tailoring and configuration are supported by different systems at different degrees. UNIX System V streams [1] enable applications to configure software modules to protocols at runtime. Haas [2] describes a horizontally oriented protocol for high speed communications (HOPS) built from simple, user selected, protocol functions. O'Malley and Peterson suggest in [3] a complex protocol graph of micro-protocols and virtual protocols. Virtual protocols direct packets through the protocol graph, with each path in the graph corresponding to one protocol configuration. However, none of the previously listed approaches supports automatic mapping of application requirements onto protocol configuration. In extended XTP (XTPX) [4] QoS parameters are mapped onto XTPX procedures and parameters to adapt the protocol machine; while in the Quality of Service Architecture (QoS-A) [5] QoS specifications from service users are classified and mapped onto profiles to tailor some mechanism of the protocol machine to be used. ADAPTIVE [6] performs a similar approach: the configuration process examines the application requirements and attempts to match them to a pre-configured transport service class. The Function-Based Communication SubSystem (F-CSS) [7] offers four pre-defined service classes to applications without special service requirements. More demanding applications can use a variety of service parameters (including QoS parameters) to specify their particular requirements. A configuration process composes specially tailored protocol machines, based on application requirements and information in a protocol resource pool. Such an automatic selection of suitable protocol configurations at runtime is difficult because of its complexity [8]. In this paper, we present an approach which efficiently solves the configuration task in Da CaPo.

A second potential problem which is generic to all approaches that use dynamic protocol configuration (which however is not of relevance for CoRA) is the overhead introduced by the flexible framework. Naturally, general approaches have increased overhead in comparison to monolithic implementation, but within Da CaPo we kept the overhead small by embedding protocols in a specially tailored runtime environment [9].

In this paper, we concentrate on the design and implementation of CoRA. The next section introduces the three layer model, the foundation of Da CaPo and section 3 briefly describes the architecture of Da CaPo. In section 4, we illustrate the basic ideas of CoRA and describe its implementation in section 5. Performance measurements of CoRA are presented in section 6. Section 7 summarizes this work and indicates future work.

2. Three layer model

Da CaPo is based on a three layer model [10] which splits communication systems into the layers A, C and T (Figure 1). End systems communicate with each other via layer T, the transport infrastructure. The transport infrastructure represents the existing and connected communication infrastructures offering end-to-end connectivity. The service of layer T is a generic service and might correspond to layer 2a, 2b, 3 or 4 services in the OSI Reference Model. In layer C the end-to-end communication support adds functionality to the T services such that at the AC-interface services are provided to run distributed applications (layer A).

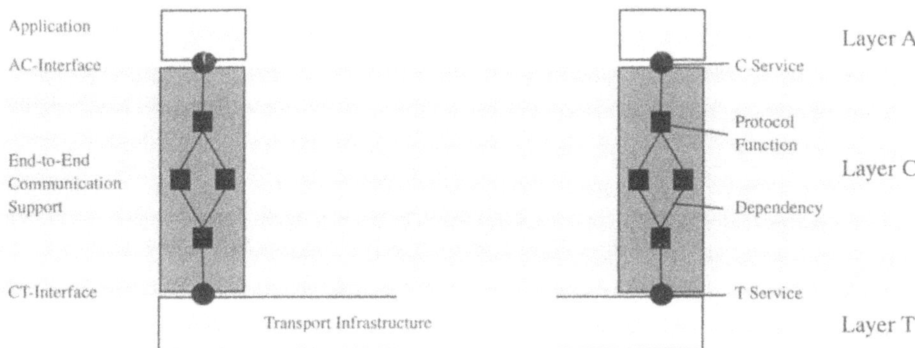

Figure 1. Three layer model

Layer C is decomposed into protocol functions instead of sub-layers. Each protocol function encapsulates a typical protocol task like error control, flow control, encryption and decryption, presentation coding and decoding, etc. Data dependencies between protocol functions, arranged on top of a T service, define a partial order on protocol functions and are specified in a protocol graph. A protocol graph is an abstract protocol specification which has to be defined by a protocol engineer. Independence between protocol functions is directly expressed in the protocol graph and indicates the possible parallel execution of these protocol functions. If multiple T services can be used, there is one protocol graph for each T service to realize a layer C service.

Protocol functions can be accomplished in multiple ways, by different protocol mechanisms, as software or hardware solutions [11]. We call the implementation of a protocol function a *module*. Modules implementing the same protocol functions are characterized by different properties, e.g., different throughput figures or different degrees of error detection and correction. To configure a protocol each protocol function in a protocol graph must be instantiated by one of its modules.

Four types of information are relevant for the configuration process: protocol graphs, available modules, module and T service properties, as well as application requirements. Protocol graphs and information on available modules for the protocol functions are stored in a local database. Furthermore, the database contains specifications of module properties describing the influence of single modules on the offered QoS. Application requirements are specified from applications within connection establishment requests.

We use a common syntax to describe the properties of modules, T services, and application requirements. We call this language *L*. All descriptions in L are based on tuples of attribute types and values or functions. The attribute types are elements of an extensible set of attribute types denoted $\mathcal{A}$. $\mathcal{A}$ contains types like "throughput", "delay", "delay jitter", and "packet loss probability". For each attribute type there is a value set $\mathcal{V}$ with an associated relation. Partially ordered value sets are associated with the relation "≥" or "≤" and unordered value sets with the relation "=". All expressions in L comprise the previously introduced basic elements. In particular, the following four types of expressions are supported:

- *T service properties* (denoted e_{ST}) are simply specified by tuples of attribute types and attribute values: $e_{ST} = <st_1, ..., st_n>$ and st_i= (type, value). The specification of T service properties is equal to common QoS specifications.

- *Module properties* (denoted e_M) specifications comprise tuples of types and functions: $e_M = <m_1, ..., m_k>$ and m_i = (type, function). The central idea of module property specification is to describe the influence of a module on offered QoS aspects with mathematical functions. 0-ary functions (i.e., constants) define the service aspect guaranteed by the module (e.g., a CRC16 module guarantees a residual error rate lower than 10^{-9}). N-ary functions describe the influence of the module on the particular attribute. For example, a CRC16 module reduces the offered throughput to a certain percentage depending on the current load of the end system and the offered throughput. Module properties are measured and evaluated before they are integrated in Da CaPo.[1]

- *Application requirements* (denoted e_{AR}) are specified by tuples of attribute types, attribute values and weight function: $e_{AR} = <ar_1, ..., ar_m>$ and ar_i = (type, value | *, weight function). Attribute values in the application requirements specify so-called *knock-out conditions* and indicate that the selected protocol must fulfill these requirements (equal to guaranteed QoS parameters). If there are no threshold values, it is denoted with the *don't care value* "*", that is equal to the least element in ordered value sets. Weight functions serve to define the relative importance of the attribute with respect to other attributes. The obligation of layer C to fulfill the objectives of the application defined by weight functions is weaker than for guaranteed QoS parameters and stronger than for best effort QoS parameters in traditional approaches. L enables us to formulate contradictory requirements (e.g., high bandwidth and low costs are required) and to deal with a wide range of application requirements, mainly introduced by new multimedia applications.

- *Layer C protocol properties* (denoted e_P) are obviously equal to C service properties. Consequently, they are specified analogous to T service properties: $e_{SC} = <sc_1, ..., sc_l>$ and sc_i= (type, value), which are equal to common QoS specifications.

A basic approach to select the best protocol configuration is to estimate the properties of all possible combinations and to compare these properties with the application requirements. The configuration algorithm must retrieve the protocol graphs of the invoked C service from the database. The instantiation of all protocol functions in a protocol graph with one of their modules creates one possible configuration. The resulting graph is called the *module graph.* To estimate the properties of a protocol configuration (denoted e_P) we start with the properties of the T service e_{ST} and calculate the influence of the next higher modules on e_{ST}. Step by step, we calculate the influence of the next higher module in the module graph on the previous results up to the highest module. The result of this process is a property description of the protocol configuration e_P. The unified representation enables a direct comparison of application requirements e_{AR} and protocol properties e_P. If a configured protocol fulfills the application requirements, i.e., does not violate any knock-out conditions, we say the protocol is in compliance with the application requirements: e_{AR} *comp* e_P. We measure the grade of correspondence of e_{AR} and e_P in the *compliance degree*: $cd(e_{AR}, e_P)$, by considering the weight functions of e_{AR}. The protocol with the highest compliance degree is the best configurable protocol with respect to the application requirements. Criterion (C1) formalizes the configuration task:

1. Currently, we are developing a tool to automatically derive the performance related aspects of module properties.

$$\text{find P such that:} \quad cd\,(e_{AR}, e_P) \rightarrow \max \quad \text{subject to:} \quad e_{AR}\,\text{comp}\,e_P \qquad (C1)$$

3. Da CaPo architecture

The realization of the three layer model is characterized by three co-operating active entities and a passive database:

- The heuristics method CoRA - which is described in detail in this paper - determines appropriate protocol configurations at runtime.
- The connection management assures that communicating peers use the same protocol for a layer C connection [12]. The connection manager negotiates with its peer a common configuration, initiates protocol establishment and release, and handles errors which cannot be treated inside single modules. Three different negotiation scenarios are maintained and can be selected by application requirements: unilateral, bilateral and combined configuration. Furthermore, the connection manager coordinates the reconfiguration of a protocol if the application requirements are no longer fulfilled.
- The resource manager provides an efficient runtime environment for Da CaPo protocols [13]. The resource manager performs the following tasks: linking and initialization of modules, packet forwarding within protocols, synchronization of parallel modules, monitoring of all protocols and available resources, and release of modules and resources. The monitoring component stores all relevant information in the local database and requests the connection manager to coordinate the protocol reconfiguration if knock-out conditions are offended.
  In contrast to other architectures, which need a process per data packet [14], Da CaPo uses only one process per protocol. Furthermore, it is possible to integrate the application into this process as the application can be designed to have the same interface as the modules. This way, the application can directly benefit from the buffer management provided by the resource manager. Thus, unnecessary copy operations are avoided. On a single processor machine the modules are sequentially executed, while on a multi-processor machine or on a machine with specialized hardware the modules are started in parallel and synchronized by the resource manager [9].
- A database stores information of general interest: protocol graphs, available modules, and module properties. Mainly, CoRA and the resource manager are handling this information.

4. Solving the configuration task

Applications request a layer C connection from Da CaPo by specifying communication partner(s)[2] and application requirements. Consequently, protocol configuration has to be done after the connect request from the application because application requirements and properties of T services which might be used are not known in advance (in particular, in heterogeneous environments). The configuration must be as efficient as possible to keep the connection establishment delay of layer C connections small. But configuration in general and protocol configuration in particular is a complex task. The set of all possible combinations depends on the number of modules per protocol function PF_i (denoted $N\text{-}PF_i$), the number of protocol functions in the protocol graph

2. The current version of Da CaPo supports only one-to-one connections, but we intend to integrate multicast connections in the near future.

PG_j (denoted N-PG_j), and the number of usable T services (denoted N-ST)[3]. Altogether, the configuration process must consider

$$N = \sum_{i=1}^{N\text{-}ST} \left(\prod_{j=1}^{N\text{-}PG_i} N\text{-}PF_j \right)$$

possibilities. Our first approach for an algorithm (we subsequently call this the *base approach*) was based on an exhaustive search of all possible configurations. The results were rather disappointing, as measurements on a NeXT workstation[4] showed that a protocol graph with only two protocol functions took 56 ms to configure, while more reasonable problem sizes immediately caused configuration times in the order of tens of seconds. Thus, the base approach is not suitable for real-time protocol configuration.

In order to decrease configuration time we developed the heuristics method CoRA. The following sections analyse three important aspects of our model with respect to the development of the configuration heuristic:

- protocol graphs and module properties,
- classification of attributes and modules, and
- integration of complex protocol mechanisms.

4.1. Protocol graphs and module properties

The configuration task requires to instantiate protocol functions with appropriate modules. In our base approach, we implicitly judge the suitability of modules by computing their influence onto the offered QoS in order to compare properties of a configuration with application requirements. In other words, only full configurations are explicitly judged and selected. This approach comprises two disadvantages:

- The exhaustive search for the best configuration simply calculates the compliance degree of all possible configurations. Obviously, there is no structure in this approach but efficient search algorithms are generally based on a structure (e.g., breadth-first search or branch-and-bound algorithms are performed on tree structures).
- Module selection is based on the comparison of protocol properties with application requirements, there is no support for explicit judgement and selection of single modules.

The central structure in the configuration task is the protocol graph. Generally, several modules may be available for each protocol function in the protocol graph. Instead of examining all module combinations we sort the modules of a protocol function according to the most relevant aspect of configuration of light-weight protocols: the module weight. In Figure 2, the left protocol graph is associated with the unordered set of available modules while the right protocol graph includes for each protocol function an ordered list of available modules.

The module weight describes the end system load introduced by this module. In current end systems CPU and available bandwidth at network interfaces are major bottlenecks in high-speed

3. The number of different T services is equal to the number of different protocol graphs.

4. Measurements were performed on a NeXTstation with Motorola 68040 processor (25 MHz).

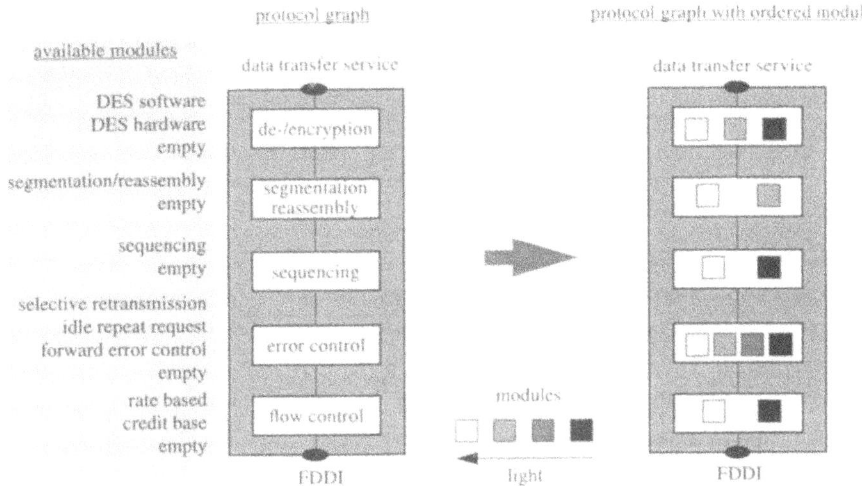

Figure 2. Sorting modules

communications, whereas memory usage is of minor importance. Consequently, we define module weight in terms of

- number of CPU cycles needed to process a packet (denoted CPACKET) independent of the packet length,
- number of CPU cycles per byte (denoted CBYTE) used to process a full packet and
- relative influence on the number of bytes to be transmitted (denoted NET_TRAFFIC). For instance, forward error control modules increase the amount of data because they introduce redundancy (i.e., NET_TRAFFIC > 1), while compression modules decrease the amount of data (i.e., NET_TRAFFIC < 1).

All factors are combined in criterion (C2) to define the module weight:

$$\text{WEIGHT} = \ln(\exp(\text{CBYTE}) + \text{CPACKET})\,\text{NET\_TRAFFIC} \qquad \text{(C2)}$$

The particular combination of CBYTE and CPACKET is used to define the module weight as independent as possible from currently used packet length. The derivation of criterion (C2) is based on several experiments with different packet length and different module properties [15].

In general, there is a linear relation between module weight and performance reduction caused by the module: the higher the module weight the higher the performance reduction. The only exception are modules storing packets for a certain time, e.g., a flow control module stores a packet for a certain time if the current packet rate is too high. The module weight is evaluated off-line and enables us to define a general rule: to configure the lightest protocol means to select the lightest module (i.e., the empty module) for each protocol function.

Nevertheless, configured protocols should not only be as light as possible; it is of major importance to satisfy knock-out conditions and weight functions of the application requirements. In several cases a module obviously offends a knock-out condition, consequently, this module can be excluded from the configuration process. For example, if the application demands a high degree of security all modules of the protocol function encryption/decryption which do not fulfill this requirement (e.g., the empty module) need not to be considered further. Each eliminated module reduces the number of possible configurations and thereby decreases configuration time.

The consideration of other knock-out conditions and weight functions which are not corresponding to the weight order of the modules is more complex. Classification of attributes and modules is an approach to reduce this complexity.

4.2. Attribute and module classification

All attribute types in $\mathcal{A}$ may be classified into five groups (see Figure 3), similar to the classification discussed in [16]. The first three groups summarize user aspects and might be considered in application requirements:

- performance related attributes like throughput and delay (denoted PERF($\mathcal{A}$)),
- reliability related attributes like bit error probability, packet loss probability, ordering, and duplication (denoted REL($\mathcal{A}$)), and
- miscellaneous attributes (denoted MISC($\mathcal{A}$)), including all important aspects for the application (e.g., costs, security, data compression, and synchronization) except performance and reliability attributes.

Attributes of the following two groups express layer C aspects and are used within layer C to determine a consistent and suitable configuration:

- resource related attributes (denoted RES($\mathcal{A}$)), including factors of module weight and availability of resources (e.g., CBYTE, CPACKET, and NET_TRAFFIC), and
- protocol related attributes (denoted PROT($\mathcal{A}$)), used to check the consitency of a protocol configuration (e.g., preconditions of modules) and to adapt modules to the current configuration (e.g., maximal transfer unit size and header description).

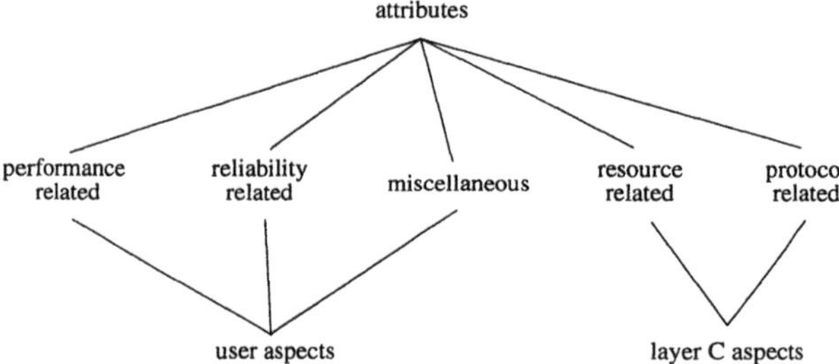

Figure 3. Attribute classification for CoRA

Obviously, all modules influence performance related and resource related attributes (except empty modules). In particular, all modules decrease the performance (except fast hardware compression modules). Consequently, the performance of T services has to be at least equal to the performance required by the application. That means, the selection of a proper T service is of central importance for the configuration of a protocol with sufficient performance. Further performance based selection of modules is influenced by the module weight (i.e., resource related attributes).

Attributes related to reliability issues and miscellaneous issues are influenced by different sets of protocol functions or modules, respectively. The reliability attributes are influenced by modules like CRC, idle repeat request, selective retransmission, and rate based flow control modules. Miscellaneous attributes are influenced by modules performing for instance presentation coding, encryption and decryption, or compression and decompression. These disjoint sets of protocol functions may be independently configured, which in turn drastically decreases the complexity of the entire configuration process. Let us consider a simple example with a protocol graph consisting of four protocol functions, each protocol function can be instantiated by four different modules. In total, there are $4^4 = 256$ possible configurations. Assuming that two protocol functions only affect reliability attributes and two protocol functions only affect miscellaneous attributes there are $4^2 + 4^2 = 32$ possibilities (i.e., only 12.5 percent of all configurations need to be investigated).

4.3. Integrating complex protocol mechanisms

In our model, a protocol function can be realized by different protocol mechanisms, i.e., protocol functions and protocol mechanisms are in a (1:n)-relation. In practice, there are several well known protocol mechanisms, each realizing multiple protocol functions, i.e., a (m:n)-relation between protocol functions and mechanisms. For example, the protocol mechanism "idle repeat request" performs the protocol functions "flow control", "packet loss detection", "packet loss correction", and "resequencing". This (m:n)-relation is contradicting to the simple abstraction hierarchy (protocol function - protocol mechanisms - modules) in our model.

To support protocol functions of any granularity and to integrate complex protocol mechanisms we extended our basic model by introducing one-level nodes (i.e., protocol functions) and two-level nodes in protocol graphs. Two-level nodes may be instantiated by sub-graphs consisting of a main protocol mechanism (which might be empty) and protocol functions. Additional protocol functions are combined with the main mechanism to specify the particular pre- and post-processing functionality. Protocol graphs as well as sub-graphs are defined by protocol engineers. Figure 4 illustrates several sub-graphs for the two-level node "reliability". The main mechanism "forward error control" requires ordered packet sequences, consequently its pre-processing part contains the protocol function "resequencing". In contrast, the main protocol mechanism "selective retransmission" supports no packet resequencing, therefore its post-processing part contains the protocol function "resequencing".

5. CoRA

All concepts discussed in the previous section are combined in the heuristic CoRA. CoRA consists of six steps:

(1) pre-decision,

(2) module elimination,

(3) T service selection,

(4) configuration,

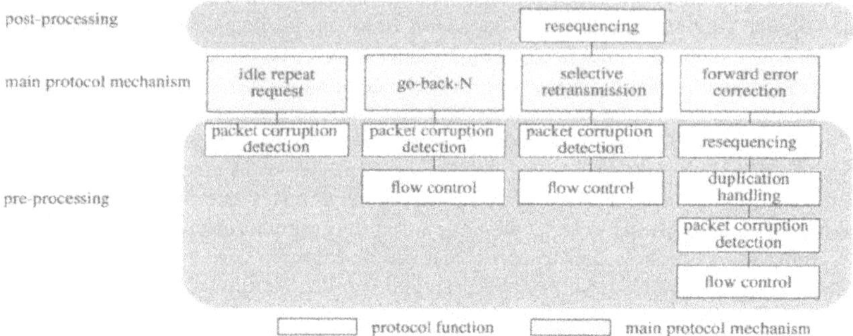

Figure 4. Sub-graphs of two-level node reliability

(5) optimization, and

(6) post-decision.

No single step guarantees to find a suitable configuration, but the combination of these steps within the modes FIRST, MINI, MEDIUM, and FULL guarantees to find at least a configuration (if it exists) which is in compliance with the application requirements. In mode FIRST, CoRA stops after finding the first configuration (step(4)) which is in compliance with the application requirements. In the other modes the steps (3), (4), and (5) are performed at least once to increase the compliance degree.

5.1. Pre-decision

The first step of CoRA is the pre-decision to determine whether it is currently appropriate to configure and establish a new protocol or not. It might be impossible to configure a suitable protocol because of high end system load or fully utilized network interfaces. Furthermore, after introducing a new layer C connection (i.e., additional end system load) all established connections should fulfill the application requirements. The decision is based on three factors:

- CPU_LOAD denotes the relative load of the CPU during the last time interval. Values of CPU_LOAD are in the range [0..1]. The value zero indicates an idle situation and value one full load on all CPUs, i.e., the higher CPU_LOAD the lower the probability to establish a new layer C connection without offending knock-out conditions of established connections.
- NET_STATE summarizes the current utilization of network interfaces (respectively, T services). Values of NET_STATE are in the range [0..1]. The value of NET_STATE is equal to zero if all network interfaces are idle and equal one if they are fully utilized. The higher NET_STATE the lower the probability to establish a new layer C connection without offending knock-out conditions of established connections.
- COMP serves to estimate the average distance between properties of all established layer C connections and the corresponding knock-out conditions. If the properties of established layer C connections are much better than demanded in the application requirements (i.e., knock-out

conditions) then the probability is high that the properties of all connections will remain in compliance with the corresponding application requirements. In contrast, if the distance between knock-out values and properties of connections is low, then the probability of offending knock-out conditions by establishing a new layer C connection is high. Values of COMP are in the range [0..1]. The higher COMP the higher the average distance and the higher the probability to introduce a new layer C connection without offending knock-out conditions of established connections.

The combination of these three factors has to be lower than a threshold value named PRE_THRESHOLD. The threshold value PRE_THRESHOLD has to be determined according to the resource allocation strategy performed in the particular end system (e.g. by a system administrator). The higher PRE_THRESHOLD the more layer C connections may be established at the same time. Criterion (C3) merges CPU_LOAD, NET_STATE and COMP according to their positive respectively negative influence:

$$\text{PRE\_THRESHOLD} > \frac{\text{CPU\_LOAD} + \text{NET\_STATE}}{2}(1 - \text{COMP}) . \tag{C3}$$

If criterion (C3) is not fulfilled CoRA terminates and indicates that it is currently impossible to configure a proper protocol and to establish the corresponding connection.

5.2. Elimination

The aim of the module elimination is to exclude as many modules as possible at the start of the protocol configuration to decrease its complexity. Candidates to be excluded are:

- Modules that are currently unavailable (e.g., hardware modules) and T services that are currently fully utilized.
- Modules whose weight is too high to run under the current end system load.
- Modules whose costs are higher than tolerated in the application requirements.
- Modules of the miscellaneous class which obviously do not fulfill the application requirements. For example, the application requires the presentation coding mechanism eXternal Data Representation (XDR). All modules - except the XDR module - can be excluded from the search.

5.3. T Service Selection

Generally, there is one protocol graph defined for each supported T service. The step T service selection serves to order the T services (i.e., protocol graphs) and to concentrate on the most promising one. Primarily, performance related attributes are considered to order the T services, because layer C protocols generally decrease the performance offered by a T service. Consequently, the performance of the T service must be at least in compliance with the application requirements to find a suitable configuration, i.e., $\text{PERF}(e_{ST}) \geq \text{PERF}(e_{AR})$. Additionally, the T service should correspond to the weight functions in the application requirements, i.e., $\text{cd}(e_{AR}, e_{ST}) \rightarrow \max$. From the systems point of view, resources should be economically allocated and the difference between application requirements and T service properties (responding to network resources to be allocated) $|e_{AR} - e_{ST}|$ should be as small as possible. Criterion (C4) combines these aspects to select a T service:

$$\text{find ST such that: } \frac{cd(e_{AR}, e_{ST})}{|e_{AR} - e_{ST}|} \rightarrow \max \qquad \text{subject to: } \frac{PERF(e_{ST})}{REDUCT} \geq PERF(e_{AR}). \qquad (C4)$$

The parameter REDUCT estimates the relative performance reduction of the layer C protocol to be configured.

5.4. Configuration

Step (4) looks for a configuration that is in compliance with the application requirements. Consequently, only knock-out conditions comprising attributes of the classes REL($\mathcal{A}$), PERF($\mathcal{A}$), and MISC($\mathcal{A}$) have to be considered. Step (4) selects a module for all nodes in one of these classes. The three classes are independently processed to decrease complexity.

The first task is to generate a one-level protocol graph out of the protocol graph determined in the previous step. Two-level nodes are expanded by selecting a sub-graph. The selection is based on the subsequently described judgement of the main protocol mechanisms (denoted MPM) in the sub-graphs. The compliance degree of the protocol mechanism MPM on top of the selected T service (ST) should be as high as possible, i.e., $cd(e_{AR}, e_{ST-MPM}) \rightarrow \max$. Furthermore, the performance of this configuration must be higher than required by the application, and the number of unresolved pre-conditions of the protocol mechanism should be small. Criterion (C5) combines these aspects to select a sub-graph:

$$\text{find MPM such that: } \frac{cd(e_{AR}, e_{ST-MPM})}{\#\text{unresolved Preconditions} + 1} \rightarrow \max$$

$$\text{subject to: } PERF(e_{ST-MPM}) \text{ comp } PERF(e_{AR}). \qquad (C5)$$

The second task is to instantiate each node in a one-level protocol graph with a module such that the resulting configuration is in compliance with the application requirements. We start with the lowest node of the protocol graph, consider application requirements and preconditions of higher nodes and try to fulfill them by selecting a suitable module (according to the weight order). If a single module cannot fulfill the requirements, we look for higher modules influencing the same attribute and examine the different combinations. If no module or module combination can be found, we start a further iteration of step configuration and generate a further one-level protocol graph. Step (4) terminates after the first configuration is found that is in compliance with the application requirements.

5.5. Optimization

The optimization step attempts to improve the protocol configuration (i.e., its compliance degree) determined in step (4). We order the tuples of the application requirements according to their importance for the application to purposefully increase the compliance degree. Two aspects have to be considered to define the importance of the application requirement tuples. First, the value of the weight function applied to the minimal required value (i.e., knock-out value) and second the distance between the knock-out value and the value offered from the currently examined protocol configuration. This protocol configuration is in compliance with the application requirements. Consequently, the lower the distance between current value and required value the higher the probability to increase the compliance degree. By placing the distance in the denominator of criterion (C6) we prefer application requirements that are weakly fulfilled[5]:

$$IMPORTANCE = \left|\frac{required\_value}{current\_value - required\_value}\right| weight(required\_value) \qquad (C6)$$

Step (5) takes the most important tuple and tries to increase the compliance degree for this attribute by examining all modules influencing this attribute. After improving the compliance degree, all tuples are ordered again and the optimization step tries to improve the most important attribute. If the optimization of one attribute is not possible, the next attribute (according to the order) will be examined. This is done until all attributes are examined and no further improvement is possible.

5.6. Post-decision

The post-decision step serves to decide whether it is possible to establish a connection for the selected protocol without decreasing the performance of existing connections too much. In contrast to step (1) the weight of the particular configuration is now known (denoted W_P) and a more precise decision can be taken. We compare the relative utilization of the end system (measured in CPU_LOAD and NET_STATE, see step (1)) with the relative protocol weight instead of the threshold value PRE_THRESHOLD. The relative protocol weight is given by the relation between the protocol weight W_P and the protocol weight W_{max}. W_{max} denotes the weight of a theoretical protocol that would fully utilize the end system, i.e., the weight of the heaviest protocol. The post-decision step allows the establishment of a new protocol if criterion (C7) holds:

$$\frac{W_P}{W_{max}} \leq \frac{CPU\_LOAD + NET\_STATE}{2} \quad \text{(C7)}$$

5.7. Modes

The stepwise approach of CoRA supports four different modes with increasing computational complexity and improved results:

- FIRST: Mode FIRST returns the first configuration which is found in step (4).
- MINI: Mode MINI operates on a one-level protocol graph, that is determined in step (4), and looks for a configuration of this one-level protocol graph with the highest compliance degree in step (5). In other words, mode MINI extends mode FIRST by additionally performing step (5).
- MEDIUM: Mode MEDIUM operates on a two-level protocol graph, that is determined in step (3), and examines all sub-graphs, in contrast to mode MINI. In other words, step (4) and (5) are performed several times.
- FULL: Mode FULL considers all two-level protocol graphs and all possible sub-graphs to find the configuration with the highest compliance degree. The steps (3), (4) and (5) are performed multiple times in mode FULL.

6. Performance evaluation

We implemented CoRA in ANSI C such that all criteria documented in this paper (C1) - (C7) might be changed easily. This enables us to later adapt CoRA based on experiences. In order to evaluate the performance of CoRA we elaborated several sample scenarios. The values used in our

5. In case that current_value is equal to required_value we simply assign a very high importance to the application requirement. By this we avoid a division by zero and indicate that there is a high probability to improve the compliance degree with this application requirement.

scenarios to specify application requirements, properties of layer T connections, and module properties are derived from the literature, our measurements and estimations. We compared the compliance degree of CoRA results with the maximum compliance degree (result of the base configuration approach) to determine the quality of CoRA results (measured in percentage of the maximum compliance degree). All measurements are performed on a Sun SparcStation 10/30.

Generally, the performance of CoRA depends on two aspects: first, the complexity of the problem (measured in number of possible configurations) and second, the structure of the particular configuration task. Obviously, the configuration task is determined by the application requirements. Restrictive knock-out conditions are applied in CoRA for module elimination and influence the performance of CoRA in two ways. In mode FIRST and mode MINI it is harder to find a configuration which is in compliance than in scenarios with non-restrictive application requirements (i.e., knock-out conditions that could be easily fulfilled). However, mode MEDIUM and mode FULL benefit from the module elimination based on restrictive knock-out conditions. This behavior is documented in the results of a sample scenario with 1,876,896 possibilities (Table 1). The scenario comprises two protocol graphs including the protocol functions "monitoring", "compression", "security", "presentation coding", and the two-level node "reliability" that are defined on the T services "IP" and "ATM/AAL5". Several modules with different properties are available for each protocol function. Application requirements 1 (AR-1) comprise no hard requirements, e.g., the performance requirements could be fulfilled by both T services. In contrast, the application requirements 2 (AR-2) include restrictive knock-out conditions on the attributes "throughput", "delay jitter", and "packet loss". The throughput and delay jitter requirements could only be fulfilled by the T service "ATM/AAL5". The base approach needs more than 12 minutes to solve the problems. Table 1 compares the configuration times of CoRA for AR-1 and AR-2 applied to the same protocol graphs and set of modules.

Table 1. Non-restrictive versus restrictive application requirements

	FIRST	**MINI**	**MEDIUM**	**FULL**
AR-1	12 ms; Quality 83%	19 ms; Quality 83%	79 ms; Quality 88%	170 ms; Quality 100%
AR-2	13 ms; Quality 91%	22 ms; Quality 98%	49 ms; Quality 100%	100 ms; Quality 100%

We examined the relation between configuration times and complexity of the configuration task with several hundred different measurements in four different basic scenarios (denoted A, B, C, and D). Each scenario is based on some protocol graphs and a set of available modules, that determine the complexity of the configuration task. Scenario A comprises 175,959 possible configurations, scenario B 1,876,896 possibilities, scenario C 7,288,848 possibilities and scenario D 58,560,768 possibilities. Within each scenario we performed measurements with different application requirements. The configuration times of the base approach directly depend on the complexity: it needs approximately 50 seconds in scenario A, 12 minutes in B, 28 minutes in C, and 3.5 hours in scenario D.

The graphs in Figure 5 represent the average configuration times for the four CoRA modes as well as the average qualities of the results (denoted Q) in the modes FIRST, MINI, MEDIUM, and FULL. The maximal standard derivation of the configuration times is nearly 50%, caused by the different application requirements. These deviations show the same behavior of CoRA than discussed in the first sample scenario (Table 1). However, the maximum standard derivation of the quality of the results is below 15%. Obviously, the scenario A with the lowest complexity needs in most modes the longest configuration times. This demonstrates that the response times of CoRA depend more on the particular problem structure (e.g., number of T services, number of two-level nodes, properties of modules and T services, and number of knock-out conditions and weight functions) than on the number of possible configurations.

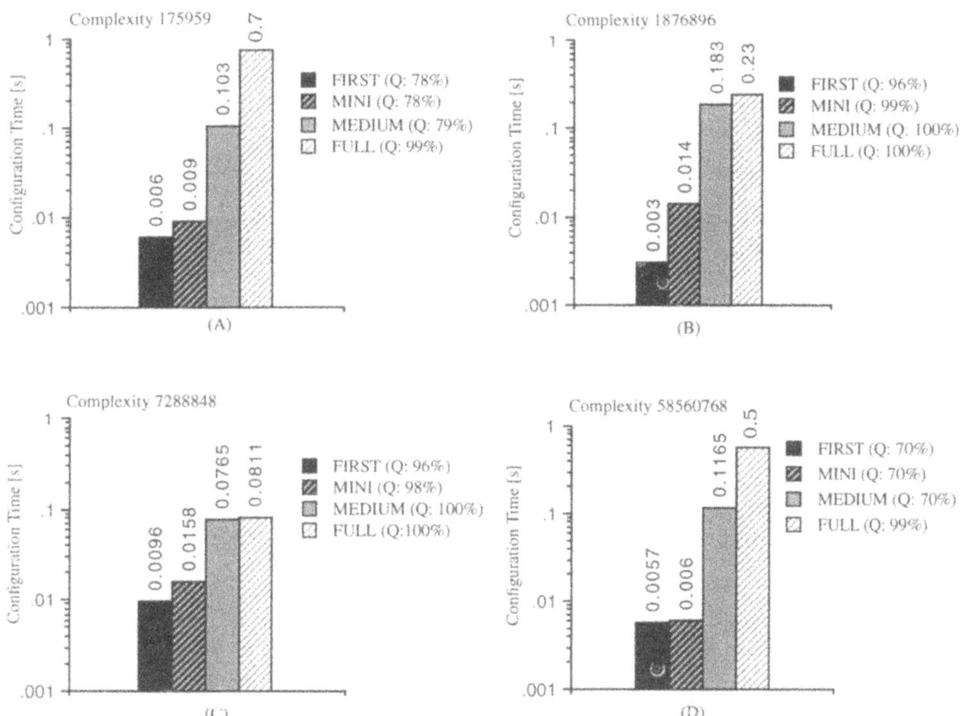

Figure 5. Average configuration times

Summarizing the results presented in Figure 5, we recognize that mode FIRST determines configurations with 70% and more of the maximum compliance degree within less than 10 milliseconds. Furthermore, the quality of results in modes MEDIUM and FULL is in most cases equal to the optimal configuration (i.e., 100% quality), while the configuration times in mode MEDIUM are less than 200 milliseconds and in mode FULL less than 500 milliseconds in most scenarios.

7. Conclusions

In this paper, we presented the concepts and implementation of the configuration heuristic CoRA applied within the Da CaPo project. The definition of module weight is used twice in CoRA; on the one hand it speeds-up the configuration process and on the other hand it supports careful resource allocation. A further speed-up of the configuration is possible by classifying attributes, protocol functions, and modules. The integration of complex protocol mechanisms enables us to study fine granular protocol mechanisms as well as complex mechanisms and to compare them in the same environment. Measurements demonstrate the high performance and quality of CoRA. CoRA enables us to configure protocols at runtime, only marginally increasing connection establishment delay with respect to connection establishment delay of fixed protocols.

Our future work is to extend the set of modules and to derive their properties. The implementation of multimedia applications on top of Da CaPo will be used to evaluate the performance and to demonstrate the feasibility of CoRA and the entire Da CaPo system in the context of "real-life" problems. Furthermore, the practical experiences should be used to improve and refine the configuration and particularly the resource allocation in CoRA.

Acknowledgements

We would like to thank the anonymous reviewers for their valuable comments. We want also to thank all the Da CaPo team and all students that made contributions to Da CaPo, namely Marcel Dasen, Peter Imhof, Alireza Olumni, Mahan Satari, Martin Vogt, Janusch Waclawczyk, Thomas Walter, and Thomas Ward. Finally, we would like to thank Arlette Gaillard for a fruitful discussion.

References

[1] Harris, D.: "Streams", in: Kochanan, S.G., Wood, P.H. (Editors): "UNIX Networking", pp. 133-170

[2] Haas, Z.: "A Communication Architecture for High-speed Networking", in: Proceedings of IEEE INFOCOMM'90, 9th Annual Joint Conference of the IEEE Computer and Communications Societies, Los Almos, California; Vol. 2, June 1990, pp. 433-441.

[3] O'Malley, S.W., Peterson, L.L.: "A Dynamic Network Architecture", in: ACM Transactions on Computer Systems, Vol. 10, No. 2, May 1992, pp. 110-143

[4] Metzler, B., Miloucheva, I.: "Specification of the Broadband Transport Protocol XTPX", CEC Deliverable Number: R2060/TUB/CIO/DS/P/001/b2, February 1992

[5] Campbell, A., Coulson, G., Gracia, F., Hutchinson, D., Leopold, H.: "Integrated Quality of Service for Multimedia Communications", in: Proceedings of IEEE INFOCOMM'93, San Francisco, March 1993

[6] Box, D. F., Schmidt, D. C., Tatsuya, S.: "ADAPTIVE - An Object-Oriented Framework for Flexible and Adaptive Communication Protocols", in: Proceedings hpn92, 4th IFIP Conference on High Performance Networking, December 1992

[7] Zitterbart, M., Stiller, B., Tantawy, A.M.: "Application-Driven Flexible Protocol Configuration", in: IEEE Journal on Selected Areas in Communications, Vol. 11, No. 4, May 1993, pp. 507-518

[8] Rzehak, R.: "In Discussion: Mrs. Zitterbarts Contribution to KiVS 93" (in German), PIK Praxis der Informationsverarbeitung und Kommunikation, 2/93

[9] Vogt, M., Plagemann, T., Plattner, B., Walter, T.: "Parallelism Aspects in Da CaPo" (in German), GI Workshop on Architecture and Implementation of High-Performance Communication Systems, Karlsruhe, January 1994

[10] Plagemann, T., Plattner, B., Vogt, M., Walter, T.,: "A Model for Dynamic Configuration of Light-Weight Protocols", in: Proceedings IEEE Third Workshop on Future Trends of Distributed Computing Systems, Taipei, Taiwan, April 1992, pp. 100-110

[11] Plagemann, T., Plattner, B., Vogt, M., Walter, T.: "Modules as Building Blocks for Protocol Configuration", Proceedings of International Conference on Network Protocols ICNP'93, San Francisco, October 1993, pp. 106-113

[12] Plagemann. T., Waclawczyk, J., Plattner, B.: "Management of Configurable Protocols for Multimedia Requirements", to appear in: Proceedings of First ISMM International Conference on Distributed Multimedia Systems and Applications, Honolulu, August 1994

[13] Vogt, M., Plagemann, T., Plattner, B., Walter, T.: "A Runtime Environment for Da CaPo", in: Proceedings of INET'93, International Networking Conference Internet Society, San Francisco, August 1993

[14] Hutchinson, N. C., Peterson, L. L.: "The x-Kernel: An Architecture for Implementing Network Protocols" in: IEEE Transactions on Software Engineering, January 1991, pp. 64-76

[15] Gotti, A: "CoRA - Configuration and Resource Allocation in Da CaPo" (in German), Diploma Thesis at Computer Engineering and Networks Laboratory, ETH Zurich, March 1994

[16] ITU - Telecommunications Standardization Sector, Study Group 7 / WP: "Quality of Service Framework (draft no. 3)", Source: SC 21/WG (N 1298), Geneva, February 1994

PART FOUR

Parallel Implementations and Error Handling

8

Measuring the Impact of Alternative Parallel Process Architectures on Communication Subsystem Performance

Douglas C. Schmidt and Tatsuya Suda
schmidt@ics.uci.edu and suda@ics.uci.edu
Department of Information and Computer Science
University of California, Irvine, California 92717*

Abstract

A communication subsystem consists of protocol functions and operating system mechanisms that support the implementation and execution of protocol stacks. To effectively parallelize a communication subsystem, careful consideration must be given to the process architecture used to structure multiple processing elements. A process architecture binds one or more processing elements with the protocol tasks and messages associated with protocol stacks in a communication subsystem. This paper outlines the two fundamental types of process architectures (task-based and message-based) and describes performance experiments conducted on three representative examples of these two types of process architectures – Layer Parallelism, which is a task-based process architecture, and Message-Parallelism and Connectional Parallelism, which are message-based process architectures. These experiments measure the impact of the process architecture on connectionless and connection-oriented protocol stacks (based upon UDP and TCP) in a shared-memory multi-processor operating system. The results from these experiments indicate that the choice of process architecture significantly affects communication subsystem performance.

1 Introduction

Advances in VLSI and fiber optic technology are shifting performance bottlenecks from the underlying networks to the communication subsystem. A communication subsystem consists of *protocol functions* (such as connection management, end-to-end flow control, remote context management, segmentation/reassembly, demultiplexing, message buffering, error protection, session control, and

*This research is supported in part by grants from the University of California MICRO program, Hughes Aircraft, Nippon Steel Information and Communication Systems Inc. (ENICOM), Hitachi Ltd., Hitachi America, and Tokyo Electric Power Company.

presentation conversions) and *operating system mechanisms* (such as process management, asynchronous event invocation, message buffering, and layer-to-layer flow control) that support the implementation and execution of communication protocol stacks composed of protocol functions.

Executing protocol functions and OS mechanisms in parallel on multi-processor platforms is a promising technique for increasing protocol processing rates and reducing latency. To significantly increase communication subsystem performance on shared memory multi-processor platforms, however, the speed-up obtained from parallelism must outweight the *context switching* and *synchronization* overhead associated with parallel processing. A context switch is triggered when an executing process relinquishes its associated processing element (PE) voluntarily or involuntarily. Depending on the underlying OS and hardware platform, performing a context switch may involve dozens to hundreds of instructions to flush register windows, memory caches, instruction pipelines, and translation look-aside buffers. Synchronization overhead arises from locking mechanisms that serialize access to shared objects (such as messages, message queues, protocol connection records, and demultiplexing tables) used when processing protocols in parallel.

A number of *process architectures* have been proposed as the basis for parallelizing communication subsystems [1, 2, 3, 4]. There are two fundamental types of process architectures: *task-based* and *message-based*. Task-based process architectures are formed by binding one or more PEs to units of protocol functionality (such as presentation layer formatting or transport layer segmentation/reassembly, acknowledgment processing, end-to-end flow control, and retransmission timer processing). In a task-based process architecture, parallelism is achieved by executing protocol tasks in separate PEs and passing data messages and control messages between the tasks/PEs. In contrast, message-based process architectures are formed by binding the PEs to data messages and control messages received from applications and network interfaces. In a message-based process architecture, parallelism is achieved by escorting multiple data messages and control messages on separate PEs through a stack of protocol tasks.

Protocol suites (such as the Internet and ISO OSI reference models) may be implemented using either task-based or message-based process architectures. However, these two types of process architectures exhibit significantly different performance characteristics that are affected by the underlying operating system and hardware platform. For instance, on shared memory multi-processor platforms, task-based process architectures often result in high data movement and context switching overhead [5]. Likewise, in a message-passing transputer multi-processor environment, message-based process architectures typically result in high levels of synchronization overhead [2].

Existing research has generally selected a single type of process architecture (either task-based or message-based) and studied it in isolation. Moreover, since different studies have been performed on different OS and hardware platforms, using different protocols and implementation techniques, it is difficult to compare the results obtained from these studies in a controlled manner. This paper describes results obtained from systematic comparisons of the performance impact of task-based and message-based process architectures. These results were obtained using an object-oriented framework that facilitates controlled experiments with alternative process architectures on shared memory multi-processor platforms [6]. The framework controls for a number of key confounding factors (such as protocol functionality, concurrency control schemes, and application traffic characteristics) in order to precisely measure the performance impact of different process architectures for parallelizing

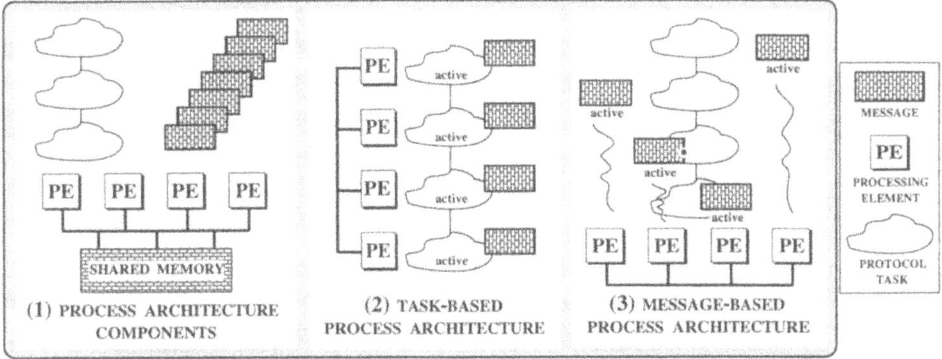

Figure 1: Basic Process Architecture Components and Interrelationships

communication protocol stacks.

This paper is organized as follows: Section 2 outlines the fundamental types of process architectures and compares related work accordingly; Section 3 describes the design and implementation of the protocol stacks and process architectures used in the experiments reported in Section 4; and Section 5 presents concluding remarks.

2 Alternative Process Architectures

Figure 1 (1) illustrates the basic elements that form the foundation of a process architecture:

- *Control messages and data messages* – which are sent and received from one or more applications and network devices
- *Protocol processing tasks* – which are the units of protocol functionality that process the control messages and data messages
- *Processing elements* (PEs) – which execute protocol tasks

There are two fundamental types of process architectures (*task-based* and *message-based*) that structure these basic elements differently. Task-based process architectures bind one or more PEs to protocol processing tasks. In this architecture, tasks are the active elements, whereas messages processed by the tasks are the passive elements (shown in Figure 1 (2)). Conversely, message-based process architectures bind the PEs to the control messages and data messages received from applications and network interfaces. In this architecture, messages are the active elements and tasks are the passive elements (shown in Figure 1 (3)).

The remainder of this section briefly examines several alternative process architectures in each category.

2.1 Task-based Process Architectures

Task-based process architectures associate processes[1] with clusters of one or more protocol tasks. Two representative examples of task-based process architectures are the *Layer Parallelism* and *Functional Parallelism* process architectures. The primary difference between these two process architectures involves the granularity of the protocol processing tasks. Layers are more "coarse-grained" than functions since they cluster multiple protocol tasks together to form a composite service (such as the end-to-end transport service provided by the OSI transport layer).

Layer Parallelism associates a separate process with each layer (*e.g.,* the presentation, transport, and network layers) in a protocol stack. Certain protocol header and data fields in the outgoing and incoming messages may be processed in parallel as they flow through a pipeline of protocol stack layers. Buffering and flow control are generally necessary since processing activities in each layer may execute at different rates.

Functional Parallelism associates a separate process with each protocol function (such as header composition, acknowledgement, retransmission, segmentation, reassembly, and routing). These protocol functions execute in parallel and communicate by passing control messages and data messages to each other.

In general, implementing pipelined task-based process architectures is relatively straightforward. Task-based process architectures map directly onto conventional layered communication models using well-structured "producer/consumer" designs. Moreover, minimal synchronization mechanisms are necessary *within* a layer or function since parallel processing is typically serialized at a service access point (such as the transport layer or application layer interface). However, as shown in Section 4, task-based process architectures are susceptible to high context switching overhead on shared memory platforms. This problem is exacerbated when the number of protocol tasks exceeds the number of PEs, due to the context switching performed when transferring messages between protocol tasks.

2.2 Message-based Process Architectures

Message-based process architectures associate processes with messages rather than protocol layers or functions. Two common examples of message-based process architectures are *Connectional Parallelism* and *Message Parallelism.* The primary difference between these approaches involves the granularity at which messages are demultiplexed onto processes. Connectional Parallelism demultiplexes all messages bound for the same connection onto the same process, whereas Message Parallelism demultiplexes messages onto any available process.

Connectional Parallelism uses a separate process to handle the messages associated with each open connection. Within a connection, a series of protocol processing tasks are invoked sequentially on each message as it flows through a protocol stack. Outgoing messages generally borrow the thread of control from the application process and use it to escort messages down a protocol stack. For

[1] In this paper, the term "process" is used to refer to a series of instructions executing within an address space; this address space may be shared with other processes. Different terminology (such as lightweight processes [6] or threads [7]) has also been used to denote the same basic concepts. Our use of the term process is consistent with the definition adopted in [8].

incoming messages, a network interface or packet filter typically performs demultiplexing operations to determine the correct process for each message.

Message Parallelism associates a separate process with every incoming or outgoing message. A process receives a message from an application or network interface and escorts the message through the protocol processing tasks in the protocol stack. As with Connectional Parallelism, outgoing messages generally borrow the thread of control from the application that initiated the message transfer.

In general, a large degree of potential parallelism exists with the message-based process architectures. The degree of parallelism depends on characteristics that change dynamically (such as messages or connections), rather than on the relatively static characteristics (such as the number of layers or protocol functions) that are associated with task-based process architectures. Depending on other communication subsystem characteristics (such as memory and bus bandwidth), this dynamism may enable message-based process architectures to effectively use a larger number of PEs.

2.3 Related Work

A number of studies have investigated the performance characteristics of task-based process architectures developed to run on either message passing or shared memory platforms. [5] measures the impact of several implementations of the transport and session layers in the OSI reference model using an ADA-like rendezvous-style of Layer Parallelism in a nonuniform access shared memory environment. [9] measures the performance of a Functional Parallelism process architecture for presentation layer and transport layer functionality on a shared memory multi-processor. [10] measures the performance of a de-layered, function-oriented transport system [11] using Functional Parallelism on a message passing transputer multi-processor platform. An earlier study [2] measured the performance of the OSI transport layer and network layer in a similar transputer environment. [12] also uses a multi-processor transputer platform to measure the performance of several data-link layer protocols.

Other studies have investigated message-based process architectures. All these studies utilize shared memory platforms. [13] measured the performance of the TCP, UDP, and IP protocols using a Message Parallelism process architecture on a uniprocessor platform running the *x*-kernel. [1] measures the impact of synchronization on Message Parallelism implementations of TCP and UDP transport protocols built within a multi-processor version of the *x*-kernel. [8] measures the performance of the Nonet transport protocol on a multi-processor version of Plan 9 STREAMS developed using Message Parallelism. [3] measures the performance of the OSI protocol stack, focusing primarily on the presentation and transport layers using Message Parallelism. [14] measures the performance of the TCP/IP protocol stack using Connectional Parallelism in a multi-processor version of System V STREAMS.

The work presented in this paper extends existing work by measuring a number of task-based and message-based process architectures in a controlled environment. Our experiments consider the impact of both synchronization and context switching overhead. In addition to measuring data link, network, and transport layer performance, our experiments also investigate presentation layer performance. The presentation layer is widely considered to be one of the primary bottlenecks in high-performance communication subsystems.

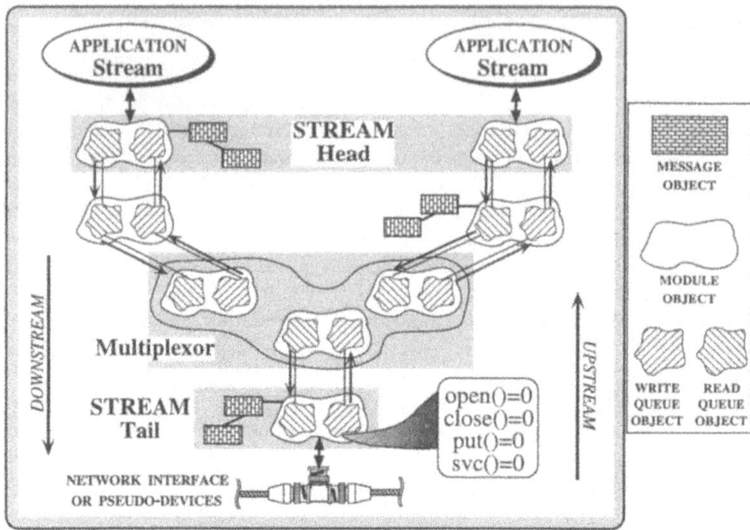

Figure 2: Components in the ADAPTIVE Service eXecutive Framework

3 Structure of the Experiments

This section describes the object-oriented framework, communication protocols, and process architectures we developed and used in the performance experiments reported in Section 4.

3.1 The ADAPTIVE Service eXecutive Framework

The communication protocols and process architectures in this study were developed using components provided by the ADAPTIVE Server eXecutive (`ASX`) framework [15]. The `ASX` framework is an integrated set of object-oriented components that facilitate experimentation with task-based and message-based process architectures on shared memory multi-processor platforms.

Components in the `ASX` are responsible for coordinating one or more *Streams*. A Stream is an object used to configure and execute protocol-specific functionality in the `ASX` framework run-time environment. As illustrated in Figure 2, a Stream contains a series of inter-connected `Modules` that may be linked together by developers at installation-time or by applications at run-time. `Modules` are objects that developers use to decompose the architecture of a protocol stack into a series of inter-connected, functionally distinct layers. Each layer implements a cluster of related protocol-specific functions (such as an end-to-end transport service, a presentation layer formatting service, or a real-time PBX signal routing service). Every `Module` contains a pair of `Queue` objects that partition a layer into its constituent read-side and write-side protocol-specific processing functionality.

Any layer that performs multiplexing and demultiplexing of message objects between related Streams may be developed using a `Multiplexor` object. A `Multiplexor` is a C++ template-based container class that provides mechanisms to route messages between `Modules` in a collection of related Streams. A complete Stream is represented as an inter-connected series of `Module` objects that communicate by exchanging messages with adjacent objects. `Modules` and `Multiplexors` may be joined together in essentially arbitrary configurations in order to satisfy application requirements and enhance component reuse.

The ASX framework employs a number of object-oriented design techniques (such as design patterns [16] and hierarchical decomposition) and C++ language features (such as inheritance, dynamic binding, and parameterized types). These design techniques and language features enable developers to incorporate protocol-specific functionality into a Stream without modifying the protocol-independent framework components. For example, incorporating a new level of protocol functionality into a Stream at installation-time or at run-time involves the following steps:

1. Inheriting from the `Queue` interface and selectively overriding several methods (described below) in the `Queue` subclass to implement protocol-specific functionality
2. Allocating a new `Module` that contains two instances (one for the read-side and one for the write-side) of the protocol-specific `Queue` subclass
3. Inserting the `Module` into a Stream object at the appropriate level (*e.g.*, the transport layer, network layer, data-link layer, etc.)

The ASX framework incorporates concepts from several other modular communication frameworks including System V STREAMS [17], the *x*-kernel [13], and the Conduit [18] (a survey of these and other communication frameworks appears in [19]). These frameworks all contain features that support the flexible configuration of communication subsystems by inter-connecting building-block protocol components. These frameworks encourage the development of standard reusable protocol components by decoupling protocol-specific processing functionality from the surrounding framework infrastructure. In addition to supplying building-block protocol and service components, the ASX framework also extends the existing communication frameworks by providing additional components that decouple protocol functionality from the following configuration decisions:

- The type of locking mechanisms used to synchronize access to shared objects
- The use of message-based and task-based process architectures
- The use of kernel-level vs. user-level execution agents

3.2 Communication Protocols

Two types of protocol stacks are used in the experiments. One protocol stack is based on the connectionless UDP transport protocol. The other protocol stack is based on the connection-oriented TCP transport protocol. The protocol stacks contain the data-link, network, transport, and presentation layers. The presentation layer is included in the experiments since it represents a major bottleneck in high-performance communication subsystems, due primarily to the large amount of data movement overhead it often incurs.

Both the connectionless and connection-oriented protocol stacks were developed by specializing reusable components in the ASX framework via inheritance and parameterized types. Inheritance and parameterized types are used to hold the protocol functionality constant while systematically varying the process architecture. Each layer in a protocol stack is implemented as a `Module` whose read-side and write-side both inherit interfaces and implementations from the `Queue` class described in [15]. The necessary synchronization and demultiplexing mechanisms are parameterized using C++ template arguments that are instantiated based on the type of process architecture being tested.

Data-link layer processing in each protocol stack is performed by the `DLP Module`. This `Module` transforms network packets received from a network interface into the canonical message format used internally by the Stream components.[2] The network and transport layer components of the protocol stacks are based on the IP, UDP, and TCP implementation in the BSD 4.3 Reno release. The 4.3 Reno TCP implementation contains the TCP header prediction enhancements, as well as the slow start algorithm and congestion avoidance features. The UDP and TCP transport protocols are configured into the ASX framework via the `UDP` and `TCP Modules`. Network layer processing is performed by the `IP Module`. This `Module` performs routing and segmentation/reassembly of Internet Protocol (IP) packets.

Presentation layer functionality is implemented in the `XDR Module` using marshalling routines produced by the ONC eXternal Data Representation (XDR) stub generator (`rpcgen`). The ONC XDR stub generator automatically translates a set of type specifications into marshalling routines. These routines encode/decode implicitly-typed messages before/after they are exchanged among hosts that may possess heterogeneous processor byte-orders. The ONC presentation layer conversion mechanisms consist of a type specification language (XDR) and a set of library routines that implement the appropriate encoding and decoding rules for built-in integral types (*e.g.*, char, short, int, and long) and real types (*e.g.*, float and double). In addition, these library routines may be combined to produce marshalling routines for arbitrary user-defined composite types (such as record/structures, unions, arrays, and pointers). Messages exchanged via XDR are implicitly-typed, which improves marshalling performance at the expense of run-time flexibility. The `XDR` functions selected for both the connectionless and connection-oriented protocol stacks convert incoming and outgoing messages into and from variable-sized arrays of structures containing both integral and real values. This conversion processing involves byte-order conversions, as well as dynamic memory allocation and deallocation.

3.3 Process Architectures

The remainder of this section outlines the structure of connectionless and connection-oriented protocol stacks developed using task-based and message-based process architectures.

3.3.1 Structure of the Task-based Process Architecture

[2]Preliminary tests using the widely-available `ttcp` benchmarking tool indicated that the PE, bus, and memory performance of the SunOS multi-processor platform used in the experiments was capable of processing messages through the protocol stack at a much faster rate than the 10 Mbps Ethernet network interface was capable of handling. Therefore, for our process architecture experiments, the network interface was simulated with a single-copy pseudo-device driver operating in loop-back mode.

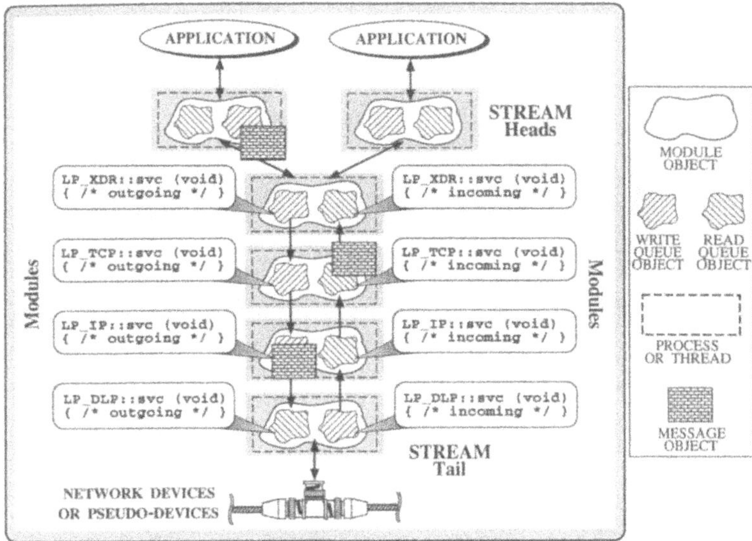

Figure 3: Layer Parallelism

• **Layer Parallelism:** Figure 3 illustrates the ASX framework components that implement a Layer Parallelism process architecture for the TCP-based connection-oriented and UDP-based connectionless protocol stacks. Protocol-specific processing at each protocol layer is performed via the `Queue::svc` method. This method is invoked by a daemon process associated with the `Module` that implements the protocol layer (*e.g.*, `LP_XDR`, `LP_TCP`, `LP_IP`, and `LP_DLP`). These daemon processes cooperate in a producer/consumer manner, operating on the header and data fields of messages corresponding to their particular protocol layer in parallel. Each `svc` method performs its protocol functions before passing the message to an adjacent `Module` that runs asynchronously in a separate daemon process. Since daemon processes all share a common address space, messages are not copied when passed between adjacent `Modules`. However, moving messages between processes may invalidate per-PE data caches.

The connectionless and connection-oriented Layer Parallelism process architecture protocol stacks are designed in a similar manner. The primary difference is that the objects in the connectionless transport layer `Module` implement the simpler UDP functionality. UDP does not generate acknowledgements, keep track of round-trip time estimates, or manage congestion windows, etc.

3.3.2 Structure of the Message-based Process Architectures

• **Connectional Parallelism:** The protocol stack depicted in Figure 4 (1) illustrates an ASX-based implementation of the Connectional Parallelism process architecture. Each connection is associated

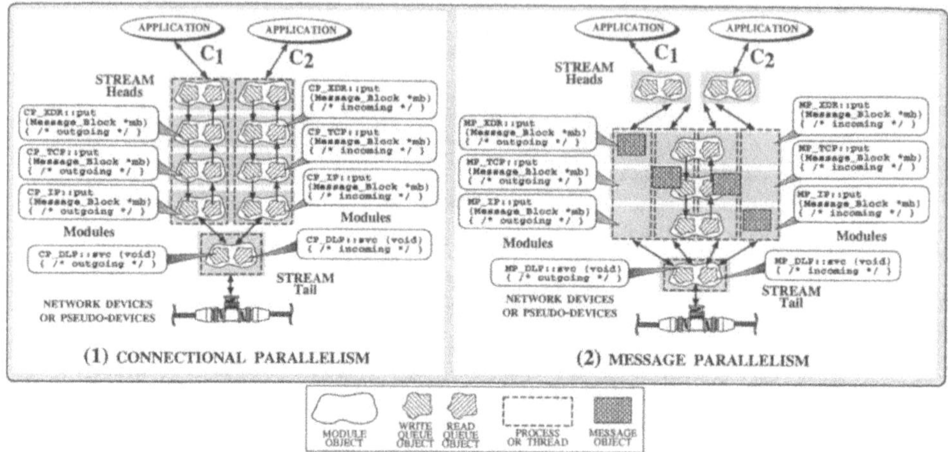

Figure 4: Message-based Process Architectures

with a separate process that performs the data-link, network, transport, and presentation layer functionality for that connection. Protocol tasks are divided into four inter-connected `Modules`, corresponding to the data-link, network, transport, and presentation layers in the ISO OSI communication model. Data-link processing is performed in the `CP_DLP Module`. This `Module` uses its read-side `svc` method to (1) transform network messages into the canonical internal message format that is processed by higher-level components in a Stream and (2) demultiplex incoming messages onto the appropriate transport layer connection.[3] Once a message has been demultiplexed onto a connection, all that connection's context information is directly accessible within the address space of the associated process. This is beneficial since (1) pointers to messages may be passed between protocol layers via simple procedure calls (rather than using more complicated and costly interprocess communication mechanisms used for Layer Parallelism process architecture), (2) cache affinity properties may be preserved since messages are processed largely within a single PE cache, and (3) minimal internal locking is required within a connection. Therefore, a process may operate on its connection's messages without incurring additional demultiplexing, synchronization, and context switching overhead. The `CP_IP`, `CP_TCP`, and `CP_XDR Modules` all perform their processing synchronously in their respective `put` methods.

• **Message Parallelism:** Figure 4 (2) illustrates a message-based process architecture for the connection-oriented protocol stack. When an incoming message arrives, it is handled by the `MP_DLP::svc` method, which manages a pool of pre-spawned threads. Each message is associated with a sep-

[3] The connection-oriented implementation of Connectional Parallelism performs "eager demultiplexing" via a packet filter at the data-link layer.

arate thread that escorts the message synchronously through a series of inter-connected `Queues` in a Stream. Each layer of the protocol stack performs its protocol functions and then makes an upcall to the next adjacent layer in the protocol stack by invoking the `Queue::put` method in that layer. The `put` method executes the protocol tasks associated with its layer. For instance, the `MP_TCP::put` method utilizes mutual exclusion (mutex) objects that serialize access to per-connection control blocks as separate messages from the same connection ascend the protocol stack in parallel.

The connectionless message-based protocol stack is structured in a similar manner, though it performs the simpler set of UDP functionality. Unlike the `MP_TCP::put` method, the `MP_UDP::put` method handles each message concurrently and independently, without explicitly preserving inter-message ordering. This reduces the number of synchronization operations required to locate and update shared resources, which improves performance.

4 Communication Subsystem Performance Experiment Results

This section describes experiments that measure the performance impact of different combinations of the protocol stacks and process architectures described above. The multi-processor platform and the measurement tools used in the experiments are also discussed.

4.1 Multi-processor Platform

All experiments were conducted on an otherwise idle Sun 690MP SPARCserver, which contains 4 SPARC 40 MHz processing elements (PEs), each capable of performing at 28 MIPs. The memory bandwidth of the SPARCserver platform was measured at approximately 150 Mbits/sec, which represents an upper limit on protocol processing throughput. Protocol processing throughput is also significantly affected by context switching and synchronization overhead exhibited by the different task-based and message-based process architectures. The costs of context switching and synchronization overhead in the SPARCserver platform are described below.

The operating system used for the experiments is release 5.3 of SunOS, which provides a multi-threaded kernel that allows multiple system calls and device interrupts to execute in parallel [6]. All the process architectures in these experiments execute protocol tasks in separate *unbound* threads multiplexed over 1, 2, 3, or 4 SunOS *lightweight processes* (LWPs) within a process. SunOS 5.3 maps each LWP directly onto a separate kernel thread. Since kernel threads are the units of PE scheduling and execution in SunOS, this mapping enables multiple LWPs (each executing protocol processing tasks in an unbound thread) to run in parallel on the SPARCserver's PEs.

Rescheduling and synchronizing a SunOS LWP involves a kernel-level context switch. The time required to perform a context switch between two LWPs was measured to be approximately 30 *u*secs. During this time, the OS performs system-related overhead (such as flushing register windows, instruction and data caches, instruction pipelines, and translation lookaside buffers) on the PE and therefore does not process protocol tasks. Measurements also revealed that it requires approximately 2 *u*secs to acquire or release a `Mutex` object implemented using a SunOS spin-lock. Likewise, measurements indicated that approximately 90 *u*secs are required to synchronize two LWPs using

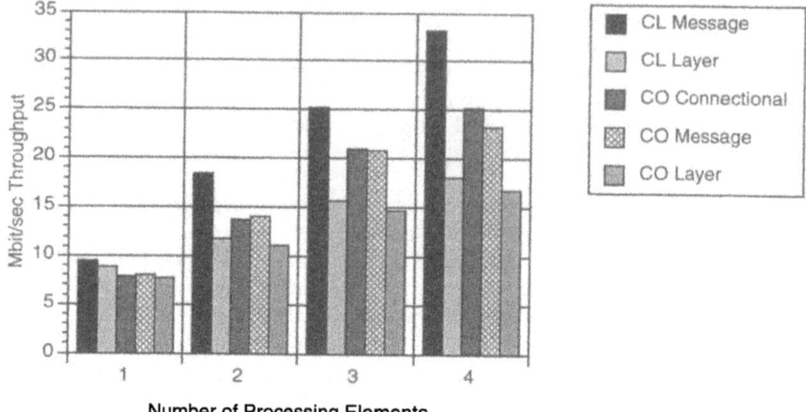

Figure 5: Process Architecture Throughput

`Condition` objects implemented using SunOS sleep-locks. The larger amount of overhead for the `Condition` object operations compared with the `Mutex` object operations occurs from the more complex locking algorithms involved, as well as the additional context switching incurred by SunOS sleep-locks.

4.2 Measurement Results

This section presents results obtained by measuring the data reception portion of the connection-oriented and connectionless protocol stacks implemented using the Layer Parallelism task-based process architecture and the Connectional Parallelism and Message Parallelism message-based process architectures. Three types of measurements were obtained for each combination of process architecture and protocol stack: *total throughput*, *context switching overhead*, and *synchronization overhead*.

Total throughput was measured by holding the protocol functionality, application traffic patterns, and network interfaces constant and systematically varying the process architecture to determine the resulting performance impact. Each benchmarking session consisted of transmitting 10,000 4 Kbyte messages through an extended version of the widely available `ttcp` protocol benchmarking tool. The original `ttcp` tool measures the processing resources and overall user and system time required to transfer data between a transmitter process and a receiver process communicating via TCP or UDP. The flow of data is uni-directional, with the transmitter flooding the receiver with a user-specified number of data buffers. Various sender and receiver parameters (such as the number of data buffers transmitted and the size of data buffers and protocol windows) may be selected at run-time.

The version of `ttcp` used in our experiments was enhanced to allow a user-specified number of communicating applications to be measured simultaneously. This feature measured the impact of multiple connections on the performance of process architectures (the connection-oriented process

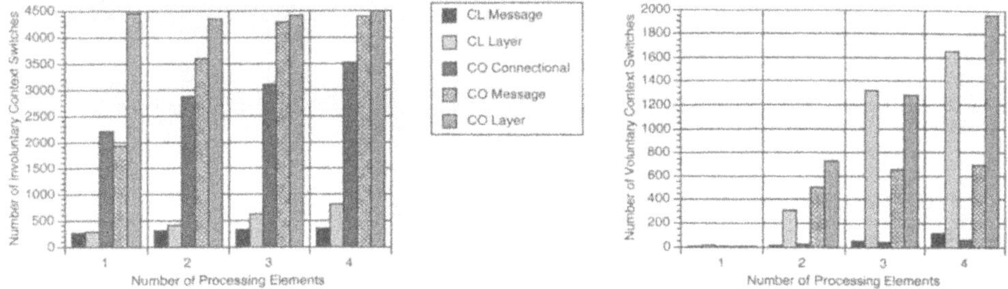

Figure 6: Process Architecture Context Switching Overhead

architecture tests were run using 4 connections). The `ttcp` tool was also modified to use the `ASX`-based protocol stacks configured according to the process architectures described in Section 4.2. To measure the impact of parallelism on throughput, each test was run using 1, 2, 3, and 4 PEs. Furthermore, each test was performed multiple times to detect the amount of spurious interference incurred from other internal OS tasks (the variance between test runs proved to be insignificant).

Context switching and synchronization measurements were obtained to help explain differences in the throughput results. These metrics were obtained from the SunOS 5.3 `/proc` file system, which records the number of voluntary and involuntary context switches incurred by threads in a process, as well as the amount of time spent waiting to obtain and release locks on `Mutex` and `Condition` objects.

Figure 5 illustrates throughput (measured in Mbits/sec) as a function of the number of PEs for the task-based and message-based process architectures used to implement the connection-oriented (CO) and connectionless (CL) protocol stacks.[4] The results in this figure indicate that increasing the number of PEs improves throughput for all the process architectures. However, the message-based process architectures significantly outperformed their task-based counterparts as the number of PEs increased from 1 to 4. For example, the performance of the connection-oriented task-based process architecture was only slightly better using 4 PEs (approximately 16 Mbits/sec, or 1.92 milliseconds per-message processing time) than the message-based process architecture was using 2 PEs (14 Mbits/sec, or 2.3 milliseconds per-message processing time). Moreover, if a larger number of PEs had been available, it appears likely that the performance improvement gained from parallel processing in the task-based process architectures would have leveled off sooner than the message-based tests due to the higher rate of growth for context switching and synchronization shown in Figure 6 and Figure 7.

The Connection Parallelism process architecture exhibited the highest levels of throughput for the connection-oriented protocol stacks when the number of PEs equaled the number of connections. The major limitation with Connectional Parallelism, however, is that it only utilizes parallelism to improve *aggregate* end-system performance since each individual connection still executes sequentially. In

[4]The Connectional Parallelism process architecture does not support the connectionless protocol stack.

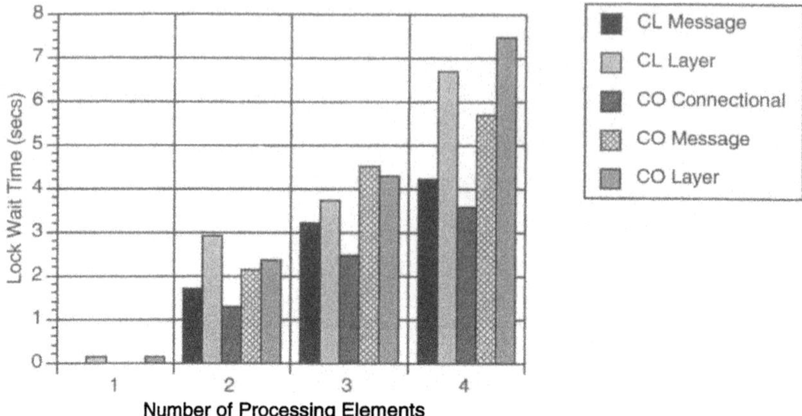

Figure 7: Process Architecture Locking Overhead

contrast, Message Parallelism also utilizes multiple PEs effectively for a single connection.

Figure 6 illustrates the number of *involuntary* and *voluntary* context switches incurred by the process architectures measured in this study. An involuntary context switch occurs when the OS kernel preempts a running thread. For example, the OS preempts running threads periodically when their LWP time-slice expires in order to schedule other threads to execute. A voluntary context switch is triggered when a thread puts itself to sleep until certain resources (such as I/O devices or synchronization locks) become available. For example, when a protocol task attempts to acquire a resource that may not become available immediately (such as obtaining a message from an empty list of messages in a `Queue`), the protocol task puts itself to sleep by invoking the `wait` method of a `Condition` object. This action causes the OS kernel to preempt the current thread and perform a context switch to another thread that is capable of executing protocol tasks immediately.

As shown in Figure 6, The Layer Parallelism task-based process architectures exhibited slightly higher levels of involuntary context switching than the message-based process architectures. This is due mostly to the fact that the Layer Parallelism tests required more time to process the 10,000 messages and were therefore pre-empted a greater number of times. Furthermore, the task-based process architectures also incurred significantly more voluntary context switches, which accounts for the substantial improvement in overall throughput exhibited by the message-based process architectures. The primary reason for the increased context switching is that the locking mechanisms used by the message-based process architectures utilize adaptive spin-locks (which rarely trigger a context switch), rather than the sleep-locks used by task-based process architectures (which *do* trigger a context switch). Note that the Connectional Parallelism process architecture incurred the least amount of context switching for the connection-oriented protocol stacks.

Figure 7 indicates the amount of execution time that the `/proc` metrics reported as being devoted to waiting to acquire and release locks in the connectionless and connection-oriented benchmark programs. As with context switching benchmarks, the message-oriented process architectures incurred considerably less synchronization overhead, particularly when 4 PEs were used. As with context switching, the spin-locks used by message-based process architecture reduce the amount of time spent synchronizing, in comparison with the sleep-locks used by the task-based process architectures.

5 Concluding Remarks

Despite an increase in the availability of operating system and hardware platforms that support networking and parallel processing, developing communication subsystems that effectively utilize parallel processing remains a complex and challenging task. A key aspect of communication subsystem performance involves the type of process architecture selected to structure parallel processing of protocol tasks. Measurement results reported in this paper indicate that task-based process architectures incur much higher levels of context switching and synchronization overhead on a shared memory platform, which significantly reduces performance. Conversely, the message-based process architectures (particularly Connectional Parallelism) incur much less context switching and synchronization, and therefore exhibit higher performance.

The ASX framework contributed to these performance experiments by helping to decouple the protocol-specific functionality from the underlying of process architecture. This decoupling increased reuse and simplified development, configuration, and experimentation with parallel protocol stacks. Components in the ASX framework are freely available via anonymous ftp from `ics.uci.edu` in the file `gnu/C++_wrappers.tar.Z`. This distribution contains complete source code, documentation, and example test drivers for the C++ components. Components in the ASX framework have been ported to both UNIX and Windows NT. The ASX framework is currently being used in a number of commercial products including the AT&T Q.port ATM signaling software product, the Ericsson EOS family of PBX monitoring applications, and the network management portion of the Motorola Iridium mobile communications system.

References

[1] Mats Bjorkman and Per Gunningberg, "Locking Strategies in Multiprocessor Implementations of Protocols," in *SIGCOMM Symposium on Communications Architectures and Protocols*, (San Francisco, California), ACM, 1993.

[2] M. Zitterbart, "High-Speed Transport Components," *IEEE Network Magazine*, pp. 54–63, January 1991.

[3] M. Goldberg, G. Neufeld, and M. Ito, "A Parallel Approach to OSI Connection-Oriented Protocols," in *Proceedings of the* 3rd *IFIP Workshop on Protocols for High-Speed Networks*, (Stockholm, Sweden), May 1992.

[4] J. Jain, M. Schwartz, and T. Bashkow, "Transport Protocol Processing at GBPS Rates," in *SIGCOMM Symposium on Communications Architectures and Protocols*, (Philadelphia, PA), pp. 188–199, ACM, Sept. 1990.

[5] C. M. Woodside and R. G. Franks, "Alternative software architectures for parallel protocol execution with synchronous ipc," *IEEE/ACM Transactions on Networking*, vol. 1, Apr. 1993.

[6] J. Eykholt, S. Kleiman, S. Barton, R. Faulkner, A. Shivalingiah, M. Smith, D. Stein, J. Voll, M. Weeks, and D. Williams, "Beyond Multiprocessing... Multithreading the SunOS Kernel," in *Summer USENIX Conference*, (San Antonio, Texas), June 1992.

[7] A. Tevanian, R. Rashid, D. Golub, D. Black, E. Cooper, and M. Young, "Mach Threads and the Unix Kernel: The Battle for Control," in *USENIX Summer Conference*, USENIX, August 1987.

[8] D. Presotto, "Multiprocessor Streams for Plan 9," in *United Kingdom UNIX User Group Summer Proceedings*, (London, England), Jan. 1993.

[9] B. Lindgren, B. Krupczak, M. Ammar, and K. Schwan, "Parallelism and Configurability in High Performance Protocol Architectures," in *Proceedings of the Second Workshop on the Architecture and Implementation of High Performance Communication Subsystems*, (Williamsburg, Virgina), IEEE, September 1993.

[10] T. Braun and M. Zitterbart, "Parallel Transport System Design," in *Proceedings of the 4th IFIP Conference on High Performance Networking*, (Belgium), IFIP, 1993.

[11] M. Zitterbart, B. Stiller, and A. Tantawy, "A Model for High-Performance Communication Subsystems," *IEEE Journal on Selected Areas in Communication*, vol. 11, pp. 507–519, May 1993.

[12] D. Giarrizzo, M. Kaiserswerth, T. Wicki, and R. Williamson, "High-Speed Parallel Protocol Implementations," in *Proceedings of the 1st International Workshop on High-Speed Networks*, pp. 165–180, May 1989.

[13] N. C. Hutchinson and L. L. Peterson, "The *x*-kernel: An Architecture for Implementing Network Protocols," *IEEE Transactions on Software Engineering*, vol. 17, pp. 64–76, January 1991.

[14] S. Saxena, J. K. Peacock, F. Yang, V. Verma, and M. Krishnan, "Pitfalls in Multithreading SVR4 STREAMS and other Weightless Processes," in *Winter USENIX Conference*, (San Diego, CA), pp. 85–106, Jan. 1993.

[15] D. C. Schmidt and T. Suda, "The ADAPTIVE Service eXecutive: an Object-Oriented Architecture for Configuring Concurrent Distributed Applications," in *The proceedings of the 8th International Working Conference on Upper Layer Protocols, Architectures, and Applications*, (Barcelona, Spain), IFIP, June 1994.

[16] D. C. Schmidt, "Reactor: An Object Behavioral Pattern for Concurrent Event Demultiplexing and Dispatching," in 1st *Annual Conference on the Pattern Languages of Programs*, (Monticello, Illinois), ACM, August 1994.

[17] D. Ritchie, "A Stream Input–Output System," *AT&T Bell Labs Technical Journal*, vol. 63, pp. 311–324, Oct. 1984.

[18] J. M. Zweig, "The Conduit: a Communication Abstraction in C++," in *USENIX C++ Conference Proceedings*, pp. 191–203, USENIX Association, April 1990.

[19] D. C. Schmidt and T. Suda, "Transport System Architecture Services for High-Performance Communications Systems," *IEEE Journal on Selected Areas in Communication*, vol. 11, pp. 489–506, May 1993.

9

A Modular VLSI Implementation Architecture for Communication Subsystems

Torsten Braun, Jochen Schiller, Martina Zitterbart
University of Karlsruhe, Institute of Telematics, Zirkel 2, 76128 Karlsruhe
Phone: +49 721 608-[3982,4003,4026], Fax: +49 721 388097
Email: [braun,schiller,zit]@telematik.informatik.uni-karlsruhe.de

Abstract

Implementation platforms for integrated service communication subsystems providing high performance capabilities are increasingly required. In order to provide high performance, dedicated VLSI components can be used for time-critical processing tasks such as retransmission support or memory management. A modular VLSI implementation architecture designed with specialized components allows for service flexibility. The components can be parametrized and may be selected individually dependent on the service required by the application. The implementation architecture is not limited to a certain protocol and allows the implementation of high-speed protocols with fixed size packet headers. The paper describes the architecture in general and dedicated parts in more detail. Preliminary performance values are presented and compared with measurements of typical software implementations.

Keyword Codes: B.4.1; C.1.2; C.5.4
Keywords: Data Communication Devices; Multiple Data Stream Architectures; VLSI Systems

1 Introduction

Emerging applications, mostly, require both high performance as well as support of a wide variety of communication services. For example, audio, video, and data transmission may require different services. Networks, (e.g., ATM-based networks) are able to fulfill the application requirements by providing data rates exceeding a gigabit per second and supporting different kinds of services. However, current communication subsystems (including higher layer protocols) are not able to deliver the available network performance to the applications.

Different approaches [1], [2], [3], [4] on implementing high performance communication subsystems have been undertaken during the last few years: software optimization, parallel processing, hardware support and dedicated VLSI components. Some of the approaches deal with efficient implementations of standard protocols such as OSI TP4 or TCP. Others developed protocols especially suited for advanced implementation environments.

The use of dedicated VLSI components is mostly limited to very simple communication protocols, only (e.g., [5], [6]). In this paper, we present a VLSI architecture that is especially targeted towards more complex protocols. Connection oriented protocols with advanced protocol mechanisms can be implemented on this architecture. Specific support for processing intensive functions is provided. For example, selective retransmission is provided by a dedicated VLSI component.

The architecture is highly independent of the specific protocol to be implemented and, thus, forms a general implementation platform for high-performance communication protocols.

This paper is organized as follows: Section 2 gives an overview of the VLSI implementation architecture. The extended memory management unit (E-MMU) is described in Section 3. The E-MMU also supports segmentation and reassembly. The connection processor that performs the core of the presented VLSI architecture is discussed in Section 0. Section 5 discusses the functionality of a dedicated connection component processor for managing retransmission and presents some preliminary performance results of the discussed component. Section 6 summarizes the paper and points out some future directions.

2 Modular VLSI Implementation Architecture

The modular VLSI architecture presented in this paper is designed around multiple so-called *Connection Processors (CP)* as shown in Figure 1. The architecture distinguishes between the receive and send part in order to allow concurrent processing and, thus, to increase system throughput. In addition to multiple *CP*s, a memory unit and a memory management unit are implemented for the send and the receive side, respectively.

The VLSI implementation architecture forms the connection between the network unit, which, e.g., handles the layers below the media access interface in 802.x LANs or ATM and lower layers in B-ISDN, and a host such as a workstation. Therefore, the application component in Figure 1 includes the host and an appropriate interface (e.g., DMA interface) for data transfer between host memory and send/receive RAM. To avoid buffering of data within the send/receive memory, the user data may also be copied directly from the network component into the workstation memory.

The basic architecture is similar to [6]. The *CP*s are responsible for implementing the transport protocol of the communication subsystem. As long as enough free processing power of the *CP*s is available, a connection can be mapped onto a *CP*. However, parallelism in protocol implementation is not the major issue addressed in this paper. Proper embedding in a subsystem architecture including system functions, such as memory management, is extremely important to high performance communication subsystems.

In order to process packets at extremely high rates as it may be needed, e.g., in servers inside gigabit networks, we propose the implementation of various VLSI components.

The architecture (cf. Figure 1) consists of two data memories, for the receive side (*receive RAM*) and for the send side (*send RAM*), respectively. They are managed by specialized memory management units (*E-MMU*s). The network unit, the applications, and the *CP*s can access the data memories via the *E-MMU*s. The use of pointers avoids data copies during protocol processing. Operations on the memory, such as segmentation/reassembly and allocation/deallocation, are completely handled by the *E-MMU*s. Applications may read and write data via the *E-MMU*s.

A key feature of the architecture is the replication of identical *CP*s similar to [6]. They include registers, arithmetic logical units (ALUs), timers, and other components required in a protocol implementation. The main purpose of the remaining components (*management*, *A_MUX*, *N_MUX*, *N_DMX*) is the distribution and collection of relevant information. Received data is divided into user data and a protocol header by the *E-MMU*. User data is written into the *receive RAM*. The protocol header together with a reference to the user data is delivered to the *N_DMX* that forwards it to an appropriate *CP*.

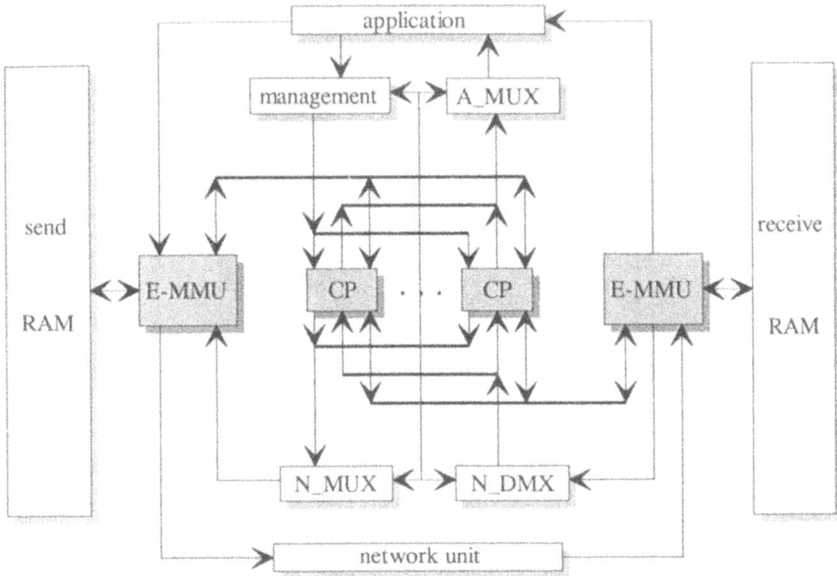

Figure 1: Modular VLSI Implementation Architecture

The *N_MUX* collects protocol headers and references to user data to be sent to the network. The *CP*s cooperate with an application component via *management* and *A_MUX*. The manager maps a request for connection establishment onto a *CP*, which then is responsible for handling that connection properly. The mapping information is distributed to the *A_MUX*, *N_MUX*, and *N_DMX*.

The four components *management*, *A_MUX*, *N_MUX*, and *N_DMX* are connected to the *CP*s via dedicated busses. The connections between *CP*s and *E-MMU*s are used for issuing *CP*-commands to the *E-MMU*s and receiving corresponding responses. In the subsequent section, the main components of the architecture, the *E-MMU* and the *CP*, are explained in some detail.

3 Extended Memory Management Unit

Memory management is one of the most time-critical system tasks of a communication subsystem. For efficient memory management, special data structures to allow simple implementation in hardware are required. These data structures have to be designed to reduce data movement to a minimum, especially in the case of segmentation and reassembly. Because several external components (application - connection processor - network unit) use the same data structures for memory management, management and access to those data structures have to be easily controlled to avoid any inconsistencies and conflicts. Conflicts may be caused by multiple processing units concurrently accessing common data structures.

In our approach, buffers and segments are the basic units for memory management. A buffer is a certain portion of consecutive, but not necessarily completely filled memory (Figure 2). A segment consists of one or more buffers.

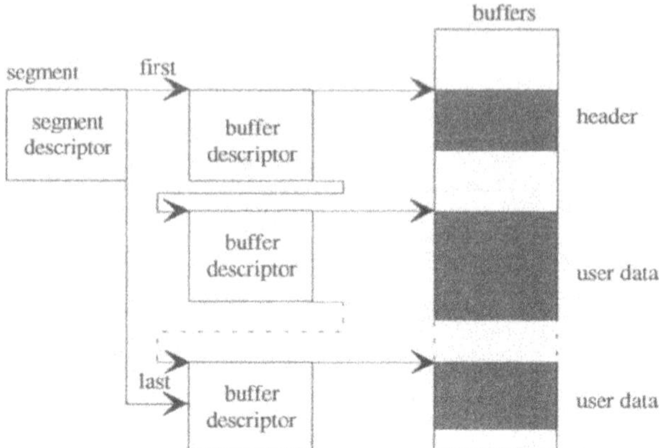

Figure 2: Data Structures for Efficient Memory Management

So-called descriptors are used to describe the content of segments and buffers. They carry information such as memory addresses, reference counters, length values etc. The buffer descriptor contains information to describe the beginning and the end of the buffer, as well as the start and length of the buffer's data. The segment descriptor describes the segment by the first and the last buffer, and the total length of data over all buffers. Because of segmentation (e.g., if the packet to segment consists of a single buffer) a buffer must be able to be part of several segments simultaneously. Therefore, the segment descriptor needs information about the beginning of its own buffer data in a buffer shared with other segments. Also, the buffer descriptor has a reference counter of segments to which it belongs. The segments and buffer descriptors themselves are collected in simply linked lists.

The internal architecture of the *E-MMU* is shown in Figure 3. The various components are interconnected with a crossbar dependent on the operations triggered by external components, an application, the connection processors, or the network unit, respectively. External components are interconnected via special interface components to the *E-MMU*. The data memory and the descriptor memory are also interconnected by such an interface.

The separation of header and user data is reflected in the E-MMU-architecture. User data must be allocated from the data memory. The header memory is usually located in the connection processor to enable direct and fast access of the connection processor to the protocol headers. Headers are passed to a demultiplexer component (N_DMX), that delivers incoming headers to the appropriate connection processor. Outgoing headers are received from a multiplexer component (N_MUX) and passed to the network unit. However, the architecture would also be able to manage a separated header memory to store the protocol headers of the complete transport subsystem.

Because user data usually have a variable length, buffer management may be very complex. To reduce the amount of wasted memory and to allow fast operations, we decided to implement the buddy algorithm for managing user data. The data structures required for implementing the algorithm are stored in an internal memory of the data MMU. The buddy algorithm provides data buffers with a length of 2^k bytes. If a data buffer with a length of m is required, a buffer with the size of 2^p is allocated with $2^{p-1} < m \leq 2^p$. If there is no appropriate buffer with the size of 2^p, an appropriate buffer is generated by dividing a buffer with the length of 2^{p+1} into two buffers of size 2^p. The fact that the buffer addresses differ in a certain bit only simplifies buffer release. If both pieces of size 2^p have to be released, they can easily combined to a buffer of size 2^{p+1}.

The buddy algorithm is useful to provide fast and efficient memory management. Furthermore, it reduces wasted memory, which is a problem when single fixed size buffers should be used, e.g., for large AAL PDUs, and it avoids fragmentation of the memory.

The descriptors reside in the descriptor memory. Because the size of a descriptor is fixed in contrast to a portion of user data, its management is much more simple than the management of user data. The descriptor memory is managed by the descriptor MMU using a stack, that contains a set of memory addresses, in an internal memory. Each memory address refers to a fixed size memory block. A memory address is taken from the stack to allocate a descriptor, while the address is pushed to the stack for release.

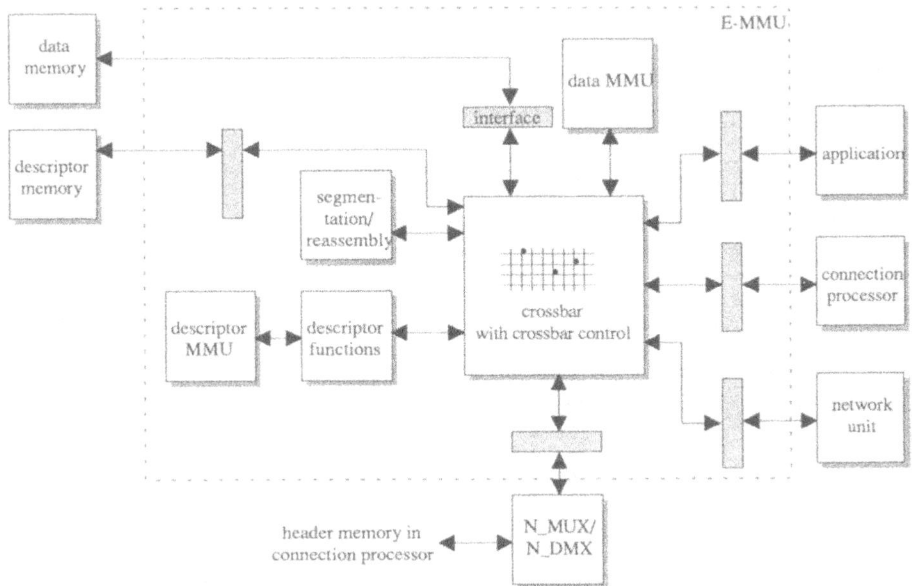

Figure 3: Architecture of the Extended Memory Management Unit (E-MMU)

If an external component, such as the network unit, allocates a data memory block, several processing steps have to be performed. First, the desired memory size must be allocated by the data MMU. Second, a buffer descriptor must be allocated by the descriptor MMU. To initialize the buffer descriptor, the address of the allocated data memory must be written into a reserved field of the buffer descriptor. Functions for initialization, allocation, and release are implemented within the descriptor function block.

Another advantage of the described data structures is the facility to efficiently support segmentation and reassembly procedures. Segmentation and reassembly can be performed arbitrarily often without copying of user data by manipulating the segment and buffer descriptors only. For segmentation, the references within the descriptors have to be adapted and new descriptors must be allocated to generate segments. For reassembly, the segments have to be combined and, thus, segment descriptors must be released.

All these actions must be initiated by the crossbar control. The crossconnect also interconnects the internal components to perform the operation required by an external component. Generally, the architecture is based on a modular decomposition of independent internal components. The components can collaborate with each other and are able to operate in parallel. The crossbar allows concurrent interconnections among the internal components and, thus, operations simultaneously triggered by external components can be performed concurrently. Therefore, the actions resulting from several concurrent operations must be coordinated to avoid inconsistencies.

The E-MMU provides the following operations to write or read to the user data memory or the descriptor memory, to allocate and release segment and buffer descriptors, to allocate and release a certain number of bytes of the data memory, and to support segmentation/reassembly functions:

command	input parameter	output parameter	comment
memory_read	type, priority, offset, address	memory word	read from memory
memory_write	type, priority, offset, address, data		write to memory
get_segment_descriptor		address	allocate segment descriptor
free_segment_descriptor	address		release segment descriptor
get_buffer_descriptor		address	allocate buffer descriptor
free_buffer_descriptor	address		release buffer descriptor
free_descriptors	address		release all descriptors of a segment
get_data_memory	number	address	allocate data buffer
free_data_memory	address		release data buffer
segment	address, segment_size	address_list	segmentation support by manipulation of the descriptors
reassemble	address_list	address	reassembly support by manipulation of the descriptors

Although, the E-MMU provides an interface to manipulate user data, access to the user data by the protocol processors should be avoided. Necessary data manipulation functions such as checksum calculation or presentation conversion functions should be performed while copying the data between the network interface and the send/receive memory or while copying between the application component and the send/receive memory, respectively.

4 Connection Processor

The main components inside a *Connection Processor* are the *FSM* processing units (*FSM_i*) that perform all protocol functions. The FSMs of a formal protocol specification can be mapped onto these processing units (e.g., protocol and interface FSMs of PATROCLOS or single FSMs of standard protocols). All *FSM* processing units work independently; they communicate via asynchronous signals, thus, enabling concurrency among different protocol functions. *FSMs* may use global or local accessible *timers*. In case of global accessible timers, multiple *FSMs* can issue commands (e.g., start, restart) to such timer components. In addition, there is a global accessible *ALU* to manipulate registers that store variables used by more than one *FSM* (e.g., c.Cntl in XTP, a flag set by three different FSMs if sending a control packet is desired). *FSMs* issue a command to the *ALU* that performs the required operation on the data and may return a result (e.g., in case of comparisons). To guarantee consistency, *FSMs* can access the registers through the *ALU* only. A specialized connecting element, e.g., a cross-connect, interconnects all components inside the connection processor and external components (E-MMU, MUX, DMX).

Due to its modular design, this architecture can be adapted to different protocols. For example, XTP needs global variables and global accessible timers and, therefore, a global accessible ALU with registers and global timer unit(s) are required. PATROCLOS needs local variables and local timers only and, thus, the global components are not required.

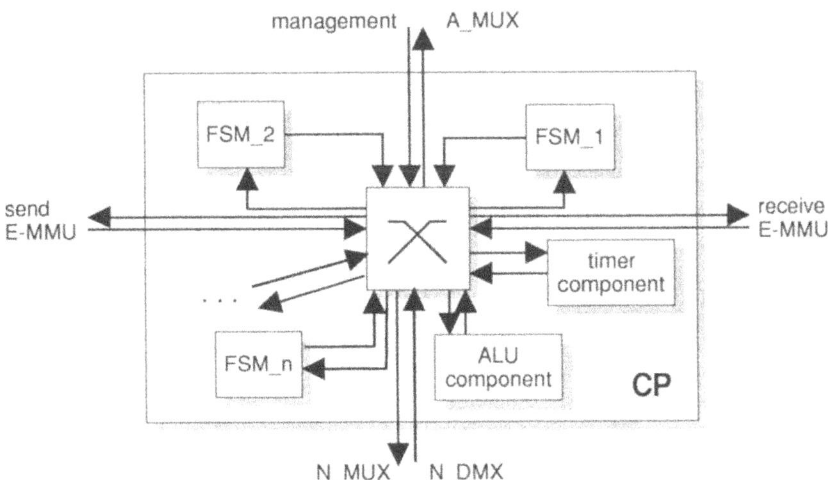

Figure 4: Internal Structure of a Connection Processor

Components (e.g., FSMs) internal to the *CP* are interconnected via a connecting element. To provide a maximum of flexibility they are designed with identical interfaces. The input signals consist of a 32 bit data bus d_i, a request line r_i, and an acknowledge line a_i. Valid data is signaled via $r_i = 1$. The component shows its readiness to receive the data via $a_i = 1$. After transmitting all data to the component, the sender pulls down r_i to 0. The output interface consists of corresponding signals: a 32 bit data bus d_o, a request line r_o, and an acknowledge line a_o. The protocol to transmit data to another component corresponds to the input interface.

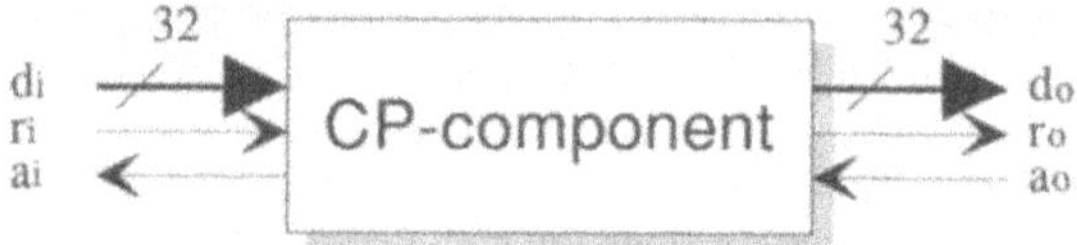

Figure 5: General CP Component Design

4.1 Connecting Component

The connecting component may provide two separated unidirectional connections to every component inside the processor and to the external units. Two different connections should not interfere each other due to blocking. Therefore, one could realize the connecting component via a crossbar switch. Dedicated or seldom used connections could be multiplexed. Due to the internal design of the connected components, the connecting component itself does not need storage components such as queues.

4.2 Finite State Machine

Every communication protocol specification consists of one or more finite state machine(s). In this paper, we mainly address protocols, which are specified as a set of multiple highly independent FSMs. Examples are XTP [2], SNR [7] and Patroclos [8], [9]. Each of these protocols is able to be implemented on the VLSI implementation architecture by implementing the appropriate FSM processing components.

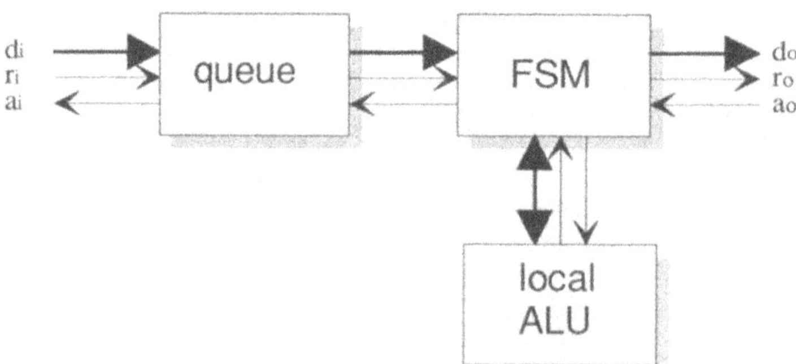

Figure 6: Finite State Machine Component

An *FSM* processing component (cf. Figure 6) forms the basis for an implementation of parallel FSMs. Each FSM processing unit consists of local registers, a local ALU, and a control unit. A local ALU is customized to individual requirements of the corresponding *FSM*. It can be as simple as an adder or as complex as a management unit for dynamic lists. To decouple *FSMs* from other units inside a *Connection Processor*, each *FSM* utilizes a separate input queue. Commands to an *FSM* will be inserted into this queue, thus providing asynchronous communication among the *FSMs*.

4.3 Timer Component

Several protocol functions, such as retransmission, reassembly, and connection management, require timer operations. In most implementations architectures, timers are supported by the operating system. However, this support often forms a major performance bottleneck [10]. Therefore, the modular VLSI implementation architecture comprises a dedicated *timer component*. According to the performance needs either this component can be scaled or several identical components can be implemented.

The *timer component* (cf. Figure 4) manages a dynamic list of timers. Several commands exist to manipulate the list:

command	input parameter	output parameter	comment
create	conn id, time out, receiver	timer_id	create a new timer entry in the timer list; return the timer identification
set	conn id, timer id, time_out		set an existing timer to a new time-out value
reset	conn_id, timer_id		reset an existing timer to its original time-out value
delete	conn_id, timer_id		delete an existing timer
read	conn_id, timer_id		read the actual value of an existing timer
alarm		timer_id	time-out has occurred

The parameters are defined as follows. *conn_id* indicates the unique identification of the connection the timer belongs to, *time_out* is a 32 bit value indicating the time-out value. The appropriate receiver(s) of an alarm are listed in the *receiver* vector. The *timer_ids* are managed by the global accessible *timer* itself. The resolution of the global accessible *timer* is 100 ns, the maximum time-out value is, therefore, greater than 7 minutes.

4.4 ALU Component

A global *ALU* (cf. Figure 4) is similar to an ALU of a general purpose processor. It implements standard operations, such as OR, XOR, ADD, SUB, SHR, SHL, and MULT. Multiplication (MULT) is the most complex operation of such an ALU, however, it is needed for several protocol functions. The ALU behaves like a coprocessor and works asynchronously. It consists of an input *queue* and the *ALU* itself (cf. Figure 7). The ALU comprises several *registers*. To guarantee data consistency, global data is stored in the *registers* and can be accessed via the *ALU* only.

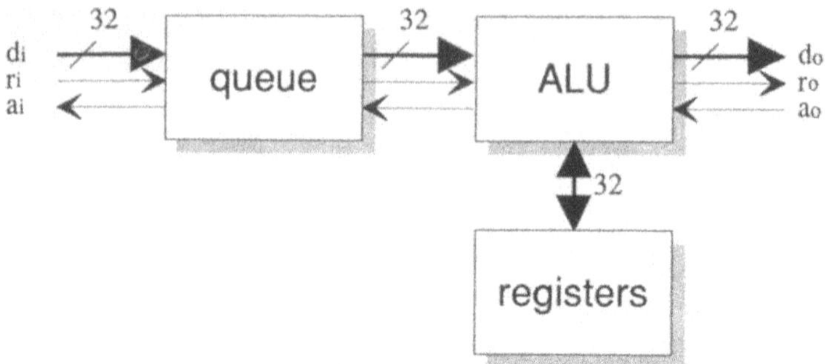

Figure 7: ALU Component

5 VLSI for Retransmission Support

One of the most complex ALU in a *Connection Processor* is retransmission support. It manages information for retransmission and, therefore, can be used as a *local ALU* for a retransmission FSM.

Enhanced transport protocols, such as XTP, support selective retransmission due to the unacceptable overhead of a go-back-n mechanism. Information needed for selective retransmission are gaps representing spans of bytes not yet correctly transmitted. The sender has to retransmit the bytes indicated by a gap if the service guarantees correct and complete transmission.

The processing overhead associated with handling retransmissions and the required data structures may be extremely high compared to other protocol functions. As an example some data about an XTP implementation on Digital Alpha Computers (150 MHz, 6.66 ns cycle time). The function to insert a new gap in a list needs in the best case 824 4-byte commands and takes, therefore, 5.4 µs. The best case occurs if the new entry can be inserted at the beginning of the list. If the new entry has to be inserted after the first 10 entries it needs 4054 commands or 27.03 µs due to the search operations in the list. These calculations assume that the processor is not interrupted during execution of this function and all data is stored in the fast processor cache. XTP was implemented using C without special inline assembly code. Note that the protocol has been implemented without any code optimizations on C or assembly language level.

If data is sent at a rate of 1 Gbit/s this results in more than 122,000 1024-byte packets per second. If the retransmission of each packet has to be controlled this results in a new entry in the list in less than 10 µs.

Clearly, retransmission support forms a time critical task. Therefore, we are implementing dedicated VLSI support for this task. The retransmission support presented in the following can handle negative selective, positive selective, and positive cumulative acknowledgement and can be used for gaps in received data to support the acknowledgement function or for gaps in transmitted data to support the retransmission mechanism. The ALU has a set of commands to set, delete, insert, and read gaps.

5.1 Logical Representation of Data

A dynamic linked list stores gaps of transmitted data in the following representation: *[seq_no_1, seq_no_2]* with *seq_no_1* and *seq_no_2* representing the beginning and ending of a gap. These gaps are connected via linked lists (cf. Figure 8).

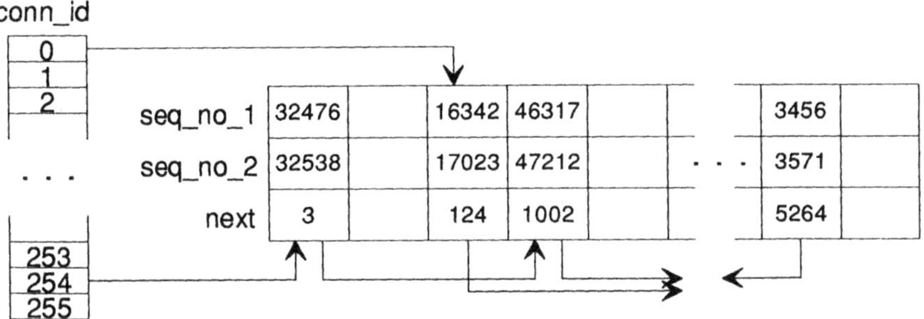

Figure 8: Logical Structure of Linked Lists for Retransmission Support

5.2 Operations of the Retransmission ALU

The following table shows the operations supported by the specialized ALU.

operation	input parameters	output parameters	comment
`init_list`	`conn_id, seq_no`		initializes a new list for the connection *conn_id* with the initial sequence number *seq_no*, sets the error flag if *conn_id* is already in use
`close_list`	`conn_id`		closes the list for connection *conn_id*, sets the error flag if the list does not exist
`set_high_ack`	`conn_id, seq_no`		sets the *high_ack* register to the value of *seq_no*; sequence numbers less than *high_ack* have been already acknowledged
`shift_high_ack`	`conn_id, length`		shifts the *high_ack* register to *high_ack* + *length*
`set_high_seq`	`conn_id, seq_no`		sets the *high_seq* register to the value of *seq_no*; *high_seq* represents the highest sequence number in use
`shift_high_seq`	`conn_id, length`		shifts the *high_seq* register to *high_seq* + *length*
`set_gap_1`	`conn_id, seq_no, length`		inserts new entry (*seq_no, seq_no* + *length*); overlapping entries are automatically joined or deleted, respectively

set_gap_2	conn_id, seq_no_1, seq_no_2		analogous to *set_gap_1*, but the new entry is of the form (*seq_no_1, seq_no_2*)
del_gap_1	conn_id, seq_no, length		deletes an existing entry, a part of an existing entry, or several existing entries, the deleted part is of the form (*seq_no, seq_no + length*); if necessary an entry is automatically divided into two new entries
del_gap_2	conn_id, seq_no_1, seq_no_2		analogous to *delete_gap_1*, but the deleted part is of the form (*seq_no_1, seq_no_2*)
read_reg	conn_id, reg_id	cont	reads the contents *cont* of the register *reg_id* (e.g. high_ack, high_seq, number_of_gaps)
get_gap_1	conn_id, ptr	seq_no, length, next	reads the entry *ptr* points to; if *ptr* = 0, the first gap is read out, if *next* = 0 the entry represented by (*seq_no, length*) is the last one, otherwise *next* point always to the next entry of the list
get_gap_2	conn_id, ptr	seq_no_1, seq_no_2, next	analogous to *get_gap_1*, but the entry is represented by (*seq_no_1, seq_no_2*)

conn_id $\in$ [0, 255]; *seq_no, seq_no_1, seq_no_2, length, cont* $\in$ [0, 2^{32}-1]; *reg_id* $\in$ [0, 15]; *ptr, next* $\in$ [0, 2^{16}-1]

Every operation sets the error flag if it failed due to memory overflow or violation of several conditions, such as *high_ack* $\leq$ *seq_no* $\leq$ *high_seq* and other range checking.

5.3 Implementation Architecture for Retransmission Support

Figure 9 shows an overview of the internal structure of the ALU. The retransmission ALU consists of 5 memory banks (A through E) that store sequence numbers representing gaps (memory A and B), pointers of linked lists (memory C), and state information, such as connection identification, register number of an anchor element, and other flags indicating the state of a connection (memory D and E). The I/O-bus connects the input/output-port (32 bit) of the retransmission ALU with the 5 register banks (Ai through Ei, $0 \leq i \leq 3$). From these registers data can be transferred to the memory.

Two simple ALUs (*ALU A*, 32 bit and *ALU D*, 8 bit) perform operations like OR, XOR, AND, ADD, SUB, and NEG. Two specialized 32 bit modulo 2^{32} comparators (*comp A* and *comp B*) perform fast comparisons needed for list operations. The ALUs and the comparators can work concurrently if no data dependencies exist.

For the command `set_gap_2`, for example, first of all the command itself and the connection identification are read from the I/O-bus into the registers E and D, respectively. In the next two cycles the central control unit reads the sequence numbers (`seq_no_1`, seq_no_2) into the registers A and B, respectively. After reading the complete command and several range checking operations the loop for searching the right position to insert the new entry in the list starts. Therefore, the first entry of the appropriate list is loaded into the registers A and B, respectively, and compared with the new entry. The loop terminates if the new entry fits, otherwise, the next entry is loaded and compared.

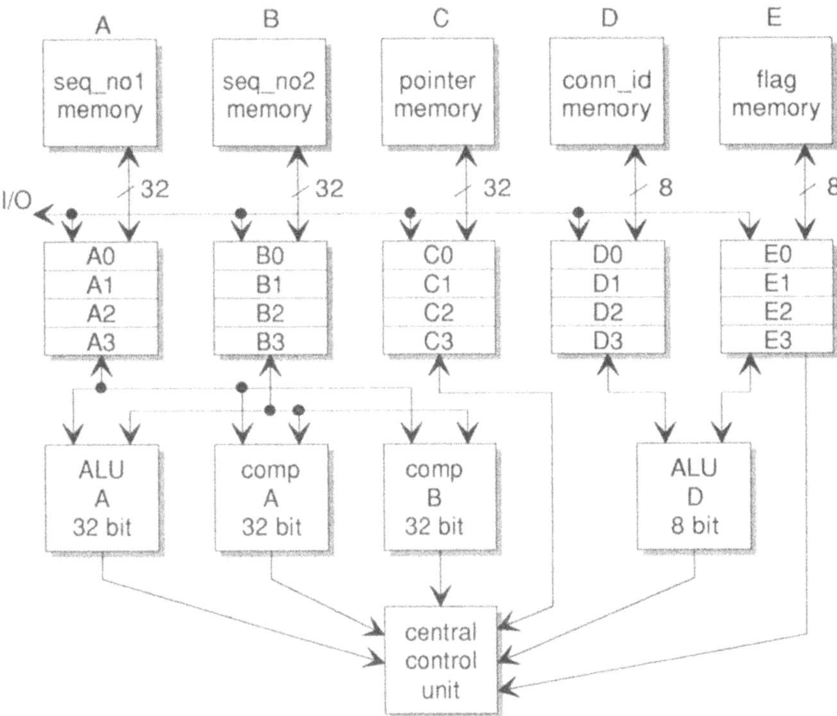

Figure 9: Structure of the Retransmission ALU

5.4 Microcode Examples of the Retransmission ALU

To provide a maximum of flexibility all functions of the *local ALU* are translated into a sequence of microcode operations. These operations are specially adapted to the implementation architecture shown in Figure 9. The *central control unit* controls the microprogram via a special micro sequencer.

Example microcode operations of the *local ALU* are listed in the following table.

operations	comment
`RMOVE S, D`	move a complete row of entries from the registers or RAM into the registers or RAM. S, D $\in$ {Ri, RAM; $0 \le i \le 3$}, Rn = (An, Bn, Cn, Dn, En), S $\neq$ D

`AMOVE I/O, D`	move data from the I/O-bus into the register D; $D \in \{Ai, Bi; 0 \le i \le 3\}$
`ACLR D`	clear register D; $D \in \{Ai, Bi; 0 \le i \le 3\}$
`AINC S, D`	S + 1 -> D; $S, D \in \{Ai, Bi; 0 \le i \le 3\}$
`ADEC S, D`	S - 1 -> D; $S, D \in \{Ai, Bi; 0 \le i \le 3\}$
`AMOVE S, D`	S -> D; $S, D \in \{Ai, Bi; 0 \le i \le 3\}$
`AADD S, D`	S + D -> D; $S, D \in \{Ai, Bi; 0 \le i \le 3\}$
`ASUB S, D`	S - D -> D; $S, D \in \{Ai, Bi; 0 \le i \le 3\}$

In addition, there are the microcode operations of the *ALU D* and complex comparison operations of the two comparators *comp A* and *comp B*. The operations of the ALUs and the comparators are always executed in parallel in one clock cycle.

5.5 Implementation Detail: A 32 bit modulo 2^{32} comparator

Due to the complexity of the complete design, only a selected part is presented in detail. This section shows the design of a 32 bit modulo 2^{32} comparator (cf. *comp A* and *comp B* in Figure 9). Such a comparator is needed to compare three 32 bit values at the same time. This comparison is done while searching for the right position in the retransmission list to insert a new entry or to delete an existing entry. Due to the modulo 2^{32} arithmetic and the need for fast search operations in the dynamic lists common comparators with only two inputs cannot be used.

Four functions are calculated at the same time:

$$\text{le\_le}(a, b, c) = \text{true} \Leftrightarrow ((a \le c) \wedge (a \le b \wedge b \le c)) \vee ((a > c) \wedge (a \le b \vee b \le c))$$
$$\text{le\_lt}(a, b, c) = \text{true} \Leftrightarrow ((a < c) \wedge (a \le b \wedge b < c)) \vee ((a > c) \wedge (a \le b \vee b < c))$$
$$\text{lt\_le}(a, b, c) = \text{true} \Leftrightarrow ((a < c) \wedge (a < b \wedge b \le c)) \vee ((a > c) \wedge (a < b \vee b \le c))$$
$$\text{lt\_lt}(a, b, c) = \text{true} \Leftrightarrow ((a < c) \wedge (a < b \wedge b < c)) \vee ((a > c) \wedge (a < b \vee b < c))$$

$a, b, c \in \mathbb{N} \bmod 2^{32}$; le: less or equal; lt: less than

The complete architecture and its components are described with the standardized hardware description language VHDL [11]. This allows for simulation and synthesis based on the same language.

```
library IEEE;
  use IEEE.std_logic_1164.all;
  use IEEE.std_logic_arith.all;

entity MOD_KOMPA is
  port ( A, B, C : IN std_logic_vector (31 downto 0);
         le_le, le_lt, lt_le, lt_lt: OUT boolean);
end MOD_KOMPA;

architecture BEHAVIORAL of MOD_KOMPA is
  begin
    process (A,B,C)
      variable h1, h2, h3, h4, h5, h6 : boolean;
      begin
        h1 := A < B; h2 := A <= B;
        h3 := A < C; h4 := A > C;
        h5 := B < C; h6 := B <= C;
        le_le <= (NOT(h4) AND (h2 AND h6)) OR (h4 AND (h6 OR h2));
```

```
        le_lt <= (h3      AND (h2 AND h5)) OR (h4 AND (h5 OR h2));
        lt_le <= (h3      AND (h1 AND h6)) OR (h4 AND (h6 OR h1));
        lt_lt <= (h3      AND (h1 AND h5)) OR (h4 AND (h5 OR h1));
    end process;
end BEHAVIORAL;
```

The above listed VHDL description was synthesized into a gate level description using a high level synthesis tool. The area of the design is 1084 gates and the estimation of the critical path 9.6 ns. The implementation of only the *le_le* function on an Alpha processor needs 20 4-byte commands which results with 6.6 ns cycle time (150 MHz) in a duration of more than 132 ns.

Figure 9: Gate Level Schematic of a 32 bit modulo 2^{32} Comparator

6 Summary and Future Work

Within this paper, a modular implementation platform for communication protocols has been presented. It is based on VLSI components dedicated to specific processing tasks. Especially, certain system and support functions can be implemented extremely efficiently by the use of dedicated hardware. Detailed examples covering memory management and retransmission support have been discussed.

However, not only high performance but specifically service integration is required for forthcoming communication subsystems. The modularity of the presented design especially achieves this requirement. Individual components may be selected based on the requested application service. Moreover, they can be parameterized. Since they provide identical interfaces and since they are interconnected via a connectivity element, a high degree of flexibility can be achieved.

The presented architecture thus may be viewed as a hardware implementation of the function-based communication subsystem F-CSS presented in [12]. In addition, the presented architecture forms a key part of a framework that eventually should cover efficient and flexible automated protocol implementations from protocol specifications. Currently, the implementation of various components of the VLSI architecture are under study. A multicast extension of the local ALU for retransmission support is under development.

Acknowledgement

The authors gratefully acknowledge Claudia Schmidt for providing the instruction counts for the XTP implementation.

References

[1] Ito, M.; Takeuchi, L.; Neufeld, G.; *Evaluation of a Multiprocessing Approach for OSI Protocol Processing;* Proceedings of the First International Conference on Computer Communications and Networks, San Diego, CA, USA, June 8-10, 1992

[2] Strayer, W.T.; Dempsey, B.J.; Weaver, A.C.; *XTP: The Xpress Transfer Protocol;* Addison-Wesley Publishing Company, 1992

[3] Feldmeier, D.C.; *An Overview of the TP++ Transport Protocol; in:* Tantawy A.N. (ed.): High Performance Communication, Kluwer Academic Publishers, 1994

[4] Sterbenz, J.P.G.; Parulkar, G.M.; *AXON Host-Network Interface Architecture for Gigabit Communications;* in: Johnson, M. J. (ed.): Protocols for High-Speed Networks, II, North-Holland, 1991, pp. 211-236

[5] Krishnakumar, A.S.; Kneuer, J.G.; Shaw, A.J.; *HIPOD: An Architecture for High-Speed Protocol Implementations*; in: Danthine, A.; Spaniol, O. (eds.): High Performance Networking, IV, IFIP, North-Holland, 1993, pp. 383-396

[6] Balraj, T.; Yemini, Y.; *Putting the Transport Layer on VLSI - the PROMPT Protocol Chip;* in: Pehrson, B.; Gunningberg, P.; Pink, S. (eds.): Protocols for High-Speed Networks, III, 1992, North-Holland, pp. 19-34

[7] Sabnani, K.; Netravali, A.; Roome, W.; *Design and Implementation of a High Speed Transport Protocol;* IEEE Transactions on Communications, Vol. 38, No. 11, November 1990, pp. 2010-2024

[8] Braun, T.; *A Parallel Transport Subsystem for Cell-Based High-Speed Networks;* Ph.D. Thesis (in German), University of Karlsruhe, Germany, VDI-Verlag, Düsseldorf, 1993

[9] Braun, T.; Zitterbart, M.; *Parallel Transport System Design;* in: Danthine, A.; Spaniol, O. (eds.): High Performance Networking, IV, IFIP, North-Holland, 1993, pp. 397-412

[10] Varghese, G; Lauck, T.; *Hashed and Hierarchical Timing Wheels: Data Structures for the Efficient Implementation of a Timer Facility;* ACM, 1987, pp. 25-38

[11] IEEE; *Standard VHDL Language Reference Manual;* IEEE Std 1076-1987

[12] Zitterbart, M.; Stiller, B.; Tantawy, A.; *A Model for Flexible High-Performance Communication Subsystems;* IEEE-JSAC, May 1993, pp. 507-518

10

Multicopy ARQ Strategies for Heterogeneous Networks

M. Aghadavoodi Jolfaei, U. Quernheim

Department of Computer Science (Info IV), Aachen University of Technology
Ahornstraße 55, D-52074 Aachen, Germany
Tel.: +49 241 80 21413, Fax: +49 241 8888220
E-mail: masoud@informatik.rwth-aachen.de

Abstract
On systems consisting of packet switched networks (e.g. LANs, HSLANs and ATM/BISDN) which are linked over satellite, there are packet losses due to buffer overflow at bridges and routers as well as bit errors on the satellite link. The resulting high error probability requires error correction methods, which are simple and effective in terms of throughput and memory requirements. For packet switched communications, quite a number of ARQ (Automatic Repeat Request) protocols have been designed which are able to cope with stringent memory requirements, but they either lack safety at high block error rates, or they require a large bandwidth due to low throughput. In this paper, we will introduce a simple and effective strategy (called Stutter-XOR or SXOR strategy) to increase throughput of exisiting protocols. In contrast to existing hybrid schemes (combinations of ARQ and forward error correction), it does not only correct bit errors, but also handles block losses. This is achieved by sending additional blocks, so called XOR blocks, which are created by combining other blocks using the XOR (modulo 2 addition) operation. We developed several variants of this strategy. Two of them will be evaluated here by means of analysis and simulation.

Keyword Codes: C.2.1; C.2.2
Keywords: Computer-Communication Networks, General; Network Architectures and Design; Network Protocols

1. INTRODUCTION

The way in which the retransmissions of erroneous blocks are carried out is the point in which the various ARQ methods differ. With continuous ARQ strategies, the sender continuously transmits new blocks, and the receiver accepts each error-free block and positively acknowledges it by sending an ACK message. On receipt of an erroneous block, the receiver negatively acknowledges the block by sending a NACK. When a NACK arrives at the sender, the transmission of new blocks is interrupted . In case of the Go-Back-N strategy, the sender does not only retransmit the negatively acknowledged block, but also all the following blocks. These are the blocks wich have been transmitted during one round trip delay after the erroneous block. In case of the selective repeat strategy, the sender retransmits only the negatively acknowledged block. These strategies are also used for point-to-multipoint communications. In this case, blocks are considered to be correct (positively acknowledged) if they are correctly received by all receivers. Accordingly, the sender decides about retransmission of a block after receipt of acknowledgements from all K receivers.

Another way to correct block errors is the use of forward error correction (FEC), which attempts to enable the receiver to correct block errors by appending redundant information to the data blocks. This method can be applied bit-wise as well as to whole blocks. While traditional FEC schemes (such as BCH or Viterbi codes) aim at the correction of bit errors, the

Stutter-XOR strategy (SXOR strategy) is able to correct whole blocks. This is based on the observation, that in packet switched networks often more packet losses are caused by buffer overflows at bridges, routers and receivers than by bit errors. This is one reason why current transport protocols (such as ISO-OSI-TP4, TCP/IP, XTP) utilize ARQ methods, which correct erroneous blocks by repeating them.

In the last years, few investigations have been done on methods for correcting packet losses [Bier 92], [LaBE 91], [OhKi 91] and these only cope with delay sensitive transmissions (e.g. video or speach transmissions), where the delay caused by repetitions of lost packets is not acceptable. The SXOR strategy can be used as a pure forward error correction method as well as a hybrid scheme in combination with existing ARQ strategies. Here we consider a variant of the stutter (or multicopy) selective repeat strategie, the Weldon scheme, combined with the SXOR strategy.

In this introduction we will first outline the basic communication model we are dealing with. In chapter 2, we will review an existing ARQ protocol, the Weldon strategy [Wel 82]. The SXOR strategy and its applications are introduced in chapter 3. In chapter 4, an analysis of the different variants of the SXOR strategy is conducted, the results of which are presented in chapter 5. Finally, chapter 6 presents an overview of the achieved results.

The time between the transmission of the first bit of the block and the arrival of the corresponding acknowledgement at the sender is called the round trip delay. While the round trip delay of a satellite link remains constant, this may not be the case in other parts of a network, where alternate routing may occur. Here we assume that the round trip delay remains constant for the whole communication path from the sending station to the receiver(s). The round trip delay can also be expressed as the maximum number S of blocks the sender can transmit until the acknowledgement for the first block arrives. For simplification, we assume an error free feedback channel (with a propagation delay equal to that of the satellite link) for the acknowledgements, i.e. the acknowledgements do not contribute to the link load and are received correctly after one round trip time.

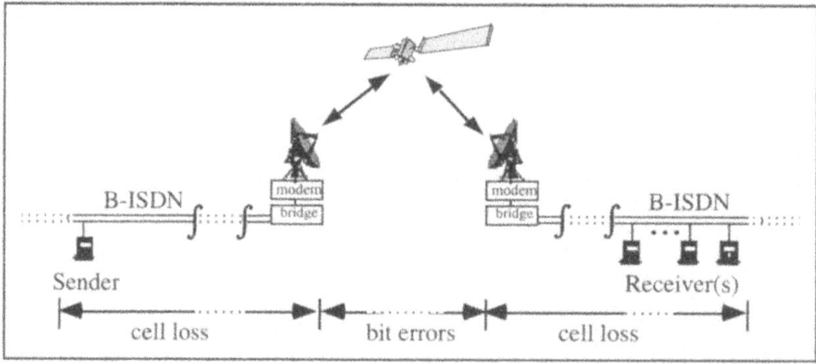

Figure 1: Connecting B-ISDN networks via satellite.

The transmitted blocks are safeguarded by a a set of parity bits, e. g. CRC (Cyclic Redundancy Check). The CRC bits enable the receiver to detect transmission errors. A block error occurs when there is at least one bit error in the block. In this case, the CRC bits do not match the data bits of the block. Undetected block errors are not considered. It is assumed that block errors occur independently of each other [AgKM 92]. A block is affected by an error with probability p, and unaffected with 1-p. Given a bit error probability of p_b, and with independent bit errors the resulting block error rate is $p = 1 - (1-p_b)^L$, where L is the block length in bits.

As a scenario we consider a heterogenous network consisting of high speed and satellite links (see figure 1). While on the terrestrial links cell losses may occur, the satellite link is dominated by bit errors. We assume that in case of an ATM network a block consists of k cells. A block is said to be erroneous, if it has at least one cell missing or contains at least one erroneous bit.

2. WELDON STRATEGY

In [Wel 82] , E.J. Weldon proposes a more advanced multicopy scheme. It is assumed that the receiver has a limited buffer capacity of q·S blocks, where S again denotes the maximum number of blocks which may be sent during one round trip time and q is a fixed integer. This means that there is room for q complete round trip times in the receiver buffer. When the sender receives a NACK for a specific block, it retransmits this block not just once but $n_1 \geq 1$ times. In case that each of the retransmitted blocks is disturbed again, the sender now retransmits the block $n_2 > n_1$ times and so on, up to n_q, where $n_1, ..., n_q$ are user determined values. When stage q is reached, a maximum of q·S blocks (new and retransmitted ones) may have been sent and stored in the receiver buffer (due to limited load, retransmissions and erroneous blocks which are discarded, the actual number will be lower) and the sender considers a receiver buffer overflow. Therefore, on further retransmission attempts the sender retransmits n_q copies of the disturbed message and one copy of each of the S subsequent messages, which are supposed to have been lost due to the buffer overflow. Figure 2 shows an example with q = 2, $n_1 = 2$ and $n_2 = 3$.

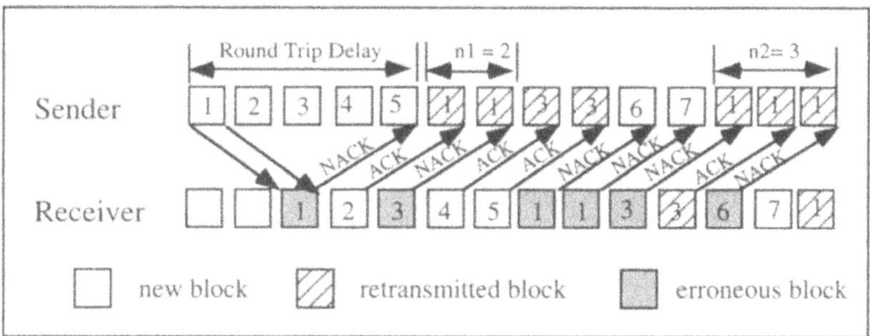

Figure 2: Weldon selective repeat ARQ strategy.

With q = 1 and $n_1 = 1$, the Weldon scheme resembles the ordinary selective repeat scheme. Values larger than q = 1 are only useful for excessively high block error rates.

3. STUTTER-XOR STRATEGY

The various ARQ schemes differ in the number of retransmissions executed to correct a block error. Multicopy schemes employ a vast number of retransmissions to obtain a high security level, but on the expense of throughput. With our new SXOR strategy, we aim to reduce the number of retransmissions necessary to correct a certain number of block errors using existing ARQ schemes. This is achieved by modulo-2-addition (bit-by-bit logical exclusive OR, ⊕) of several retransmitted blocks. The strategy works as follows:

We define a so called XOR window size n. As long as no errors occur, new blocks are transmitted continuously. If, however, one or more NACKs are received at the sending station, it will open an XOR window, in which blocks occur as either new blocks, retransmitted blocks, or XOR blocks, that is, blocks which are modulo-2 sums of the other blocks in the current window. We will term blocks other than XOR blocks basic blocks from now on. The XOR window limits the number basic blocks which can be combined to an XOR block. The receiver which is waiting for a particular block it has not received correctly yet (say, block number B), can retrieve a correct version of B by either waiting for a plain retransmission of B, or by taking advantage of an XOR block which contains B and a number of other blocks it has received correctly.

If, e.g., the receiver has correctly received blocks A,C, and D, but not B and E, it may retrieve B out of the XOR block $X = A \oplus B \oplus C \oplus D$ by calculating $X \oplus A \oplus C \oplus D$. Substituting X and reordering yields $A \oplus A \oplus C \oplus C \oplus D \oplus D \oplus B = B$, as the sum of a block with itself equals the all zero block. An XOR block containing more than one erroneous blocks cannot be used to correct either one; the block $Y = B \oplus C \oplus E$ with neither B nor E being correctly received would be useless (unless either B or E could be reconstructed from another block). It is important to note that the process of calculating the sum of received blocks can either be conducted on the fly with no extra buffer (compared to non-XOR protocols) by immediately XORing each correctly received block of the XOR window into the receiver input buffer in which incoming channel bits are assembled to blocks, or with one extra buffer especially dedicated to XORing.

3.1 Stutter-XOR Selective Repeat

If a NACK is received for a particular block B, the system enters XOR mode for one XOR window. Every group of n basic blocks is succeeded by an XOR block of the n basic blocks with the first basic block being the retransmission of B. There may be more retransmission phases (several XOR windows), if more NACKs should occur, just like ordinary selective repeat.

Fig. 3 gives an example for SXOR selective repeat. Let n = 3 be the XOR window size, and may the sender have transmitted basic blocks 1, 2, 3, 4 and 5, when it receives a NACK for block 1. Now, an XOR window starts, and n = 3 blocks are transmitted: first, the retransmission of 1 and new block 6. A NACK for block 3 arrives at this time and therefore this block is transmitted again rather than new block 7. After this, the window is closed and the XOR block $X = 1 \oplus 6 \oplus 3$ is transmitted. Now, for these four blocks (window and XOR block), there are 16 possible combinations of correct and incorrect blocks. 5 of them (those with 3 or 4 correctly received blocks) yield correct versions of 1, 3, and 6: 3, 6, X correct, 1, 6, X correct, 1, 3, X correct, 1, 3, 6 correct, and 1,3, 6, X correct. For the other 9 combinations (two, one, or no block correct), the XOR block cannot be exploited.

For both ordinary and XOR selective repeat, the exchange of acknowledgements is slightly altered in comparison with ordinary selective repeat. ACKs are emitted immediately upon correct reception, while for NACKs, there are two possible implementations. They are either immediately transmitted, or they are deferred until the end of the window. In the former case, the sender has to deduce from the received acknowledgements, whether or not the receiver has been able to correct the erroneous block. This also means, that there has to be an ACK or NACK for the XOR block. In the latter case, the first NACK is retained until either the XOR window is over (and if the block could be corrected using the XOR block, it is ACKed after all), or until the second NACK. In case of more than one error during one window, the XOR block is useless and erroneous blocks can be NACKed immediately to enable a quicker response by the sender.

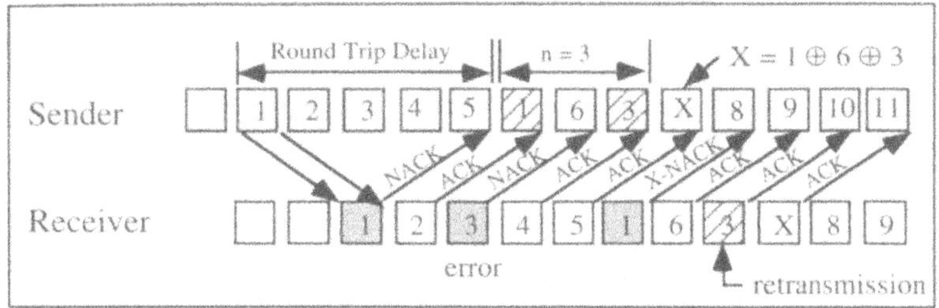

Figure 3: Example for the SXOR selective repeat scheme.

3.2 Stutter-XOR Weldon

As with ordinary Weldon, retransmissions occur in q stages, with n_i retransmissions on stage i. But now, a complete XOR window rather than a single block is retransmitted n_i times. That means, new basic blocks occur during the retransmissions, as well as XOR blocks. An XOR window consists of the block to be retransmitted plus n-1 other basic blocks, succeeded by an XOR block of the whole window (analogous to SXOR selective repeat).

An example for this strategy could look like the one in fig. 4. We let n = 2 and $n_1 = 2$: Blocks 1, 2, 3, 4 and 5 are transmitted, then a NACK for 1 is received. Stage q = 1 is entered, which means that there are $n_1 = 2$ transmissions of an XOR window of length n = 2.

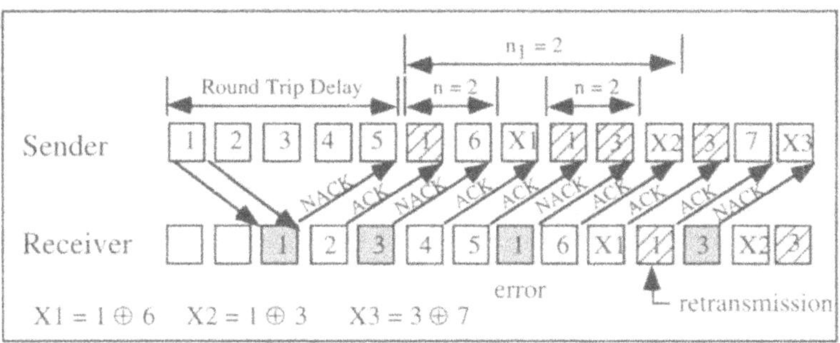

Figure 4: SXOR Weldon scheme.

The first window consists of blocks 1 and 6, followed by X1 = 1 ⊕ 6. The second window consists of blocks 1, and 3, since in the meantime, a NACK for block 3 has been received (otherwise, block 7 would have been sent), succeeded by X2 = 1 ⊕ 3. Stage q = 2 would yield n_2 XOR windows, each with 2 basic blocks and one XOR block of the whole window and so on.

4. ANALYSIS OF THE SXOR STRATEGY

For throughput analysis, we have considered two methods, which where presented by Weldon [Wel 82] and Lin/Yu [YuL 81] respectively. As simulations and analytical evaluations have shown, the Weldon method achieves a good approximation of the simulational results despite of its simplicity, but only up to a certain block error probability. The extended Lin/Yu method is more difficult, but approximates the simulational results well even at high block error probabilities. In the following we will exemplary demonstrate the Weldon method. We assume that all blocks are of equal length. The additional overhead, which is needed to contain the sequence numbers of the blocks participating an XOR block, is neglected here.

Assuming a saturated sender (no idle times), throughput T is defined as follows:

$$T = \frac{1}{E[X]} \qquad (1)$$

Here random variable X describes the number of transmissions of a representative block B. The expectation value E[X] then denotes the mean number of transmission attempts of each block.

Definition (1) must be generalized for our analysis of the SXOR strategy. The reason for this can be demonstrated considering the situation in figure 3 as an example:

Transmissions		6th	7th	8th	9th
Block Numbers		1	6	3	$X=1\oplus6\oplus3$
ACK Scenarios	1)	NACK	ACK	ACK	ACK
	2)	ACK	NACK	ACK	ACK
	3)	ACK	ACK	NACK	ACK

Here, for single block errors in the window, block X is required in 1/3 of the cases to either decode basic block 1, 3, or 6. Therefore, we assign to this block a weight of 1/3 rather than 1 when counting the number of transmissions for each of the three basic block numbers. If we sum up the total weight, three times 1 for the basic blocks and three times 1/3 for block X altogether yield the correct number of 4 block weights. In general, when we are restricting our scope to one representative block B, we assign a weight of 1/n to an XOR block consisting of n basic blocks .

For the analysis we assume a receiver buffer size of S blocks, where is denotes the number of blocks that can be sent during one round trip delay. It is assumed, that at the beginning of the transmission all receiver buffers are empty, and that the probability of block B being correctly received on the first transmission attempt is given by

$$P(X=1) = 1 - p \qquad (2)$$

In case the first transmission attempt is not successful, L copies of block B are retransmitted and all correctly transmitted succeeding blocks can be stored in the receiver buffer.The probability, that this transmission turn is successful is

$$P(X=1+L) = p\, Q(p,L) \qquad (3)$$

where Q(p, L) is the probability that one of the L copies of the block is successfully transmitted. The number of blocks sent per repetition is determined by the strategy used. The Weldon strategy simply transmits L copies of the erroneous block. Here, with a probability of $Q(p,L) = 1 - p^L$ at least one copy is correctly transmitted. With the SXOR-SR strategy, L is composed of the basic block retransmitted and its share of the XOR block, that is $L = 1 + \frac{1}{n}$.
The probability that block B can be reconstructed at this attempt is given by

$$Q(p,L) = (1-p)^{n+1} + p(1-p)^n \qquad (4)$$

This is the sum of the probabilities, that either all basic blocks of the SXOR window are being received correctly, or that exactly the observed basic block of the window is erroneous and all others are correct. In a similar manner, terms for L and Q(p,L) of the other strategies are derived.

- Selective Repeat:
  $L = 1$
  $Q(p,1) = 1-p$
- Weldon (q=1):
  $L = n_1$
  $Q(p,n_1) = 1- p^{n_1}$
- SXOR Selective Repeat:
  $L=1+\frac{1}{n}$
  $Q_{XOR\text{-}SR} = Q(p,1+\frac{1}{n}) = (1-p)^{n+1} + p(1-p)^n$
- SXOR-Weldon:
  $L = n_1(1+\frac{1}{n})$
  $Q(p,n_1(1+\frac{1}{n})) = 1-(1-Q_{XOR\text{-}SR})^{n_1}$

If a third transmission attempt is necessary, 1+L blocks have already been transmitted during the first two attempts; now we transmit additional L+(S-1) blocks, because we assume that (S-1) new blocks have been sent between the first and second transmission of B and have filled up the receiver buffer. When the third transmission of B arrives at the receiver, (S-1) more blocks have been sent meanwhile, but have been discarded due to buffer overflow, so they also have to be retransmitted. Therefore the probability that this transmission attempt is successful is given by

$$P(X=1+L+L+S-1) = p\,(1-Q(p,L))\,Q(p,L) \tag{5}$$

However, this is a pessimistic assumption: the maximum number of (S-1) blocks are only lost, if (S-1) blocks have been successfully transmitted. While this assumption works well for low block error probabilities, there is a growing deviation from the real throughput at higher block error probabilities. This is also true at further transmission attempts. For the m-th retransmission ($m \geq 2$), we get

$$P(X=1+L+m(L+S-1)) = p\,(1-Q(p,L))^{m-1}\,Q(p,L) \tag{6}$$

The sum of the weighted counts yields the expected value of the number of transmission attempts of the representative block B:

$$\begin{aligned} E[X] &= (1-p)+\sum_{i=1}^{\infty} (1+L+(i-1)(L+S-1))p(1-Q(p,L))^{i-1}\,Q(p,L) \\ &= 1 - p(S-1) + \frac{p(L+S-1)}{Q(p,L)} \end{aligned} \tag{7}$$

Setting L=1 and Q(p, L)=1-p in formula (7) leads to the mean number of transmission attempts needed to correctly transmit a block with the selective repeat strategy [MiSh 81], [Weld 82]:

$$E[X] = \frac{1+p^2(S-1)}{(1-p)} \qquad (8)$$

The throughput of the Weldon strategy for q=1 and n_1 can be determined by substituting $L=n_1$ and $Q(p,n_1) = 1- p^{n_1}$. This yields to the result for the Weldon strategy (q=1) as described in [Weld 82]:

$$E[X] = 1+n_1p+\frac{(n_1+S-1)p^{1+n_1}}{1-p^{n_1}} \qquad (9)$$

6. RESULTS

Using the selective repeat strategy, the sender retransmits only negatively acknowledged blocks. The receiver rejects blocks which are either hit by a transmission error or cannot be saved in the receiver buffer. The simulation implements the following method for buffering: Be i_0 the oldest block not yet correctly received by the receiver. Then buffer space is reserved for all blocks with a sequence number $i < i_0+S$. This space cannot be occupied by newer blocks. In this manner it is assured, that there is room in the buffer when these blocks, which are necessary for conveyance in correct order, are correctly received [YuL 81].

Figures 5/6 and 7/8 show results of analysis and simulation in comparison for point-to-point communication. At high block error probabilities, the analysis is too pessimistic. The simulated system has a lower buffer overflow probability, because erroneous blocks are not being stored in the receiver buffer. With increasing block error rate, this effect becomes predominant, so that analysis and simulation deviate more and more. Because of the growing number of receivers, this tendency is even stronger with point-to-multipoint communications; for this reason, we present only simulational results for this case.

For point-to-point communication and with an XOR window of n = 2, XOR selective repeat shows the best throughput results above a certain block error probability. A higher window size means that the probability of more than one block error in the window rises. With multiple block errors, the additional XOR block becomes useless, so the results for large window sizes approach those of the selective repeat strategy.

Figures 7 and 8 show a comparison of the selective repeat, Weldon, SXOR selective repeat and SXOR-Weldon strategies for point-to-point communication. Selective repeat achieves the lowest throughput of all schemes. Moreover, the throughput of the SXOR-SR strategy (n=2) is slightly lower than the one achieved by the Weldon strategy (q=1, n_1=2). Caused by the higher number of repetitions, the SXOR-Weldon strategy is superior to Weldon and SXOR-SR only for high block error probabilities. Since SXOR-Weldon with n_1=1 resembles SXOR-SR, it can be concluded that an adaptive scheme over n_1 would achieve the best results.

In contrast to point-to-point communications (fig. 5-8), with a higher number of receivers larger XOR window sizes prove to be useful. When different blocks are correctly received by different receivers, the XOR windows used for repetitions will frequently contain blocks, which have already been correctly received by some receivers at a former transmission attempt. These blocks can be used to decode other blocks from the XOR block, even if they are hit by an error in the current transmission. In this case, even the distortion of more than one block does not make the XOR block useless. On the other side, with a growing XOR window size the overhead caused by the additional XOR blocks is decreased. Figure 9 shows the simulational results of the SR, Weldon and SXOR-SR strategies for point-to-multipoint communications with 50 receivers.

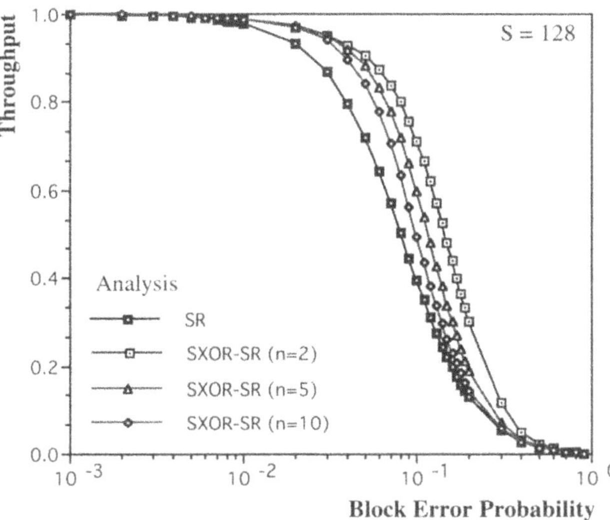

Figure 5: Analysis of the selective repeat and SXOR-SR strategies for various XOR window sizes (n = 2, 5 and 10, point-to-point communication)

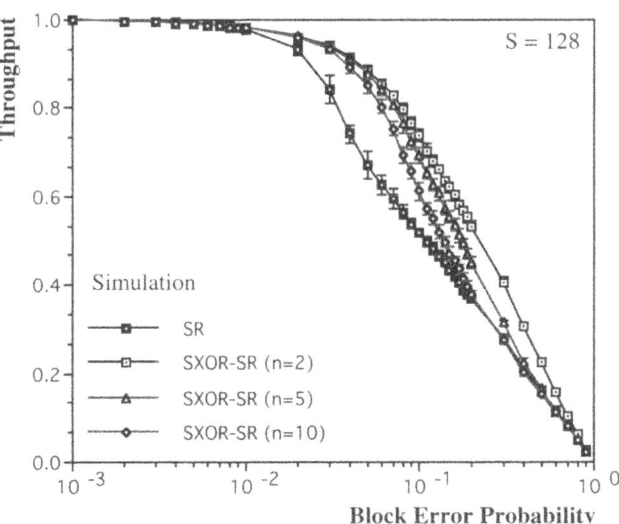

Figure 6: Simulation of the selective repeat and SXOR-SR strategies for various XOR window sizes (n = 2, 5 and 10, confidence level 99.9%)

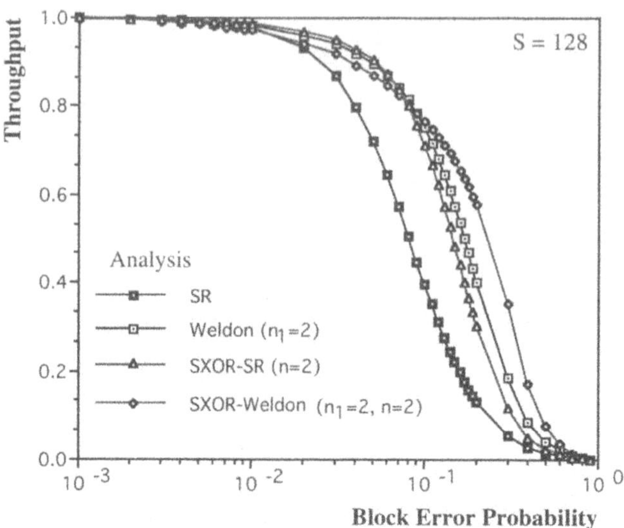

Figure 7: Analytic throughput comparison of selective repeat, Weldon, SXOR-SR and SXOR-Weldon for point-to-point communication

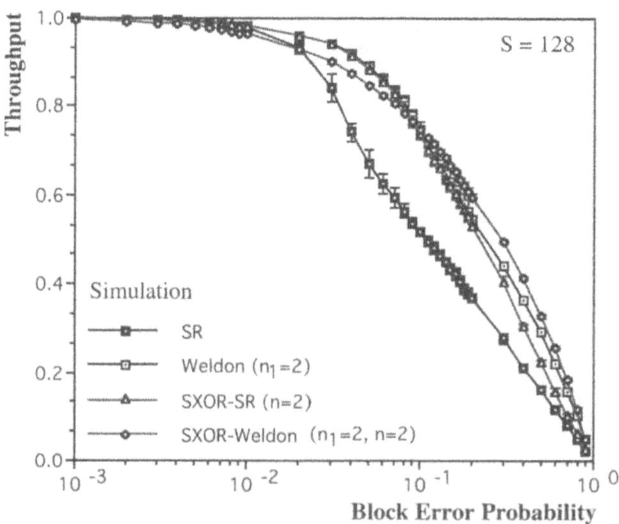

Figure 8: Simulative throughput comparison of selective repeat, Weldon, SXOR-SR and SXOR-Weldon strategies for point-to-point communication (confidence level 99.9 %)

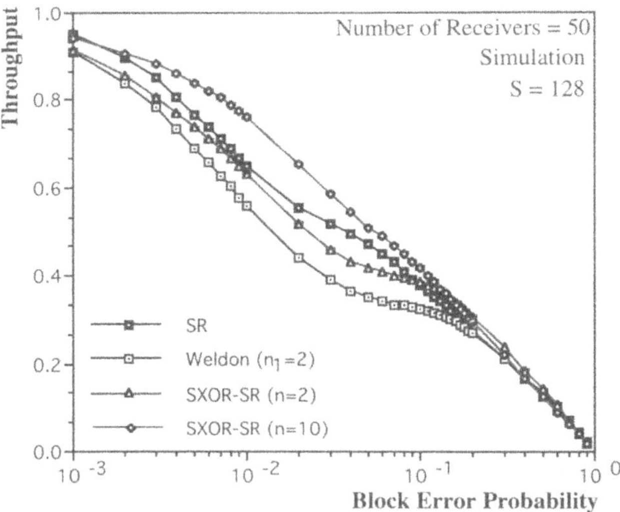

Figure 9: Simulative throughput comparison of selctive repeat, Weldon and SXOR-SR strategies for point-to-multipoint communication (confidence level 99.9 %)

As Figures 10 and 11 demonstrate, with a larger number of receivers multicopy schemes (like Weldon and SXOR Weldon), which transmit a higher number of redundant blocks, show a significantly lower throughput than selective repeat or SXOR-SR.

Figure 11 also shows that at higher block error probabilities and a small number of receivers (<5) the use of Weldon- or SXOR-Weldon is preferable. In this case, an adaptive scheme over n_1 and n would achieve the best results, since SXOR-Weldon with $n_1 = 1$ resembles SXOR selective repeat.

The SXOR strategy can be used advantageous on channels where both bit errors and packet losses due to buffer overflow occur. This is demonstrated in figure 12, where the results are shown as a function of the bit error probability with an additional constant block loss probability P(block-loss) of 0.01. As further simulations have shown, block losses in this range may occur at heavy system load [Naze 92]. Block losses are assumed to be uniformly distributed and independent of each other. The above reviewed strategies can also be combined with traditional FEC schemes. Here we used a (1023,1013) BCH encoding [LiCo 83], which is able to correct 1 bit errors. The FEC encoding is applied to all data blocks, so the number of blocks which can be sent during one round trip delay (S) is slightly decreased (S'). Furthermore, we assume an additional overhead of 8 bits for the SXOR strategy.

Now the throughput is no longer defined in terms of the number of correctly transmitted block per time unit, but in terms of data bits, so that the differing block lengths of the various strategies do not interfere. Throughput efficiency defined as follows:

$$T_{eff} = \frac{\text{Data (bits)}}{\text{Data (bits)} + \text{Overhead (bits)}} \frac{1}{E[X]} \qquad (10)$$

As the results show, in this scenario the SXOR-SR scheme is clearly superior to the other strategies, both with and without additional FEC encoding.

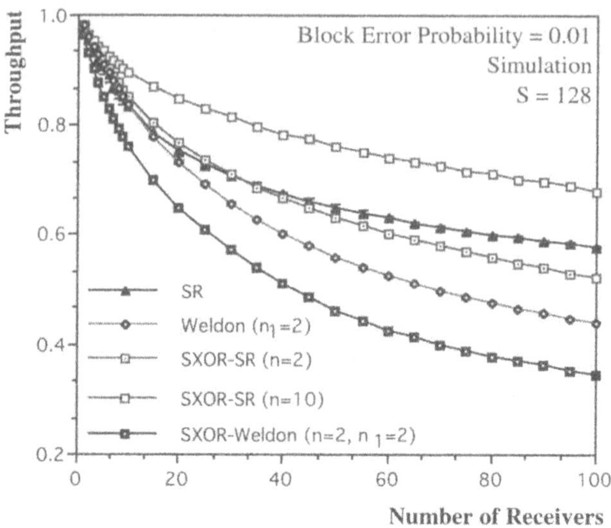

Figure 10: Simulative throughput comparison of the SR, SXOR-SR and SXOR-Weldon strategies for various numbers of receivers (confidence level 99.9 %)

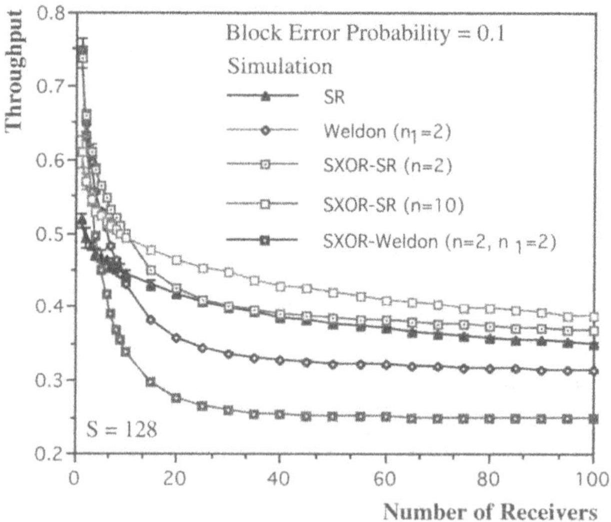

Figure 11: Simulative throughput comparison of the SR, SXOR-SR and SXOR-Weldon strategies for various numbers of receivers (confidence level 99.9 %)

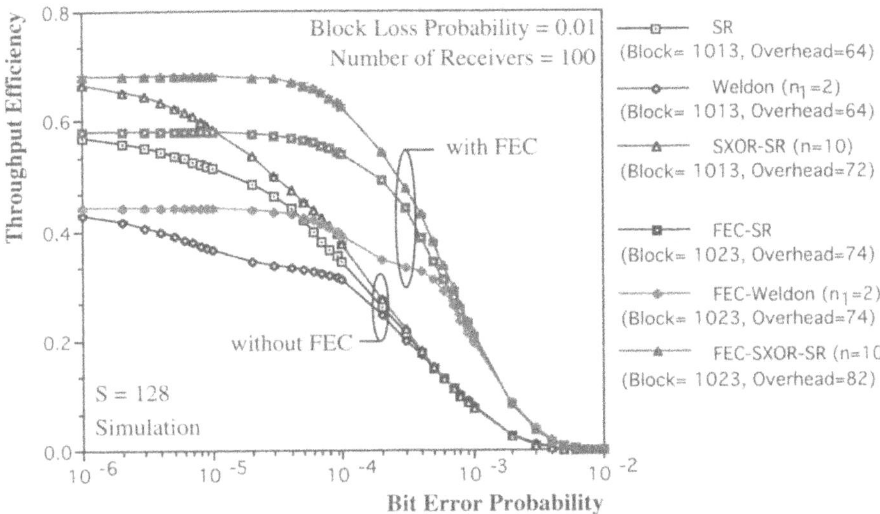

Figure 12: Simulative throughput comparison of the SR, Weldon and SXOR-SR strategies with and without FEC for 100 receivers (confidence level 99.9 %)

7. CONCLUSION

We have shown that by combining several blocks into XOR blocks the throughput for existing ARQ protocols can be improved. It can be applied to existing ARQ protocols and will generally yield better performance, especially on transmission channels where both block losses and bit errors occur. For ordinary selective repeat, the improvement is considerable. The Stutter-XOR strategy can even improve the well known Weldon strategy, which has been considered to be most efficient for high block error rates yet. Also, FEC channels can take advantage of XORing. The Stutter-XOR strategy can be implemented very efficiently in software; no fast hardware is required, as for FEC decoders.

REFERENCES

[AgKM 92] M. Aghadavoodi Jolfaei, D. Kreuer, O. Maly, U. Quernheim: "Two Time Variant Models for Satellite Channels", Proceeding Supercom/ICC 92, Chicago, USA, pp. 314.31.1- 314.3.5.

[AgMM 93] M. Aghadavoodi Jolfaei, S. C. Martin, J. Mattfeldt: "A New Efficient Selective Repeat Protocol for Point-To-Multipoint Communication"; Proceedings ICC '93, pp. 1113-1117, Genf, Swiss, May 93.

[BeF 64] R. J. Benice, Jr. A. H. Frey: "An Analysis of Retransmission System"; IEEE Trans. Commun. Technol., 135-145, Dec. 1964.

[Bier 92] E. W. Biersack: "A Simulation Study of Forward Error Correction in ATM Networks"; Computer Communications Review, 22(1):36-47, January 1992.

[BrM 86] H. Bruneel, M. Moeneclaey: "On the Throughput Performance of Some Continuos ARQ Strategies with Repeated transmissions"; IEEE Trans, Commun., 244-249, Mar. 1986.

[Bus 72] H. O. Burton, D. D. Sullivan: "Errors and Error Control"; Proc. IEEE, pp. 1293-1303, Nov. 72.

[CaE 86] S. R. Chandran, S. Lin: "A Selective-Repeat ARQ Scheme fot Point-to-Multipoint Communications and It´s Throughput Analysis"; Proc. ACM SIGCOM Conference, pp. 292-301, Stowe, VT, Aug. 1984.

[Dav 70] M. A. David: "Order Statistics"; New York: Wiley, 1970.

[GoJ 84] I. S. Gopal, J. M. Jaffe: "Point-to-Multipoint Communication Over Broadcast Links"; IEEE Trans. Commun., Com-3, pp. 1034-1044, Sep. 1984.

[LaBE 91] B. Lamparter, O. Böhrer, W. Effelsberg: "Vorausschauende Fehlerkorrektur für multimediale Datenströme"; GI/ITG-Tagungsband Verteilte Multimedia-Systeme, University Stuttgart, Feb. 1991.

[LiCo 83] S. Lin, D.J. Costello: "Error Control Coding: Fundamentals and Applications"; Prentice Hall, Englewood Cliffs, N.J. 07632, USA, 1983.

[MiSh 81] M. J. Miller, Shu Lin: "The Analysis of Some Selective-Repeat ARQ Schemes with Finite Receiver Buffer"; IEEE Trans. Comm., vol. COM-29, pp. 1307-1315, Sep. 1981.

[Mor 87] J. M. Morris: "On Another Go-Back-N ARQ Technique for High Error Rate Condition"; IEEE Trans. Commun., pp. 187-189, Jan. 1978.

[Naze 93b] M.-R. Nazeman: "Leistungsbewertung unterschiedlicher Protokolle zur Ankopplung von Ethernet-LANs über Satellit"; Diploma thesis at Aachen University of Technology (Info. IV), June 1993.

[OhKi 91] H. Ohata, T. Kitami: "A Cell Loss Recovery Method Using FEC in ATM Networks"; IEEE Journal on Selected Areas in Communications, Vol. 9, No. 9, pp. 1471-1483, December 1991.

[Sas 75] A. R. K. Sastry: "Improving Automatic Repeat-Request (ARQ) Performance on Satellite Channels under High Error Rate Conditions"; IEEE Trans. Commun. Electron., pp. 224-231, Apr. 1975.

[SaSc 85] K. Sabnani, M. Schwartz: "Multidestination Protocols for Satellite Broadcast Channels"; IEEE Transactions on Communications, Vol. COM-33, No. 3, March 1985.

[Toba 80] F. A. Tobagi: "Multiaccess Protocols in Packet Communication Systems"; IEEE Transactions on Communications, Vol. COM-28, No. 4, pp. 468-488, April 1980.

[ToMi 87] D. Towsley, S. Mithal: "A Selective-Repeat ARQ Protocol for a Point-to-Multipoint Channel"; Proceedings INFOCOM, pp. 521-526, San Franscisco, CA, March 1987.

[Tows 79] D. Towsley: "The Stutter Go-Back-N ARQ Protocol"; IEEE Trans. Comm., vol. COM-27, pp. 869-875, Jun. 1979.

[WaSi 88] J. L. Wang, J. A. Silvester: "Optimal Adaptive ARQ Protocols for Point-to-Multipoint Communication"; IEEE INFOCOM' 88, pp. 704-713, LA, 1988.

[Wel 82] E. J. Weldon: "An Improved Selective Repeat ARQ Strategy"; IEEE Trans. Comm., vol. COM-30, pp. 480-486, Mar. 1982.

[YuL 81] P. S. Yu, S. Lin: "An Efficient Selective Repeat ARQ Scheme for Satellite Channels and Its Throughput Analysis"; IEEE Trans. Commun., pp. 353-363, Mar. 1981.

[YuL 82] P. S. Yu, S. Lin: "A Hybrid ARQ Scheme with Parity Retransmission for Error Control of Satellite Channels"; IEEE Transactions on Communications, Vol. COM-30, No. 7, Juli 1982.

PART FIVE

Protocols

11

The Design of BTOP - An ATM Bulk Transfer Protocol

Liam Casey

BNR, PO Box 3511, Station C, Ottawa, Canada K1Y 4H7.
liam@bnr.ca

Abstract

Broadband networks have the raw speed to achieve large data transfers that appear instantaneous to users (i.e. that take less than a second). To realize such speeds in practice requires new protocols. BTOP (Bulk Transfer Oriented Protocol) is such a protocol, designed to be simple and efficient. Not only is BTOP intended to run directly within the AAL-5 layer, BTOP is inspired by the simplicity of AAL-5. AAL-5 was developed in reaction to the perceived complexity of AAL-3 and AAL-4. BTOP's simplicity derives from it only performing those functions that the ATM layer and AAL-5 common part sublayer do not. This paper gives an overview of BTOP and addresses the issues in designing a protocol to specifically exploit the capabilities of ATM.

Keyword Codes: C.2.2;
Keywords: Network Protocols;

1. INTRODUCTION

ATM is often touted as being ideal for bulk transfer applications such as image transfer. A high resolution image can occupy 2 Mbytes or more. Frequently discussed applications for ATM based Broadband networks are medical imaging and advertising pre-press. In the medical imaging arena there is a strong desire to be able to store, retrieve and transfer high resolution medical images in support of a wide range of activities. Within hospitals, for example, the physical transfer of X-rays between the radiology department and the emergency department could be eliminated by providing ATM connectivity between a radiology image server and "electronic light boxes" in the emergency department. Teleradiology applications include remote consultations whereby an image can be jointly viewed and discussed by a specialist in one location (e.g. a tertiary care facility) and a practitioner in another (a primary care facility).

Applications for the high speed transfer of images are also seen in advertising and civil engineering, where the production of artwork may be geographically distant from clients who want to browse through it. For examples of other image transfer applications see [1].

Images are not the only possible bulk transfer items. If a bulk transfer service were in place offering 10 Mbyte transfers in less than a second, it would be used for transferring large documents and design databases. Such applications currently may not need sub-second responsiveness (waiting 30 seconds to obtain a 1 Mbyte Postscript file from the other side of the continent, using say a T1 circuit, does not add much to the 5 minutes it takes to print the document and the longer time to read it). On the other hand, waiting a minute or more between viewable images on an electronic light box would have a definite productivity impact. Hence image transfer has been chosen as the archetypal bulk transfer application for ATM.

Sub-second bulk transfers cannot be performed with the wide area circuits available to users to-day. But it is not a foregone conclusion that they will be feasible even when an ATM based Broadband ISDN infrastructure is in place. High bandwidth is a necessary, but not a sufficient condition. A careful choice of protocol stacks is required to match ATM's bandwidth. Existing protocols, designed when 9.6 kbps was considered a very high transmission speed and 20 kbytes was the size of a computer's main memory, are not necessarily up to the task. We have designed BTOP (Bulk Transfer Oriented Protocol) explicitly for sub-second bulk transfers over ATM. The overriding design goal of BTOP has been simplicity, so that its execution speed can match ATM's transfer speed.

In the next section we summarize ATM protocol architecture. As an ATM cell is too small to deal with, individual cells are aggregated into AAL-5 PDUs (packets) with a maximum size of 64 Kbytes. In section 3 we first define Application Data Units which can consist of multiple AAL-5 PDUs. We then describe how BTOP transfers an Application Data Unit as a back-to-back sequence of AAL-5 PDUs, acknowledged by a single cell at the end.

Sections 4 and 5 summarize the protocol performance and the underlying assumptions we have made of ATM, such as a very low cell loss ratio. In section 6 we discuss some of the design guidelines we adopted. ATM provides an unique set of services and we show that BTOP's simplicity arises from the design decision not to duplicate services in higher layer protocols. Finally we look at implementation issues, then briefly contrast BTOP with existing protocols.

BTOP is in the definition phase. Our purpose here is to describe the design considerations of a truly high speed protocol. Not all the supporting assumptions we make will be true of initial, limited, ATM deployments. But, unless issues of protocol throughput are addressed, the practicability of new, data transfer intensive applications, justifying ATM's widespread adoption, is in doubt .

2. BROADBAND PROTOCOLS

The ITU-T (formerly CCITT) has been the primary source of Broadband protocols. This Standards body has determined that ATM will be the basic transport mode of Broadband ISDN. The ATM Forum exists to promote the use of ATM in both public and private networks. It bases the protocols it develops on ITU-T recommendations. Broadband protocols do not fit exactly into the ISO seven layer OSI stack. There are two general areas of difference: the 3-plane model and the unspecified upper layers.

2.1 The three planes of BISDN

The ITU-T has adopted a 3 plane model for protocol stacks using ATM, depicted below in Figure 1 [2]. The ATM Forum has also adopted this model [3].

Briefly the three planes are:

U-plane: The User plane is where applications usually work. It contains the protocols for the transfer of normal user data. BTOP is in the U-plane.

C-plane: The Control plane contains the signalling protocols for the "out of band" establishment of connections. In BISDN, ATM connections, called VCCs (Virtual Channel Connections), are established using the C-plane Q.2931 (formerly Q.93B)/ATM Forum UNI Signalling [3] protocol stacks.

M-plane: The Management plane manages the exchange of information between the C-plane and the U-plane and generally provides management functionality. For example, detecting and reporting physical link failures is an M-plane responsibility

For more details see [3] or a text such as [4].

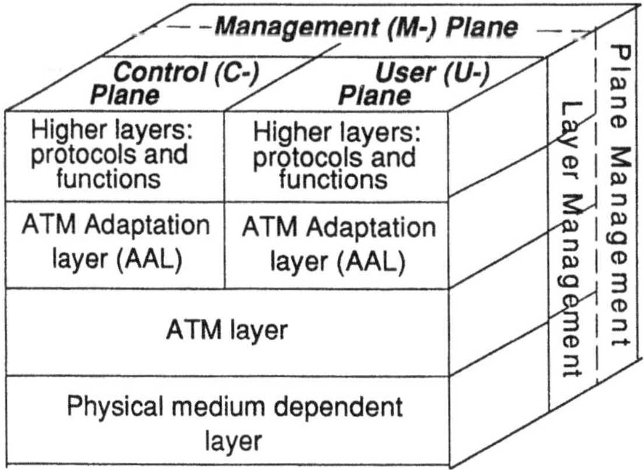

Figure 1: B-ISDN Protocol Model for ATM

Designing a protocol within the context of the three plane model is considerably simpler than designing classic, single stack protocols. For example, although BTOP is connection oriented, unlike existing transport protocols such as TCP [5], it does not have its own connection establishment phase. Applications using BTOP are responsible for establishing the VCC to be used. They may do this using C-plane signalling facilities. Alternatively applications may use permanent virtual connections (PVCs) pre-established using M-plane facilities. The latter option may be appropriate for "dumb" terminals such as electronic light boxes.

2.2 Layering - where BTOP fits.

As shown above in figure 1, there are several layers defined for both the C-plane and U-plane protocols. A particular protocol stack will consist of a Physical Medium Dependent layer, an ATM layer, an ATM Adaptation layer and higher layers. The ITU-T has defined Q.2931 as a higher layer in the C-plane but has not defined U-plane higher layers. No correspondence between the top of the ATM Adaptation layer and a position in the OSI stack has been stated. This is not surprising since the purpose of the Adaptation layer is to adapt the service offered by the ATM layer, viz. the basic relaying of cells, into the service required by the Upper layer. Thus different kinds of Upper layers require different Adaptation layers.

The ITU-T has specified 4 ATM adaptation layers, AAL-1, AAL-2, AAL-3/4 and AAL-5, each for use in providing different services. AAL-5 is an ATM Adaptation layer intended for the transfer of multi-cell data packets (called PDUs henceforth). As depicted below in figure 2, AAL-5 is itself divided into sublayers.

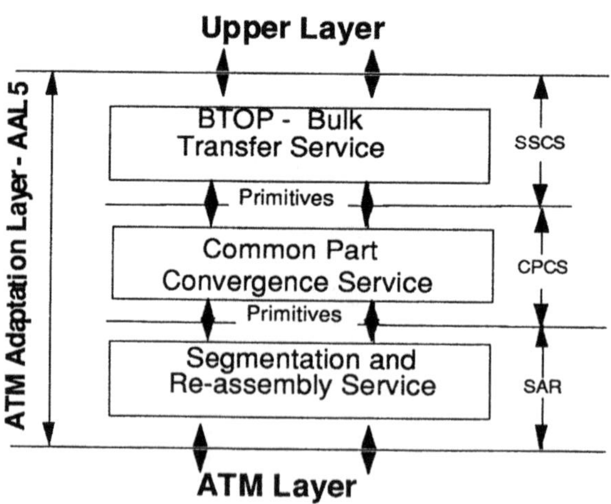

Figure 2: BTOP positioned as AAL-5 sublayer

Looking from the bottom up (as when cells are received) the sublayers and their functions are:

SAR — The Segmentation and Re-assembly Service. On the reception side this sublayer takes the contents of cells delivered by the ATM layer and concatenates them into bigger PDUs.

CPCS — The Common Part Convergence Service. On reception this sublayer takes the concatenated cell PDU (called AAL-5 CPCS PDU or basic AAL-5 PDU) and checks that it has been received correctly. It does this by examining a trailer that is carried in the last 8 bytes of the last cell of the PDU (cf. figure 3). This trailer was generated by CPCS on the sending side. The CPCS function is common to all classes of AAL-5.

SSCS: — The Service Specific Convergence Service. The functionality of this sublayer depends on the nature of the Upper layer. It implements all other services that specific Upper layers assume will be provided.

The ITU recommendation I.363 [6] leaves the SSCS unspecified - different SSCS are to be defined for different services. BTOP is the protocol of a Bulk Transfer (BT) service. Bulk Transfer is a "specific" Service Specific Convergence Sublayer of AAL-5. Other, already defined SSCSs include the null one (when the common CPCS provides all the services needed by the Upper layer) and SSCOP, which has been defined to support signalling applications.

In the case of BTOP the Upper layer is the Application layer (unless compression or special encoding are required in which case a Presentation layer would be interposed between Application and BTOP). The Bulk Transfer Convergence Sublayer transfers large Application layer PDU's without requiring any intervening layers between it and the Application.

We attach great importance to placing BTOP within the context of BISDN planes and protocol levels. Only by examining the total picture of the services each plane and level offers can we minimize the functionality that we need to provide in BTOP. For example, BTOP does not provide any flow control because this is provided by the ABR service of the ATM layer (see section 6).

3. OVERVIEW OF BTOP

3.1 Application Data Units

A Bulk Transfer is the transmission of a large Application Data Unit from a Source application to a Destination application. An Application Data Unit is a chunk of data that makes sense at the application level. The length of transport PDUs (packets) in most networks exhibits a bimodal distribution. There are short packets, carrying either protocol information (e.g. acks) or application control (e.g. a get file request), and there are maximum size packets. A sequence of maximum size packets are used to effect a large data transfer, such as a file or an image. What is really being transferred is an Application Data Unit, although existing Transport protocols do not recognize any unit above the transport PDU.

In [7] an Application Data Unit is defined as an aggregate of data that an application can process out of order. Our definition is similar: the aggregate of data that an application processes as a complete unit, in a single processing step. An image, for example, would be considered one Application Data Unit. For the applications most likely to use BTOP, the transfer of an Application Data Unit will be triggered by some form of user request (e.g. pushing a button to view the next set of X-rays).

Although Application Data Units could be of any size, for practical reasons we have limited them to 16 Mbytes. 16 Mbytes meets current needs, for example an uncompressed 2000 x 2000, 24 bit colour image occupies 12 Mbytes. A protocol to deal with multiple Application Data Units could be defined but we have chosen not to elaborate upon it in this paper (c.f. NETBLT [8] which deals with multiple bursts).

3.2 Source-Destination Associations

Before any transfers can take place there has to be a VCC established between the Source and Destination. In general we assume that the Source and Destination applications will have some form of association established between them (with its own protocol stack) to signal between themselves what information is to be transferred and when the transfer is to occur.

The nature of the signalling association is not of particular concern to BTOP. It could be "user to user" signalling in the C-plane. If in the U-plane, the protocol would probably run on a different VCC.

BTOP is not intended for the unsolicited transfer of indeterminate amounts of data. We assume that the Destination knows the size of an Application Data Unit before it is transmitted. In general we suppose that there will have been a signalling exchange between Source and Destination prior to invoking Bulk Transfer service. This enables the Destination to prepared itself for the reception of the Application Data Unit.

One particular scenario has Bulk Transfer taking place during the service phase of a remote procedure call (RPC) made from the Destination, invoking the Source. If a protocol such as ROOP [9] were used, then it might be possible that the same VCC could be used for both RPC "signalling" and bulk transfer.

3.3 Protocol and BT-PDU types.

BTOP packets (BT-PDUs) come in two classes, data and control. Both classes of BT-PDU have the format of AAL-5 CPCS PDUs. The class of BT-PDU is indicated by the CPI (Common Part Indicator) field in the CPCS-PDU Trailer (see figure 3). Four CPI values will have to be reserved to indicate different data and control types. Control BT-PDUs carry no payload and so have a Length of zero and fit into a single ATM cell. The use of the CPCS-UU (CPCS User to User Indication) is described below.

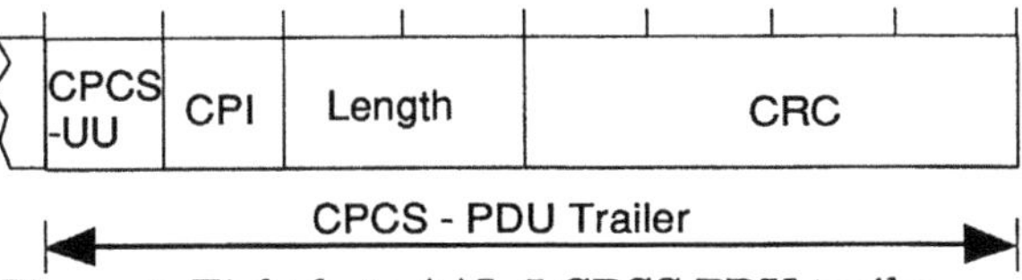

Figure 3: Eight byte AAL-5 CPCS PDU trailer

When a source BT entity has an Application Data Unit to send, it segments it into DATA PDUs. Each DATA PDU, except usually the last, is maximally sized for the VCC being used between the Source and the Destination. The DATA PDUs are each given a decrementing sequence number, carried in the CPCS-UU field. The last DATA PDU has a UU value of zero. All DATA PDUs are passed down to the ATM layer to be transmitted back-to-back. That is, the entire Application Data Unit is transmitted as a single burst of cells, subject only to traffic management policies applied at the ATM layer (see below). The entire burst is acknowledged with a single Received OK control PDU (1 cell) after the last DATA PDU been received. Figure 4 depicts the normal BTOP message sequence when the Application Data Unit occupies n DATA PDUs.

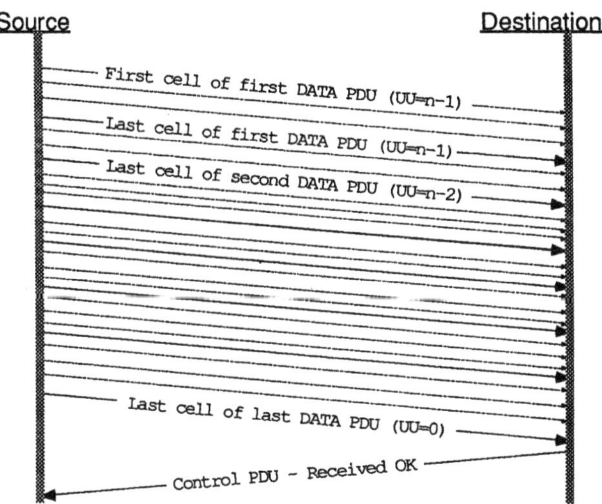

Figure 4: Cell sequence chart for error free Bulk Transfer

If there is a missing or corrupt cell in a data PDU then the destination uses a RESEND control PDU to request that the entire data PDU be retransmitted. The RESEND control PDU is a single cell AAL-5 PDU containing the sequence number of the corrupted data PDU in the UU field. (Note that under the assumptions of cell loss rate described below, losing two or more DATA PDUs in a burst will be extremely rare - the overhead of sending a separate RESEND PDUs for each lost PDU is not significant). The Source is expected to respond with the missing data PDU. Re-transmitted data PDUs have a different CPI value (RDATA) from original DATA PDUs, to avoid potential confusion with delayed responses.

As discussed later (in the section on Implementation Issues) rare occurrences, such as the loss of the Received OK control PDU can be protected against by timers. For full details of the BTOP protocol see [10].

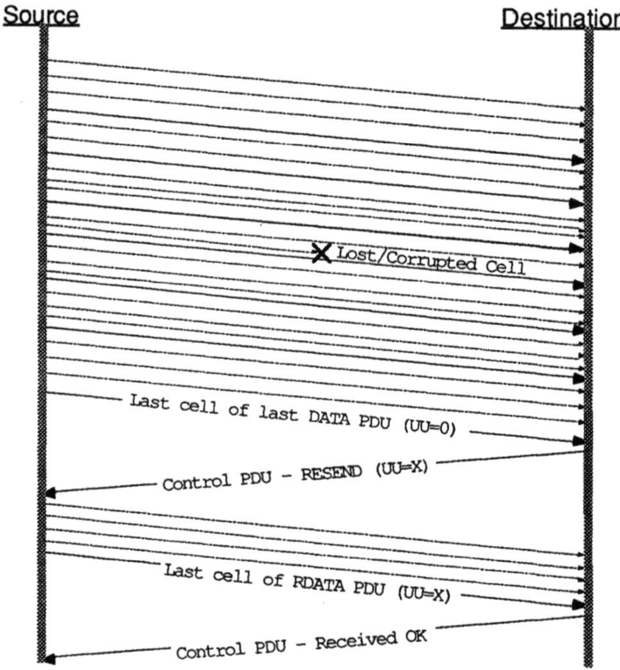

Figure 5: Cell sequence chart showing resent data PDU

3.4 BTOP APIs

Application Programming Interfaces (APIs) could be developed for applications to access BTOP's Bulk Transfer service. A basic API would offer Bulk_Send and Bulk_Recieve primitives, for the Source and Destination respectively. This basic API would be applicable where there are in-memory buffers to hold the entire burst. As noted above the maximum size of Application Data Unit that BTOP treats as a single burst is 16Mbytes. 16Mbytes is not beyond the capacity of today's workstations to store directly in primary memory. Simple devices, such as electronic light boxes in medical imaging applications, might use the Bulk_Recieve primitive to transfer an incoming image directly from the network into a display frame buffer.

The basic API for Bulk Transfer service might look like:

```
type Ind_Code is (OK, NoVCC, BufferTooBig, ...etc.);

type Buf_Address is new System.Address;

type Buf_Length  is range 1..16M;          -- In bytes

type Up_Call is access procedure(Result : Ind_Code);

procedure Bulk_Send_Request
                (B_Handle :In  Buf_Address;
                 B_Size   :In  Buf_Length;
                 Result   :Out Ind_Code);

procedure Bulk_Recieve_Request
                (B_Handle :In  Buf_Address;
                 B_Size   :In  Buf_Length;
                 Indication  : Up_Call);
```

The basic Bulk Transfer service copies data from the Source buffer onto the ATM network as cells. At the Destination the data is moved into the buffer specified by the Destination as it arrives. A call-back routine is used to notify the Destination application of the completion of the transfer. The Source application is assumed to block while waiting for the transfer to complete, although an asynchronous version of Bulk_Send could be provided also. If any retransmissions are needed, the Bulk Transfer service re-transmits the required data directly from the buffer without any intervention from the application.

Since BTOP processes an Application Data Unit in the predictable access pattern of first byte to last byte, there is no impediment to an implementation using it for data that is held on secondary storage and pre-loaded piecemeal into memory as needed. The BTOP protocol itself would not be affected, but extra API hooks would be needed to handle a chain of smaller buffers and to allow BTOP to re-access any data that has to be re-transmitted because of cell losses. Note though that re-transmissions will only occur after the first pass through the Application Data Unit is completed, and that the response time in providing data then is not critical to overall throughput.

4. SUPPORTING ASSUMPTIONS

BTOP relies on a number of properties particular to ATM and AAL-5. These are listed below.

4.1 End-to-End ATM Service.

Today ATM cards are available for PCs and workstations. ATM to the desktop is strongly emphasized in the work of the ATM Forum. Rather than carry extra network protocol overheads to cope with non-ATM stations accessing a network though routers and gateways, BTOP has been designed for deployment where ATM service is end-to-end.

4.2 AAL-5 CPCS Properties

The CPCS service used by the BT sublayer Source and Destination entities is the Non-assured Message Mode service [6]. It, in conjunction with the underlying ATM layer, has the following properties:

- No re-ordering: The delivery order of PDUs on a particular VCC is the same as the sending order (i.e. misordering is negligible).
- No duplicates: CPCS does not duplicate PDUs and the underlying ATM layer does not duplicate cells (i.e. the duplication rate is negligible).
- Notification of corrupt PDUs. Cells may be corrupted or lost (infrequently). CPCS detects the resulting errors in PDUs. An option in the AAL-5 specification allows CPCS to deliver PDUs that are corrupt. BTOP assumes that it is notified when a PDU has arrived (see below in the section on Implementation Issues).

4.3 VCC and Source-Destination Association management.

There is no provision in the BTOP protocol for detecting the loss of a connection or the termination of either end. Rather the Burst Transfer service relies upon M-plane notifications.

- Should either the Source or Destination fail, it is a Layer Management function to close down the VCC.
- When the VCC is closed down or fails (including underlying transport failure) it is a Layer Management function to notify both the Source and the Destination, including the Burst Transfer entities.

5. PROTOCOL CHARACTERISTICS

5.1 Burst size

The maximum payload length of an AAL-5 PDU is 65535 bytes (64 Kbytes less one byte). The size of the UU field limits the total number of DATA PDUs in a burst to 256. Thus BTOP can handle Application Data Units of just less than 16 Mbytes. 16 Mbytes provides a much larger unit of transfer than current protocols, while remaining within the capacity of current network and workstation technology.

5.2 Error Handling

BTOP is designed on the assumption that the cell loss ratio (CLR) of the underlying ATM transport is very low. Assuming a CLR of less than 1.7E-10, (c.f. [11]), fewer than one in 4 million data PDUs will be corrupted. With 256 data PDUs in a maximum sized Application Data Unit (16 Mbyte) we would expect fewer than one in every 16 thousand Bulk Transfers to encounter a cell loss on the first pass.

The error checking and recovery unit in BTOP is the AAL-5 CPCS PDU. A 32bit CRC is provided in the CPCS sublayer that protects the entire payload of the AAL-5 PDU including UU and CPI fields.

BTOP uses selective retransmission. The Source only services RESEND requests once it has transmitted the entire first pass burst. A design option was to have the Destination issue the RESEND control PDU as soon as it detected a corrupted DATA PDU. But since loss of a DATA PDU will be very infrequent, this optimisation has been ignored in favour of simplifying state machines and timer requirements. Thus the retransmission of each missing DATA PDU is requested and acknowledged serially, after the last DATA PDU has been received.

5.3 Performance

In normal operation one Received OK PDU (a single cell) acknowledges the entire transfer (i.e. up to 16 Mbytes of application data). The transaction is completed in the transfer time plus one round trip delay (RTD).

If a cell is lost in the transfer then, with the selective retransmission strategy described above, another round trip delay will be added before the transfer is completed.

With these delay characteristics BTOP is well suited for wide area transfers, as well as local area transfers. In the wide area the RTD can be a significant portion of the transfer time even for relatively large Application Data Units (e.g. a 3000km separation of Source and Destination gives a RTD of ~ 33msecs, which is more than the transfer time for 0.5 Mbytes at 150 Mb/s). In the wide area BTOP will perform significantly better than protocols that require multiple round trips.

6. DESIGN CHOICES

Not only is BTOP intended to run directly on top of the AAL-5 CPCS, BTOP is inspired by the simplicity of AAL-5. AAL-5 was developed by the computer industry in reaction to the perceived complexity of AAL-3 and AAL-4 [12]. To achieve the goals of simplicity and efficiency, BTOP's design was developed with the following three guidelines:

- focus on one function
- do not perform functions that are already done in other parts of the protocol stack
- exploit the special characteristics of ATM

BTOP is focused on just the transfer of large Application Data Units. It is proposed that other functions such as remote procedure calls use other protocols, cf. [9]. Below we look at the application of the second and third guidelines.

6.1 Exploiting Connection Orientation

An important aspect of BTOP's simplicity arises from it being connection oriented. ATM is inherently connection oriented (53 byte cells are too short to carry a full source and destination address in each one!) and BTOP exploits that fact. AAL-5 offers a very simple end-to-end connection oriented protocol. The AAL-5 connection is directly provided by the ATM layer VCC (Virtual Channel Connection) between Source and Destination.

Any connection oriented protocol has the overhead of connection establishment. As noted above for BTOP a VCC needs to be pre-established between Source and Destination using C-plane or M-plane facilities. The virtue of connection orientation is that there is a lot of information that needs to be exchanged between Source and Destination only once, not with every PDU transfer. For BTOP the following assumptions are made:

- At or before connection establishment there is agreement on the maximum size of AAL-5 PDU and on the maximum transfer rate.
- At or before connection establishment the Source and Destination have satisfied themselves as to the identities of their respective partners (i.e. authentication, if required, is performed as part of VCC setup, not on a per transfer basis). Likewise any encryption key exchange occurs at or before connection establishment.
- During connection establishment the network is informed of the class of service and QOS required (e.g. ABR).

To summarize, all fixed attributes for Bulk Transfer between Source and Destination are set before or during connection establishment. No fields are used in BT_PDUs to convey these attributes. Consequently no processing is needed to generate the fields on PDU transmission, nor to check their validity upon reception.

6.2 No multiplexing

Multiplexing is cheap at the ATM layer. There is generous provision for multiple VCCs between two end systems. Cell flows that are part of different associations between entities on the end systems can be given their own separate VCCs, each tailored to their own traffic characteristics. There is no need for multiplexing different streams of data at protocol layers above the ATM layer, and BTOP makes no provision for it. There is only one Service Access Point (SAP) for the BT Service per VCC. The single SAP is identified by the VCC. There is no sublayer addressing.

Neither is there any overlapping of transfers. For a given instance of the Bulk Transfer Service (i.e. a given VCC) there can be only one outstanding Bulk Transfer in progress at a time.

These design decisions result in there being no need for sub-address fields and transaction identifiers in BT-PDUs.

6.3 No Destination Flow Control

The basic capability of BTOP is to transfer up to 16Mbytes of data from memory to memory. Modern systems that want to perform large data transfers can easily afford to buffer the whole burst in memory. There is little point in buying high speed network interfaces if the attached system can not keep up with the line rate. Hence, while sender and receiver can agree on a maximum transmission rate when a connection is established, BTOP does not provide for destination flow control during a Bulk Transfer (although the definition of ABR might permit it - see below).

6.4 Network Flow Control via ABR

The ATM Forum's Traffic Management Subworking group is working towards defining an ATM layer service called ABR (Available Bit Rate) service. ABR is intended to be the ATM analogue of the service that a MAC layer offers in a LAN. In a LAN the available bandwidth is not permanently divided between all stations, rather each station gets the full bandwidth when it needs it. The media access control mechanism determines access opportunities when multiple stations contend for the bandwidth.

At the time of writing, the specification of the ABR service has yet to be nailed down, and no mechanism has been chosen for its implementation. However the general nature of ABR is clear. Stations will be able to use a substantial part of their access bandwidth to transmit bursts. In threatened congestion situations, ABR delays the transmission of cells rather than discarding them. Some form of protocol will exist to control the flow of cells in the ATM network. Thus, unlike CBR (Constant Bit Rate) and VBR (Variable Bit Rate) services, ABR will not give a tight guarantee of cell delay and cell throughput. ABR will however offer a cell loss rate close to that of the underlying medium, similar to CBR.

Given that ABR will be provided at the ATM layer, not only is it unnecessary to provide a flow control service within BTOP, it would be very undesirable to do so. Any flow control mechanism installed above ABR is likely to interfere with it in ways that are difficult to analyse and will certainly decrease efficiency. Hence BTOP relies on the ATM layer ABR service for flow control. The BTOP protocol itself does not do anything to regulate its rate, or avoid congestion during a burst.

Note that, under the assumption that the Destination can receive a burst at full rate, there is no impediment to using BTOP with CBR service. Of course, since CBR bandwidth is reserved for the lifetime of the VCC, either a VCC will have to be established and held just for each Bulk Transfer or bandwidth will be wasted between bursts.

6.5 Using large packets

In legacy networks the maximum size of packets is quite small, as low as 256 bytes. This arose in part from concern about the buffer space needed at source and destination, and in part from trying to optimize packet size for the likely transmission error rates. Such small packets cause excessive processing overhead and limit overall throughput.

Since ATM networks have a very low inherent transmission error rate, the maximum (payload) size of AAL-5 PDUs has been set at 65535 bytes. At this size the overhead is manageable (265 PDUs per second on a 150 Mb/s link) even if a processor has to get involved with each PDU (but see below) .

7. IMPLEMENTATION ISSUES

7.1 Hardware realization

BTOP is designed with future so-called "desktop area networks" in mind. To avoid interrupting a high performance CPU and moving data through main memory, bulk transfers may be direct from peripheral device to device (e.g. from disk controller to frame buffer). BTOP is intended to be fully realizable in hardware.

The key to hardware implementation is simplicity. Bulk Transfer across a network should be no more complex than say a chained SCSI transfer from disk to main memory. As we described above, connection orientation and the elimination of options has allowed us to produce a protocol that does not add any more fields to the basic AAL-5 PDU. BTOP needs to process just two otherwise unused AAL-5 fields.

Basic AAL-5 is already encapsulated in hardware. The state machines for BTOP should be easy to integrate (we do not expect a silicon implementation to reflect the layering of figures 1 and 2). We have eschewed high efficiency in low-runner situations in order to simplify implementation. For example the PDU re-transmission scheme, with retransmitting beginning only after the entire burst has been sent (and dealing with only one PDU at a time), was chosen so that Destination hardware can be set up for direct copying from cell buffer to final destination.

7.2 Dealing with special cell losses

If Application Data Units are going to be built up "in situ", then it is important that the Destination BTOP state machine have an accurate view of the sequence number of an incoming data PDU before the trailer arrives. This allows the calculation of the final location into which the contents of each cell of the data PDU is to be directly copied. To do this the state machine needs to be notified of the end of each data PDU even if the final (specially marked) cell is lost or corrupted.

The corruption of intermediate DATA PDUs due to loss or corruption of their final cells does not cause particular problems. When the final cell of an intermediate PDU is lost, the first or second cell of next data PDU should cause the SAR layer to indicate "Re-assembly Buffer Overflow", from whence it can be deduced that the final cell was lost. (Note that in all likelihood the following DATA PDU will not be received correctly, resulting in the retransmission of two DATA PDUs).

Loss of the final cell of the final DATA PDU is more problematic. BTOP requires that a control PDU (either Received OK or RESEND) be generated by the Destination after a burst has been received. Completion of the burst is signalled by the final cell of the last PDU. When the final cell of the last DATA PDU or a RDATA PDU is lost there are no more cells to trigger Re-assembly Buffer Overflow and so the normal notification of PDU completion would not occur. The final cell of the last PDU is a special cell in BTOP.

Likewise the cell that constitutes a control PDU is a special cell: its loss, if not provided for, will cause a Bulk Transfer to "hang".

The simplest implementation approach is to ignore the loss of these special cells and let the Bulk Transfer remain in an uncompleted state until a user intervenes and resets the ATM connection. This approach is very attractive if ATM connections meet the specified cell loss ratio of 1.7E-10 mentioned above. With such a low loss rate and performing a Bulk Transfer every second (i.e. generating two special cells per second), a connection will hang only once every 95 years. This is far less frequent than most disk controllers hang.

A more conservative approach requires timers. Dealing with timers adds expense to an implementation, especially if the timers have to be manipulated on every cell. AAL-5 allows an optional SAR re-assembly timer, to detect partially filled PDUs that are not going to be completed. Any SAR re-assembly timer value must be very loose or it will interact negatively with ABR service, which delays cells rather than lose them. Using a SAR re-assembly timer gives almost complete coverage for detecting the last DATA PDU, but it does not cover single cell control PDUs or last DATA PDUs that happen to consist of a single cell. Our recommendation is to use a subsequent application request to send or receive an Application Data Unit as an opportunity to detect a hung connection. If when a new request is received, it appears to the BT layer entity that a current transfer could be hung then it should set a long timer. Should the timer expire the BT layer entity should tidy up and then proceed to process the new request.

8. EXISTING PROTOCOLS

Existing protocols that are used for Bulk Transfer are neither tailored to, nor appropriate for, ATM. They were designed for connectionless transports with small packet sizes, high packet loss rates and low bandwidth-delay products.

TCP [5] is the protocol used by FTP. FTP is the most commonly used bulk transfer protocol today. TCP establishes its own connection on top of connectionless service (IP). In a Broadband environment, IP in turn would be carried on top of AAL-5 connections either directly or over a LAN emulation layer. This layering is very inefficient and completely unnecessary in an environment that is end-to-end ATM. There are so many protocol fields to be filled in that even a simple "ack" will not fit in a single cell. BTOP is considerably simpler.

With TCP, retransmissions are of the "go-back-n" variety. This is not efficient in Broadband networks with their high bandwidth delay product.

BTOP uses selective retransmission. TCP also has a slow start windowing scheme. This mixes up network congestion control, error recovery and destination flow control. Slow start can double or triple the transfer time of a burst in a wide area network. There are also questions as to how it will interact with ABR flow control. BTOP leaves congestion control to the underlying ABR service.

XTP [13] was designed for high speed protocol processing but is too general. In [14] Kure and Sorteberg examine the problems of running XTP over ATM. XTP is not targeted just at bulk transfers. It has both rate and flow control schemes, and a priority scheme for handling interleaved packets on the same connection (which it manages itself). All of these features are present in ATM with ABR service and so BTOP does not duplicate them. Duplicating them in XTP result in extra processing and PDU space overheads. XTP was also designed to work over a number of link layers, with concomitant protocol overhead. However both XTP and BTOP have the right flavour of retransmission scheme (i.e. retransmit missing blocks).

NETBLT [8] is explicitly targeted at the rapid transfer of a large quantity of data between computers. The main concerns of the protocol are to control the rate of the sender so as to avoid network congestion and client overrun. It is intended to run over IP and performs its own connection establishment and maintenance. It too uses selective re-transmission. In some ways BTOP can be regarded as the re-engineering of NETBLT's burst transfer for the ATM environment - discarding IP, and exploiting ATM's connection oriented services.

In [1] Turner and Peterson specifically study the transfer of images. Their work is predicated on there being a significant rate of packet loss in a network (1% or more). They show that compression/encoding schemes that allow reconstruction of missing parts of an image can be more effective than retransmitting missing packets. In BTOP we too have attempted to do an end-to-end design but we have not considered interpolated reconstruction necessary given the relatively lossless nature of ATM. BTOP is not confined to just images. In the image domain it will work with any compression scheme but is particularly appropriate for applications that are intolerant of any artefacts in images.

9. SUMMARY

ATM will provide high bandwidth over long distances. This could be exploited to give sub-second transfer times for large Application Data Units, such as an uncompressed image, but to do so will require new protocols. BTOP is such a protocol that exploits the unique properties of ATM. Existing protocols were designed for considerably less powerful networking technology and are inefficient when run over ATM networks.

BTOP has been designed to reduce the number of layers involved in performing a Bulk Transfer. It offers service directly to applications and runs on top of basic (CPCS) AAL-5. AAL-5 runs on top of the ATM layer. BTOP achieves high efficiency and low overhead by not performing any function done in another protocol layer.

BTOP operates in the context of the ATM three plane model. BTOP relies on the C-plane functionality to establish and tear down the VCC it uses. BTOP does not have any "connection establishment" phase of its own. BTOP exploits connection establishment to set up all required parameters, authentication etc. rather than performing these functions for every transfer, or, even worse, for every constituent packet. The ATM layer is assumed to provide ABR service to take care of network congestion. BTOP uses ABR, it does not compete with it.

BTOP transfers Application Data Units across an ATM network as a sequence of AAL-5 PDUs. A positive acknowledgement is returned when all PDUs have been successfully received. Selective re-transmission of an AAL-5 PDU is invoked if any of its constituent cells should go missing or be corrupted.

BTOP is the first of what we hope will be a growing number of protocols that attempt to match an application area directly with the capabilities of ATM [15]. We regard this as a more worthwhile activity than contorting ATM to handle old protocols past their prime.

REFERENCES

[1] Image Transfer: An End-to-End Design, C. Turner and L Peterson, Proc SIGCOMM '92 (Baltimore) as ACM SIGCOMM Vol 22 No 4 Oct 1992 pp 258-268.

[2] CCITT (now ITU-T) Recommendation I.121: Broadband Aspects of ISDN, IXth Plenary Assembly (Melbourne) Nov 1988.

[3] ATM User-Network Interface Specification, Version 3.0, The ATM Forum, Mountain View CA, Sept 93, (to be published by Prentice Hall).

[4] *Asynchronous Transfer Mode: Solution for Broadband ISDN*, Martin de Prycker, Ellis Horwood series in computer communication and networking, Chichester, 1991.

[5] Transmission Control Protocol - RFC 791, Defense Advanced Research Projects Agency, Arlington Va, Sep 1981

[6] AAL Type 5, Draft Recommendation text for section 6 of I.363, ITU-T SWP XVIII/8-5, Geneva Jan 1993.

[7] Architectural Considerations for a New Generation of Protocols, D Clark and D. Tennenhouse, Proc SIGCOMM '90 (Philadelphia) as ACM SIGCOMM Vol 20 No 4 Sept 1990 pp 200-208.

[8] NETBLT: A Bulk Data Transfer Protocol - RFC 998, D Clark, M Lambert and L Zhang, MIT, Mar 1987.

[9] ROOP - Remote Operations straight on top of AAL-5, L Casey, ATM_FORUM/94-0083, (Lake Tahoe) Jan 1994.

[10] BTOP Specification, L. Casey BNR CRL Internal Report 94123, Jun 1994.

[11] TA-NWT-001110: Broadband ISDN Switching System Generic Requirements, Issue 2 Bellcore, Aug 1993.

[12] The Development of ATM Standards and Technology: A Retrospective, R.Vickers, IEEE Micro Vol 13 No 6, Dec 1993, pp 62-73

[13] XTP Protocol Definition, Rev 3.1. G. Chesson et al, Protocol Engines Inc, Santa Barbara, 1988

[14] XTP over ATM, Ø. Kure & I. Sorteberg, Computer Networks and ISDN Systems Vol 26, 1993, pp 253-262.

[15] BTOP - A Bulk Transfer protocol over AAL-5, L Casey, ATM_FORUM/94-0082, (Lake Tahoe) Jan 1994.

12

High performance presentation and transport mechanisms for integrated communication subsystems

W. S. Dabbous

INRIA, 2004 Route des Lucioles, BP-93,
06902 Sophia Antipolis Cedex, FRANCE
e-mail: dabbous@sophia.inria.fr

The performance enhancement of communication protocols is an essential step toward the building of high performance communication subsystems. Integrated Layer Processing (ILP) has been proposed as an engineering principle for the optimization of communication protocol implementations. However, the applicability of this principle to complete operational communication subsystems has not been yet addressed. This is in part due to the lack of high performance implementation of presentation and transport functions. In this paper, we show to what extent "ILP based" optimizations improve the performance of some data manipulation functions (e.g. presentation encoding or checksumming) independently. We present how some of these implementation optimization techniques were applied to the ASN.1 encodings rules. We also analyze the impact of such optimization techniques according to the support hardware. A prototype implementation of the XTP protocol is also described. ILP based optimization applied to the checksum calculation algorithm show that an important performance enhancement can be achieved. These results should facilitate the support of ILP within an operational communication subsystem.

1 Introduction

The emergence of high speed and low error rate networks motivated many research work on the design of more efficient communication protocols. This includes the tuning of existing transport protocols (e.g. TCP extensions for high speed paths [Jac90], TP4 1992 revision [CCI88], and other related work [Wat87], [Col85]) and the design of a new transport protocols (e.g. XTP [PEI92]). At the presentation level, this concerns the performance optimization of the presentation coding and decoding routines either by defining new "light weight transfer syntaxes" [Hui90] or by implementation enhancements of the standard ASN.1 Basic Encoding Rules (BER) routines [Dab92], [Hui92], [Lin93].

Good implementation techniques represent one of the most important factors in determining the performance of a given protocol [Cla89]. These techniques depend on the environment more than on the protocol itself. The proposed solutions focused on the enhancement of the protocol implementation performance in a given software or a hardware environment: outboard protocol processors (e.g. [Kan88], [Coo90]), hardware protocol implementations (early work on XTP/Protocol Engine [Che89]) or parallel implementations of transport protocols [Bra92], [Rüt92], [Lap92], [Bjö93]. A detailed survey of protocol implementation optimization techniques can be found in [Dab91] and [Fel93b].

The performance of workstations has increased with the advent of modern RISC architectures but not at the same pace as the network bandwidth during past years. Furthermore, access to primary memory is relatively costly compared to cache and registers and the discrepancy between the processor and memory performance is expected to get worser. The memory access is expected to represent a bottleneck [Dru93].

Protocol processing can be divided into two parts, control functions and data manipulation functions. Example of data manipulation functions are presentation encoding, checksumming, encryption and compression. In the control part there are functions for header and connection state processing. Jacobson et al. have demonstrated that the control part processing can match gigabit network performance for the most common size of PDUs with appropriate implementations [Cla89].

However, data manipulation functions present a bottleneck [Cla90], [Gun91]. They consist of two or three phases. First a read phase where data is loaded from memory to cache or registers, then a manipulation or "processing" phase followed by a write phase for some functions, e.g. presentation encoding. For very *simple functions*, e.g. checksumming or byte swap, the time to read and write to memory dominates the processing time. For other *processing oriented functions*, like encryption and some presentation encodings, the manipulation time dominates with current processor speeds. However, the situation is expected to change with the increase of processor performance: the memory access will be the major bottleneck rather than data processing.

The data manipulation functions are spread over different layers. In a naive protocol suite implementation, the layers are mapped into distinct software or hardware entities which can be seen as atomic entities. The functions of each layer are carried out completely before the protocol data unit is passed to the next layer. This means that the optimization of each layer has to be done separately. Such ordering constraints is in conflict with efficient implementation of data manipulation functions [Wak92], [Cla90].

One could accuse the layered model (TCP/IP or OSI) for causing this conflict. In fact, the operations of multiplexing and segmentation both hide vital information that lower layers need to optimize their performance. Although it is important to distinguish between the architecture of a protocol suite and the implementation of a specific end system or a relay node, A certain degree of flexibility is needed in the way the functions are organized within the layers. One solution is to integrate all layers in a single block, in essence discarding the modularity of layers in exchange for performance. The alternative is to perform multiplexing/demultiplexing and segmentation/reassembly exactly once in the protocol stack. Another approach is to perform multiplexing/demultiplexing and reassembly exactly once in the protocol stack and allow multiplex segmentations to occur (see [Fel93a]).

Integrated Layer Processing (ILP) is an engineering principle that has been suggested for addressing the cited problem [Cla90]. The main concept behind ILP is to minimize costly memory read/write operations by combining data manipulation oriented functions within one or two processing loops instead of performing them serially as is most often done today . It is expected that the cost reduction due to this optimization will result in better overall performance as it will reduce time consuming memory access. This optimization may be applied within a single data manipulation function (*intra-function optimization*) or across several functions (*inter-function optimization*). The interest of the ILP principle (and similar software pipelining principles such as lazy message evaluation and delayed evaluation) has been discussed in [Cla90], [Gun91], [Par93], [O'M90], [Peh92] and [Abb92].

However, the examples cited in these papers correspond to the expected performance gain when some specific data manipulation functions were integrated. They do not address the problem of the design and implementation of a complete operational communication subsystem according to the ILP principle (some issues concerning the design of an entire stack according to ILP are presented in [Abb93a]). As the performance benefits of ILP are applicable when the memory read and write operations are the bottleneck, previous work reported examples with simple manipulation functions (e.g. checksumming or byte swap). However, the development of complete operational communication subsystems requires the integration of other "processing oriented" data manipulation functions (e.g. presentation encoding) in order to provide the desired service. The optimization of these processing oriented data manipulation functions will result in a double benefit: in addition to their execution at higher speeds, their integration with other manipulation functions will be facilitated leading to increased performance gain. This optimization is now both feasible and attractive due to the increase of processor speeds and of the discrepancy between processor and memory performance.

We worked on the optimization of specific processing oriented data manipulation functions: the ASN.1 presentation encoding and decoding routines. In addition, in order to experiment integration techniques with transport level functions, we performed a prototype implementation of the XTP protocol at the user level. The integration of presentation and transport level functions is more easily achievable if the transport is implemented at the user level. We also optimized the XTP checksum implementation which resulted in high speed checksum calculation. We present the results of work on the optimization of both presentation and transport level functions i.e. presentation encoding and decoding routines and the XTP checksum calculation which are two examples of intra-function ILP optimizations. The rest of the paper is organized as follows: Section 2 presents the work on the enhancement of the presentation routines. In section 3, we describe the XTP implementation and present some performance test results. Section 4 concludes the paper.

2 High speed presentation

Our work was centered around the optimization of the ASN.1 Basic Encoding Rules. The cost of the coding and decoding routines is attributed to the heavy Type-Length-Value oriented coding of ASN.1 BER. This motivated the work on "light weight" or XDR-like transfer syntaxes [Hui89] based on three design principles:

- avoid unnecessary information in the encoding,
- use fixed representation when it is possible and,

- simplify the mapping of the elements by using fixed length structures.

The first principle leads to the abandonment of the systematic "Type-Length-Value" encoding, which is replaced by new encoding rules for the ASN.1 constructs; the second and the third principles lead to a set of easily decoded *word oriented* encoding rules. The *word size*, is a parameter that characterize the syntax (16, 32, or 64 bits). The complete definition of these light weight encoding rules (Flat Tree Light Weight Syntax, FTLWS or LWS for short) is given in [Hui90]. Both BER and LWS syntaxes are supported by the MAVROS ASN.1 compiler developed at INRIA [Hui91].

We have done some experiments to compare the performance of ASN.1 BER and the LWS encoding rules. The ASN.1 specification of the data type used in these tests is:

```
IMPORTS MPDU FROM X400-1984-P1;

perf-object ::= CHOICE {
        ints[1] SEQUENCE OF INTEGER,
        strings[2] SEQUENCE OF PrintableString,
        real[3] SEQUENCE OF REAL,
        mpdus[4] SEQUENCE OF MPDU}
```

The test object is either a SEQUENCE OF INTEGER, a SEQUENCE OF REAL or a SEQUENCE OF PrintableString (mean length of 26 characters per string). In order to test the encoding rules with more complex types, we chose the MPDU type (Message Protocol Data Unit) which defines the message format (envelope and content) according to the 1984 version of the P1 Protocol (CCITT Recommendation X.411) [CCI84]

The number of the data values for each type in the performance tests is given in table 1.

Test identifier	Data Type	Number of test values
`ints`	SEQUENCE OF INTEGER	400
`real`	SEQUENCE OF REAL	200
`strings`	SEQUENCE OF PrintableString	124
`mpdus`	SET OF MPDU	18

Table 1: Number of Data values for each type

The results of performance tests accomplished on a Sun 3/60 are given in table 2.

The LWS coding routines were 1.42 to 5.85 times faster than those of the BER depending on the data types. For basic data types, the improvement is considerable (specially for the `real` type) because of the relative efficiency of the word oriented coding. However, the results for the strings and MPDU types are very disappointing, as they showed only a modest gain in performance for the light-weight transfer syntax, while resulting in a more than twice longer encoding.

After this first test series we conducted several optimizations by analyzing the coding routines for a tree structure type in order to facilitate the analysis of the "execution

Coding routines	`ints`	`reals`	`strings`	`mpdus`
BER (total)	5999	13666	7833	62330
BER (per element)	15	68.33	63.2	3462
LWS (total)	1333	2333	5499	37998
LWS (per element)	3.33	11.66	44.35	2111
Speed up	77.77 %	82.92 %	29.79 %	39.03 %

Table 2: Coding time in μs for different types.

profile". We tested a number of modifications in MAVROS ASN.1 BER coding and decoding algorithms, of which we cite:

- Inline the encodings for some types ("OCTET STRINGS"),
- Apply the "type reduction" technique in order to remove all "repeated indirections", replacing them as much as possible by direct references to the data type.
- Use "header prediction" techniques to speed up the decoding of tags and that of length fields,
- Speed up the encoding of tags and length field by using systematically the "indefinite length" encoding form.
- Use static declaration and better memory management (reduce the calls to `malloc`).

Some of these optimizations are "ILP based": minimize costly memory access operations. Some other concern the reduction of the processing operations. These optimizations were reported in the code generation programs of the ASN.1 compiler, as well as in the "run-time" library. A new series of tests were conducted again with the MPDU type on the Sun3 but also on other workstation types and the performance figures are shown in table 3.

	Coding time (μs)		Decoding time (μs)		Size (octets)		Decoding (Mbps)	
	BER	LWS	BER	LWS	BER	LWS	BER	LWS
Sun3 (`Initial`)	3462	2111	16212	13527	480	1011	0.24	0.6
Sun3 (`new`)	1861	2018	6944	6398	540	1029	0.62	1.29
SS 10/30	112	199	318	225	540	1029	13.58	36.59
Dec 5000/300	208	208	372	212	540	1029	11.6	38.8
HP 9000/715	100	100	339	155	540	1029	12.74	53.1
Dec alpha	74	55	176	92.6	540	1029	24.54	88.9

Table 3: Speed of the presentation routines.

The throughput is defined as the ratio of the code size in bits to the decoding time in seconds. In fact, this gives an indication of the speed limit in processing the incoming transport packets. The comparison of BER and LWS will be based on the decoding time instead of the decoding throughput.

Before we analyze these figures, recall that the coding procedure consists in copying the `perf-object` element structure from memory to the cache, looping on processing instructions and then storing the result (BER or LWS streamlined data) in memory. This cycle is reversed for decoding but the processing part is longer due to decoding controls and to the need for memory allocation. Therefore, if the overall decoding time is much higher than the coding time, we can deduce that the processing speed is the bottleneck. Otherwise, if the decoding and coding times are comparable, then we may say that the memory access overhead (which is the same for both routines) dominates the processing time. In addition, BER encoding requires more processing than the LWS which generates, however, a larger code size.

In the case of `initial` tests on Sun3, it is clear that the processing speed is the limiting factor for both BER and LWS (decoding slower than coding and BER slower than LWS). In this case, it is advantageous to use the LWS.

The optimizations led to a drastic reduction of the BER cost and the optimized Sun3 (`new`) figures shows that the BER coding time is lower than LWS: what is gained in processing for LWS is lost due to the higher size of the data which will transit on the bus. However, the far more CPU consuming routine (decoding) BER is still slower than the LWS.

The results are more interesting with high performance RISC workstations. On the SparcStation 10, coding BER is much more efficient than LWS which implies that the memory access dominates the processing time. However, BER decoding is still limited by the CPU performance. Both coding and decoding routines for LWS are limited by the memory access (due to the large code size stored and loaded).

Similar results are obtained on both Dec 3000 and HP workstations. Note, however, that the coding time is the same for both BER and LWS which means that both memory access and CPU bottlenecks are balanced. For decoding, the processor speed limitation is more pronounced.

The "best" results are obtained on Dec-alpha where the 64 bit bus enhances the memory access performance. The figures are "classical": BER more costly than LWS and decoding (BER or LWS) is more costly than coding. This is typically due to CPU limitation.

The maximum decoding throughput is about 24.54 Mbps for BER and 88.9 Mbps for LWS on the Dec-Alpha.

From the above results we can learn the following:

- A drastic improvement of the speed of coding and decoding routines for both BER and LWS routines can be obtained by adequate implementation optimization on high performance workstations

- The LWS is interesting if the processor speed is the bottleneck, however, a more compact syntax is desirable in order to reduce the size of the data to be transmitted on the network and the memory bus.

- Memory access is a limiting factor in most cases on RISC workstations. This confirms that integration techniques should result in increased performance on such workstations. ILP intra-function optimizations were applied and resulted in enhanced performance for BER encoding.

The optimized version of the encoding and decoding functions should facilitate the implementation of the presentation layer as a filter with "streamlined" *encoding* and *transmission* of application data units.

3 High speed transport mechanisms

We now present our work on the transport level functions. This work has two goals:

- To experiment the effect of intra-function integration mechanisms on the performance of transport functions,
- To have a substrate suitable for the building of communication subsystems with support of the ILP principle.

We performed a prototype implementation of the XTP protocol. XTP was initially designed to be implemented on specialized hardware with execution efficiency as inherent part of the design process. Therefore, many syntactical and algorithmic choices of the protocol are oriented to facilitate high speed operation over low error rate networks.

XTP provides a set of mechanisms, the application can select the desired type of service according to its own requirements. We implemented XTP in the Unix user level as a library of transport functionalities. This library is linked to the application code. The interest of a user level implementation of XTP is to provide a support for the test of the integration mechanisms. In fact, on the same processor and in the absence of scheduling overhead one could expect that the performance of both kernel and user level implementations of the same transport protocol are comparable. This is confirmed by the results reported in section 3.2.2.

In the rest of this section we will describe the XTP implementation. We will also compare the performance of user level XTP implementation with kernel supported TCP.

3.1 XTP implementation

The implementation runs on an extension of the 4.3 BSD socket interface. The extension of the kernel protocols was done in order to add the support of XTP on top of IP. The implementation architecture is depicted in figure 1.

Two functions `xtp_input` and `xtp_output` on top of IP serve as a demultiplexing layer. They process the `key` field in the XTP header. This minimal kernel support provides a secure allocation of network ports. The provision of "datagram" XTP sockets enables the application to declare "network entry points". Transmission control procedures are performed at the user level and thus can be easily configured according to the application needs. User applications have access to the XTP socket interface through the

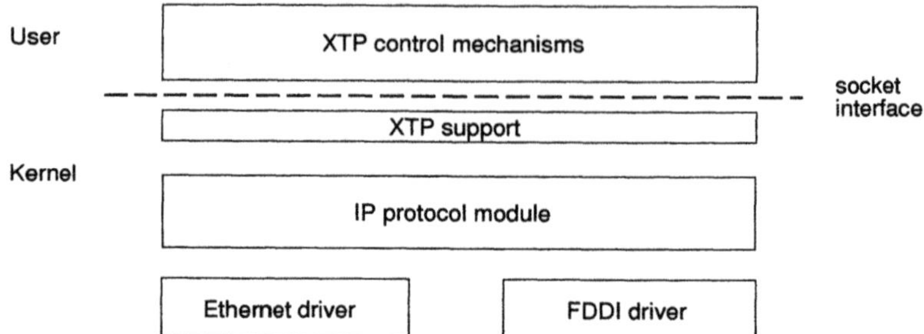

Figure 1: General architecture of the XTP implementation

standard system calls (`socket`, `bind`, `connect`, `send`, `sendto`, `writev`, `recvfrom`, `readv`, `select`).[1]

The XTP connection set up, the reliable data transfer and the connection tear down functionalities were implemented. Other XTP functionalities like rate control, route management, multicast procedures have not been implemented. A detailed description of this implementation is given[Dab93]. We present here the main implementation issues:

Context initialization An application using XTP should initialize a context before sending or receiving data. After this initialization, a field within the context structure points to a control block with default values, corresponding to default type and quality of service. The application may overwrite some or all the fields of the control block. This reflects the programmability of the XTP protocol.

Context scheduling An application may have several active contexts in the same time (several point to point associations with different application entities). The application examines the list of "active" contexts to determine the next context to be processed and the corresponding wait timer value as described in [Var87].

Buffer management The allocation of memory buffers is performed by the application. The sizes of the maximum receive or send buffers are specified in the corresponding fields in the control bloc structure. These buffers are directly used by the application to compute/process data. There is no intermediate transport level buffers. When an XTP packet is received, only the packet header is consulted. After the necessary control functions had been performed, the data is copied into the corresponding place within the user level buffer. The system calls `readv` and `writev` are used to accomplish the scatter/gather I/O. According to its own requirements (e.g when a complete frame is received for video application) the application is invited to process the message.

[1]The `writev` system call attempts to gather an array of buffers of data and send it as a single datagram (connected socket). The `readv` system call attempts to read a datagram from a socket and scatters the input data into an array of buffers.

Retransmission strategies The selective ACK-upon-request mechanism used by XTP facilitates the implementation of the two well known retransmissions strategies, goBack-N and selective repeat. Both strategies have been implemented. During the performance tests presented hereafter the receiver has adopted the selective repeat strategy.

Checksum calculation The 32-bit XTP check function called CXOR, is defined as the catenation of two 16-bit functions: XOR which is a straight "vertical" exclusive-or of each 16-bit short word in a block of information, and RXOR a rotated "spiral" exclusive-or of each 16-bit short word. We first implemented a "raw" version of the XTP checksum and then experimented the effect of ILP intra-function optimizations on the performance of the checksum algorithm. We had increased performance by a factor of 12 over the initial "raw" implementation due to these optimizations as will be shown in the section 3.2.1

3.2 Performance tests

In this section we present performance test results of our prototype XTP. Performance of the standard kernel TCP BSD implementation will be presented for reference. However, the comparison should be made very carefully: the performance of a protocol depends on the environment, implementation tunings and enhancement techniques as demonstrated in [Dab92]. Many differences exist between the choices we adopted and the TCP kernel implementation, and the XTP implementation is not as well "tuned" as the TCP kernel implementation. Therefore, it is not meaningful to make precise comparison of both protocols based on the performance figures we have. The figures give only an idea on the possible performance of XTP when implemented in the user level. Other detailed performance tests can be found in [Cab88] and [Nic91] for TCP and in [Fan93] for XTP. A user level implementation of protocol of TCP is described in [The93]. Another work on optimized TCP checksum is reported in [Kay93].

3.2.1 Performance of the checksum algorithm

The speed of checksum calculation is known to be one of the important factors that determine the performance of a transport protocol [Cla90], [Ros90]. The XTP designers proposed to perform this task in hardware [Str92]. However, the reduced flexibility of the hardware approach is a critical constraint because of the applications diversity. The following results show that it is possible to optimize the software implementation of this routine on modern workstations thus removing one of the bottlenecks for protocol performance.

The performance of TCP and ISO TP4 checksum algorithms are given in the first two rows of the table 4 for four different machine hardwares. These are byte or short word oriented calculations. The "raw" implementation of the XTP checksum as a loop of short word oriented exclusive-OR and rotate operations was too slow (maximum throughput of 13.88 Mbps on a SS-10).

We analyzed the cost of the basic operations needed for the the checksum routine in order to determine the most costly functions. The results are presented in table 5. The

Checksum	Sun 3/60	Sparc IPX	DEC 5000/200	SS-10/41
TCP	2.61 Mbps	13.3 Mbps	30.52 Mbps	31.75 Mbps
ISO	1.21 Mbps	6.50 Mbps	10.69 Mbps	12.46 Mbps
XTP (raw)	1.36 Mbps	7.02 Mbps	13.67 Mbps	13.88 Mbps

Table 4: Maximum throughput with different checksum algorithms

`rotate(x,n)` macro is defined as `n` circular shifts on the short word `x`. This is a relatively costly operation and should be saved as much as possible.

Operation	Sun 3/60	Sparc IPX	DEC 5000/200	SS-10/41
rotate(x,1)	6.24 μs	1.00 μs	0.687 μs	0.651 μs
Short XOR	4.00 μs	0.88 μs	0.445 μs	0.45 μs

Table 5: Cost of the basic checksum operations (in a loop)

We performed some hand optimizations on the checksum calculation algorithm. The results are given in table 6. A first optimization (op1) consisted in doing the XORs on words with the same RXOR rotation and saving the rotate operations until the end. This "enhancement" increased the speed slightly on Sun3, but resulted in slower operation on Sparc and Dec workstations. In fact, to implement this modification, we used an array of 16 short words to store the XOR values. As the manipulation operations on the CISC Sun3 were the bottleneck, the saving in the number of rotate operations resulted in increased performance. On the contrary, the memory access on both Dec and Sparc workstations is the bottleneck. Therefore, the load operations of the array elements decreased the performance. A second optimization (op2) consisted in declaring 16 short words instead of the array in order to save indirections (unroll the loop). This resulted in increased performance on all machine types at the cost of a slightly higher size of the checksum calculation program. The performance increase is, however, more pronounced for RISC workstations. In fact, these 16 words were defined as registers thus avoiding costly memory access in the checksum computation loop. In a third modification (op3), the number of XOR operations is divided by 2 by XORing the 16 words at the end to determine the CXOR value instead of doing it on the fly. This also increased the overall performance. In (op4) the same optimizations as (op3) are applied with long words calculations and then folding the results down to 16-bits. The results are interesting: the increase factor with regard the third enhancement vary between 2.6 for the DecStation and 5.6 for the SS10. The higher the value of this factor, the more costly is memory access relatively.

3.2.2 Throughput measures

Tests have been performed over both Ethernet and FDDI networks. We implemented a simple client/server application where the client sends raw data to the server using UDP,

Checksum	Sun 3/60	Sparc IPX	DEC 5000/200	SS-10/41
XTP (std)	1.36 Mbps	7.02 Mbps	13.67 Mbps	13.88 Mbps
XTP (op1)	1.46 Mbps	6.83 Mbps	10.93 Mbps	13.14 Mbps
XTP (op2)	2.95 Mbps	16.54 Mbps	36.96 Mbps	28.47 Mbps
XTP (op3)	5.78 Mbps	31.25 Mbps	66.06 Mbps	52.63 Mbps
XTP (op4)	22.59 Mbps	172.05 Mbps	173.53 Mbps	296.3 Mbps

Table 6: Effect of optimizations of the XTP checksum algorithm

TCP or XTP protocols. For UDP, the client will send numbered packets, the test is considered terminated when the last packet is received at the server. If this packet is lost during one of the tests, the corresponding results are ignored. For both TCP and XTP, the test is terminated when all the data is received correctly at the server. The XTP library is linked with both the client and server code. The throughput is defined in the three cases as the number of received bits divided by the time interval between the reception of the first and the last packet at the server. The tables 7, 8 and 9 give the results of the tests over both Ethernet and FDDI. The machines running the client and the server are both DECstations 5000/200.

Packet size	Ethernet	FDDI
1000 o	6.90 Mbps	7.02 Mbps
1400 o	9.57 Mbps	10.32 Mbps
2000 o	9.52 Mbps	13.22 Mbps
3000 o	9.66 Mbps	16.66 Mbps
4000 o	9.47 Mbps	29.03 Mbps
4600 o	9.32 Mbps	24.53 Mbps
9000 o	9.21 Mbps	24.66 Mbps

Table 7: Maximum throughput for UDP

One may be tempted to say that the UDP figures give the raw performance that may be obtained on either Ethernet or FDDI. However, this assertion should be taken with care: packet losses at the `ipqueue` level due the high number of packet transmitted during a test (100 to 500) reduce the throughput as defined here above. The best performance for UDP are obtained for small number of transmitted packets (15 or 20). Over Ethernet, there is no bottleneck at the UDP level: the throughput is limited by the speed of the Ethernet interface. Over FDDI, the maximum raw UDP throughput is limited either by the performance of the FDDI interface or by operating system overhead. Note that the MTU is 1460 and 4312 octets over Ethernet and FDDI respectively. Therefore, the 9000 octets packets are transmitted as 7 Ethernet packets and 3 FDDI frames and reassembled by the kernel at the UDP/IP layer.

Packet size	Ethernet	FDDI
1023 o	7.25 Mbps	8.43 Mbps
1024 o	8.35 Mbps	19.05 Mbps
1400 o	9.91 Mbps	20.74 Mbps
2000 o	9.82 Mbps	18.86 Mbps
3000 o	9.16 Mbps	18.04 Mbps
4000 o	9.00 Mbps	17.60 Mbps
5000 o	8.71 Mbps	15.73 Mbps

Table 8: Maximum throughput for TCP

ADU size	Ethernet	FDDI
2000 o	9.62 Mbps	18.01 Mbps
4000 o	9.65 Mbps	19.72 Mbps
8192 o	9.71 Mbps	20.38 Mbps
16384 o	9.77 Mbps	21.24 Mbps

Table 9: Maximum throughput for XTP

The TCP figures are more interesting: no bottleneck over Ethernet, TCP can easily saturate the 10 Mbps CSMA-CD LAN. However, the performance over FDDI are limited by the control mechanisms of TCP. The highest throughput (20.74 Mbps) is lower than the best UDP performance. The socket receive buffers have been set to 65535 octets in order to increase the window size. TCP Packets shorter than 1024 octets show poor performance even over FDDI. The limitation comes from the buffering mechanism within the kernel: small buffers of 128 octets are used, thus limiting the overall performance of the transmission. For packets longer than 1024 octets, the performance are quite similar with a tendency to decrease.

For the XTP tests, one should distinguish between two notions: ADU and NDU as they are defined in [Dav92]. ADU or *Application Data Unit* is the basic unit of data exchange between application entities. NDU or *Network Data Unit* is the unit of data processing and switching within the network. For the purpose of these tests NDUs over Ethernet are 1400 bytes long and NDUs over FDDI are 4000 bytes long. In table 9, we report the ADU sizes. Note that the results do not vary a lot with the ADU size, as the same NDU size is used. The most important result shown in this table is that XTP can saturate Ethernet and operate up to a speed of 21 Mbps over FDDI. This was made possible because of the following factors:

- minimize data copying by the use of `writev` and `readv`,
- optimize the checksum calculation by ILP intra-function optimizations,
- minimal overhead by return key based context look up in the kernel

These results show that it is possible to have good performance with user level implementation of transport protocols. The throughput figures should not be taken as an indication of the performance of XTP with all control mechanisms implemented. However, the other control mechanisms (rate control, route management) are not data manipulation oriented and may be considered as having minimal additional cost. The performance of our XTP implementation with the complete error and flow control procedures is an encouraging step in the direction of integrated implementation of protocols. Such enhanced performance transport implementation provides building blocks that will facilitate the configuration of application specific communication subsystems. Two examples of a user-space implementation of TCP can be found in [Cas94] and [The93]. However, some problems still to be solved before user level implementations provide optimal results: the execution environment should be more adapted to the support of network protocols by providing better memory management and process scheduling than Unix.

We implemented XTP as a step towards the development of new high performance communication subsystems. However, it is not clear *how* the application should configure the elementary transport and presentation "building blocks" in order to generate the complete communication subsystem. XTP can serve as a substrate for the implementation of this communication subsystem by providing efficient "building blocks". Other design studies should determine how the elementary building blocks can be grouped together in order to provide the desired service with the best performance.

4 Conclusion

In this paper, we presented our work on high performance implementation of presentation and transport functions. The experiments show that it is possible to have increased performance by proper implementation techniques. We showed specifically that an important speed up factor can be obtained by applying ILP intra-function optimizations for costly data manipulation functions. The performance enhancement of both presentation ASN.1 BER and XTP checksum algorithm show that it is possible to use such building blocks in the design and implementation of high speed integrated communication subsystems. Next step is to study how to combine elementary building blocks according to the application needs, to generate a complete communication subsystem verifying the ILP principle.

References

[Abb93a] Mark B. Abbott, Larry L. Peterson. "Increasing Network Throughput by Integrating Protocol Layers ", *IEEE/ACM Transactions on Networking*, Vol 1, No 4, October 1993.

[Abb93b] Mark B. Abbott, Larry L. Peterson. "A Language-Based Approach to Protocol Implementation", *IEEE/ACM Transactions on Networking*, Vol 1, No 1, Feb. 1993.

[Abb92] Mark B. Abbott, Larry L. Peterson. "Automatic Integration of Communication Protocol Layers", *TR 92-24*, Department of Computer Science, University of Arizona.

[Bjö93] M. Björkman and P. Gunningberg. "Locking Effects in Multiprocessor Implementation of Protocols", In *Proceedings ACM SIGCOMM'93.*

[Bra92] T. Braun and M. Zitterbart. "Parallel Transport System Design", In *Proceedings of the 4th IFIP Conference on High Performance Networking*, Liège, Belgique, December 1992.

[Cab88] Luis-Felipe Cabrera et al. "User-Process Communication Performance in Networks of Computers", *IEEE Transactions on Software Engineering*, Vol. 14, No 1, January 1988, pp. 38-53.

[Cas94] Claude Castelluccia and Walid Dabbous. "Modular Communication Subsystem Implementation using a Synchronous approach", In *Proceedings of USENIX-94, Symposium on High Speed Networking*, Oakland, CA, August 1994.

[CCI88] X.224 Recommendation, *Blue Book*, Volume VIII, Fascicle VIII.5, 1988.

[CCI84] X.411 Recommendation, *Red Book*, Volume VIII, Fascicle VIII.7, 1984.

[Che86] D. R. Cheriton, "VMTP: a transport protocol for the next generation of communication systems", In *Proceedings ACM SIGCOMM '86,* Stowe, Vermont, August 1986, pp. 406-415.

[Che89] G. Chesson, "XTP/PE Design Considerations", In *Protocols for High-Speed Networks,* H. Rudin, R. Williamson, Eds., Elsevier Science Publishers/North-Holland, May 1989.

[Cla90] David D. Clark and David L. Tennenhouse. "Architectural Considerations for a New Generation of Protocols", In *Proceedings ACM SIGCOMM '90,* September 24-27, 1990, Philadelphia, Pennsylvania, pp. 200-208.

[Cla89] David D. Clark, Van Jacobson, John Romkey, Howard Salwen. "An analysis of TCP processing overhead", *IEEE Communications Magazine*, June 1989, pp. 23-29.

[Cla87] D. Clark, M. Lambert, L. Zhang. "NETBLT: a high throughput transport protocol", *CCR*, Volume 17, Number 5, 1987, pp. 353-359.

[Col85] R. Colella, R. Aronoff, K. Mills. "Performance Improvements for ISO Transport", *Computer Communication Review,* Vol. 15, No. 5, September 1985.

[Coo90] E. C. Cooper, P. A. Steenkiste, R. A. Sansom, and B. D. Zill. "Protocol Implementation on the Nectar Communication Processor", In *Proceedings ACM SIGCOMM '90,* Philadelphia, PA, September 1990, pp. 135-144.

[Dab93] Walid Dabbous, Christian Huitema. *"XTP implementation under Unix"*, Research Report No 2102, Institut National de Recherche en Informatique et en Automatique, November 1993.

[Dab92] Walid Dabbous et al. "Applicability of the session and the presentation layers for the support of high speed applications", *Rapport Technique RT 144*, Institut National de Recherche en Informatique et en Automatique, October 1992.

[Dab91] W. Dabbous, *"Etude des protocoles de contrôle de transmission à haut débit pour les applications multimédias"*, PhD Thesis, Université de Paris-Sud, March 1991.

[Dav92] James R. Davin. *"ALF Protocol Specification"*, Internal draft, M.I.T. Laboratory for Computer Science, February 1992.

[Dru93] Peter Druschel, Mark B. Abbott, Michael A. Pagels, and Larry Peterson. "Network Subsystem Design", *IEEE network*, July 1993, pp. 8-17.

[Fan93] C. Fan, T. Luckenbach, X. Xu. "Performance Comparison and Analysis of XTP and TCP/IP over BERKOM Broadband ISDN Network", In *Proceedings of the* 12th *IEEE INFOCOM*, San Francisco, USA, 1993, pp. 1154-1161.

[Fel93a] D. C. Feldmeier. "A Data Labelling Technique for High-Performance Protocol Processing and Its Consequences", *Computer Communications Review (SIGCOMM '93)*, Vol 23, No 4, October 1993, pp. 170-181.

[Fel93b] D. C. Feldmeier. "A Survey of High Performance Protocol Implementation Techniques", *High Performance Networks – Technology and Protocols*, Ed. Ahmad Tantawy, Kluwer Academic Publishers. Boston, MA, 1993, pp. 29-50.

[Gun91] P. Gunningberg, C. Partridge, T. Sirotkin, B. Victor. "Delayed evaluation of gigabit protocols", In *Proceedings of the* 2nd *MultiG Workshop*, 1991.

[Hos93] Philipp Hoschka. "Towards tailoring protocols to application specific requirements", In *Proceedings of INFOCOM '93*, San Francisco, California, 1993, pp. 647-653.

[Hui92] C. Huitema, G. Chave. "Measuring the Performance of an ASN.1 Compiler", In *Proceedings of the IFIP WG 6.5 Conference, ULPAA '92*, Vancouver, 1992, pp.99-112.

[Hui91] Christian Huitema. *"MAVROS, Highlights on an ASN.1 compiler"*, INRIA research note, May 1991.

[Hui90] Christian Huitema. *"Definition of the Flat Tree Light Weight Syntax (FTLWS)"*, Internal Document, INRIA Sophia Antipolis, July 1990.

[Hui89] Christian Huitema, Assem Doghri. "Defining faster transfer syntaxes for the OSI Presentation Protocol", *Computer Communication Review,* Vol 19, No 5, Oct 1989, pp. 44-55.

[Jac90] V. Jacobson, R. Braden and L. Zhang. "TCP Extension for High-Speed Paths", *RFC 1185*, October 1990.

[Kan88] Hemant Kanakia, David R. Cheriton. "The VMP Network Adapter Board (NAB): High-Performance Network Communication for Multiprocessors", In *Proceedings SIGCOMM '88,* Stanford, CA, 1988, pp. 175-187.

[Kay93] J. Kay, J. Pasquale. "Measurement, Analysis and Improvement of UDP/IP Throughput for the DECstation 5000", In *Proceedings of the 1993 Winter Usenix Conference*, San Diego, USA, pp. 249-258.

[Lap92] T. F. La Porta and M. Schwartz. "A high-Speed Protocol Parallel Implementation: Design and Analysis", In *Proceedings of the* 4th *IFIP Conference on High Performance Networking*, Liège, Belgique, December 1992.

[Léo92] L. Léonard. "Enhanced Transport Service Specification". *Deliverable ULg-4*, OSI 95 project, October 1992.

[Lin93] H-A. Lin. "Estimation of the Optimal Performance of ASN.1/BER Transfer Syntax", *Computer Communication Review,* Vol. 23, No. 3, July 1993.

[Nic91] A. Nicholson et al. "High Speed Networking at Cray Research", *CCR*, Vol 21, No 1, January 1991, pp. 99-110.

[O'M90] S. W. O'Malley and L. L. Peterson. "A Highly Layered Architecture for High-Speed Networks", In *Proceedings of the IFIP Workshop on Protocols for high speed networks II*, Palo Alto, CA, 1990, pp. 141-156.

[Par93] C. Partridge and S. Pink. A Faster UDP. Submitted to *IEEE Transaction on Networking*.

[PEI92] *"XTP Protocol Definition, Revision 3.6"*, PEI 92-10, January 1992, Protocol Engines Incorporated, Santa Barbara, CA 93101, USA.

[Peh92] Bjorn Pehrson, Per Gunningberg and Stephen Pink "Distributed Multimedia Applications on Gigabit Networks", *IEEE Network Magazine*, Vol 6, No 1, January 1992, pp. 26-35.

[Ros90] Marshall T. Rose. *"The Open Book, a Practical Perspective on OSI"*, Prentice-Hall, 1990.

[Rüt92] Erich Rütsche and Matthias Kaiserwerth. "TCP/IP on the Parallel Protocol Engine", In *Proceedings of the* 4th *IFIP Conference on High Performance Networking*, Liège, Belgique, December 1992.

[Sch93] D. Schmidt, B. Stiller, T. Suda, A.N. Tantawy, and M. Zitterbart. "Language Support for Flexible Application-Tailored Protocol Configuration", *Proceedings of LCN '93*.

[Str92] W. T. Strayer, B. J. Dempsey, A. C. Weaver, *"XTP: The Xpress Transfer Protocol"*, Addison-Wesley, 1992.

[Ten89] David L. Tennenhouse. "Layered Multiplexing considered harmful", In *Proceedings of the IFIP Workshop on Protocols for high speed networks,* Zurich, Switzerland, 9-11 May, 1989.

[The93] Chandramohan A. Thekkath, Thu D. Nguyen, Evelyn Moy, and Edward D. Lazowska. "Implementing Network Protocols at User Level", In *Proceedings of the SIGCOMM'93,* San Francisco, CA, 1993, pp. 64-73.

[Var87] G. Varghese, T. Lauck. "Hashed and Hierarchical Timing Wheels: Data Structures for the Efficient Implementation of a Timer Facility", In *Proceedings of the* 11th *ACM symposium on Operating Systems Principles,* Austin, TX, November 1987.

[Wak92] Ian Wakeman, Jon Crowcroft, Zheng Wang, and Dejan Sirovica, "Layering considered harmful", *IEEE Network*, January 1992, p. 7-16.

[Wat87] Richard W. Watson, Sandy A. Mamrak. "Gaining efficiency in transport services by appropriate design and implementation choices", *ACM transactions on computer systems,* Vol 5, No. 2, May 1987, pp 97-120.

13

PATROCLOS: A Flexible and High-Performance Transport Subsystem

Torsten Braun

Institute of Telematics, University of Karlsruhe, Zirkel 2, 76128 Karlsruhe, Germany
Phone: +49 721 608-3982, Fax: +49 721 388097
E-mail: braun@telematik.informatik.uni-karlsruhe.de

ABSTRACT

Increasing application demands together with the modern network technology towards gigabit speeds require new adequate transport systems. The transport subsystem PATROCLOS (parallel transport subsystem for cell based high speed networks) is based on a modular protocol architecture with a high degree of inherent parallelism. The primary goal of PATROCLOS was to allow efficient implementations of transport oriented protocols on multiprocessor architectures combined with specialized hardware for very time critical functions. Furthermore, the protocol architecture should offer an enhanced and flexible transport service to support a wide variety of applications providing a large set of different configurable protocol functions. Other important aspects of the transport subsystem are its efficient operation on top of modern cell-based high-speed networks such as ATM.

This paper describes the PATROCLOS protocol architecture with special emphasis on the integrated protocol functions and mechanisms. The protocol architecture allows an easy integration of several kinds of transport service interfaces. Two different types of service interfaces (OSI95 [1], F-CSS interface [2]) are discussed. Moreover, the paper shows the advantages of the modular approach for parallel protocol processing by performance results of a multiprocessor implementation. The performance results show significant better throughput and speed-up values than other parallel protocol implementations.

Keyword Codes: B.4.1; C.1.2; C.2.2
Keyword: Data Communication Devices; Multiprocessors; Network Protocols

1. INTRODUCTION

PATROCLOS provides the functionalities of the layers between the media access layer interface of cell-based networks and the transport service interface in a single protocol component (functionality of OSI layers 2b-4). To support the increasing variety of applications, it is necessary that transport systems provide new service types, e.g. according to the OSI95 service definition [1], together with an enhanced set of quality-of-service parameters as proposed in [2]. Dependent on the type of service or the quality-of-service parameters, different protocol functions have to be selected. E.g., the type of retransmission (selective repeat, go-back-N, or even no retransmission) may have significant impact on the achievable delay. For that reason, future transport subsystems should provide a lot of different protocol functions, which have to be configured dependent on the application requirements and network characteristics. For example, the selection of go-back-N retransmission strategy may cause high delays in networks

with a large bandwidth-delay-product, while selective repeat strategy or forward error correction can significantly reduce the delays [3]. Therefore, configuration facilities will play an important role in future protocols to allow their adaptation to the wide variety of application requirements and network environments [2, 4].
PATROCLOS is based on a modular design concept to simplify the application-driven and network-dependent configuration. The key feature of the protocol is that it consists of a set of extended finite state machines (FSMs), where each FSM performs a dedicated protocol function such as connection management or acknowledgment. The modular approach is also useful to increase the performance of transport subsystems significantly by concurrent processing of the different FSMs. Parallel implementations of standard protocols like OSI TP4/CLNP or TCP/IP suffers from their inherent sequential structure. Therefore, parallel implementations of standard protocols use pipelining [5] or parallelism on a per-packet basis [6]. The transport subsystem PATROCLOS realizes a parallel protocol architecture with fine granularity adequate for parallel implementations and overcomes performance limitations by the integration of protocol and implementation issues. In contrast to other approaches of parallel protocols [7, 8, 9], PATROCLOS is based on a decomposition into building blocks, which are oriented on protocol functions. Therefore, the building blocks are the atomic units for implementation on a parallel implementation architecture and for configuration of the transport subsystem, simultaneously.

2. THE PATROCLOS ARCHITECTURE

The PATROCLOS approach combines a fine granularity with a protocol function oriented decomposition into basic modular building blocks to simplify protocol configuration and parallel implementation, simultaneously. Furthermore, it avoids layering to achieve a high degree of parallelism. Redundant placement of protocol functions such as segmentation and reassembly functions, which occur in layer 3 and 4 of the OSI Reference Model, in several layers can be avoided. Mainly, the characteristics of emerging cell-based networks such as ATM and DQDB influenced the PATROCLOS protocol design [10].

2.1. Parallel Protocol Architecture

The protocol architecture has been developed based on the analysis of a XTP [8] specification, which consists of a set of parallel FSMs, and the performance results of the corresponding multi-processor implementation [11]. PATROCLOS is also based on the concept of separating data transfer and transfer control functions as proposed in the integrated layer processing approach [12].
Data transfer and control functions are separated within the PATROCLOS protocol specification to allow a higher degree of parallelism. The separation of data and control path processing is also performed in the AXON approach [13] PATROCLOS has been specified with the specification language Promela [14], which has several advantages describing a protocol as a set of parallel FSMs. Those parallel FSMs are the building blocks of the protocol architecture. Together, all FSMs form a FSM system. FSMs belonging to the same PATROCLOS entity exchange internal messages for co-operation. Internal messages consist of a locally unique connection identifier and the message body. The message body may contain information such as sequence numbers, buffer state information, etc. (cf. Figure 2). All FSMs are designed as autonomous as possible to minimize interactions with other FSMs of the same entity. The communication among FSMs is uni-directional to allow a high degree of parallelism. If possible, the necessary information is exchanged periodically to reduce the communication overhead.

The modular protocol design supports an easy mapping on multiprocessor architectures, if the FSMs are implemented as processes, which can run on separate processors.
PATROCLOS consists of interface and protocol FSMs (cf. Figure 1, 2). ***Interface FSMs*** are located at the interfaces of the transport subsystem to the application and the network. They support interactions of PATROCLOS with the transport service user or with the underlying network unit and are only involved in local communications inside one entity. In contrast to the protocol FSMs, they do not have to communicate with FSMs of the peer entity. The network interface FSM (send control FSM) controls and schedules transmission of *all* protocol data units (PDUs, PATROCLOS-frames, P-frames) to the underlying network unit. Note that this FSM is missing in Figure 2. The transport service interface FSM maps service primitives to signals for the protocol FSMs, exchanges transport service data units (TSDUs) with the transport service user in both directions and is involved in performing graceful connection termination.

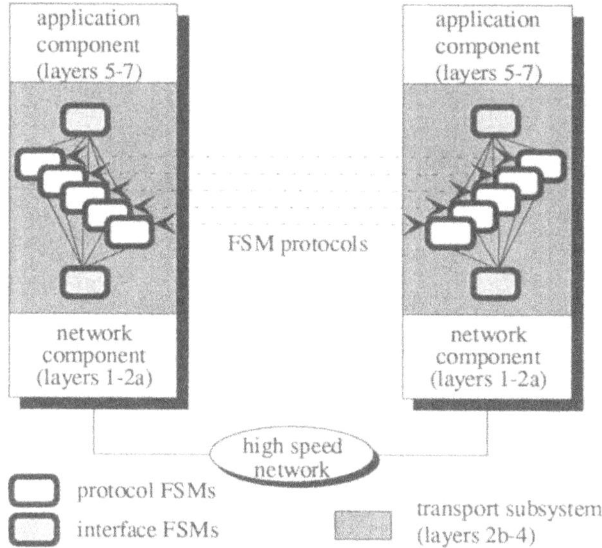

Figure 1: Protocol Architecture

The key feature of ***protocol FSMs*** is that they communicate directly by separate, so-called ***FSM protocols*** with the corresponding FSMs at the peer entity. The protocol FSMs are designed to allow parallelism between send and receive part and to decouple the connection state information exchange from user data exchange. Moreover, since a large amount of data may be traveling between communicating end systems, state exchange (e.g., acknowledgments and connection state information exchange) after dedicated requests and errors (e.g., XTP [8]) or periodic exchanges (e.g., SNR [7] and Delta-t [15]) seem to be more suitable than state exchange for each data packet. All FSM protocols use separate P-frames (external messages), individual error recovery mechanisms, and timers. P-frames containing user data are called P-data-frames or packets in contrast to the P-control-frames, which contain only control information. Each FSM protocol is performed by communicating peer FSMs, using P-data-frames containing only the necessary information for their dedicated protocol function. Multiplexing of P-frames by different FSMs is avoided to support parallelism. Together, the set of FSM protocols builds the complete PATROCLOS protocol. The decomposition into independent FSMs permits parallel execution and has the benefit of a highly modular system, which can be configured easily depending on the requirements of the applications and the characteristics of the networks as suggested in [2]. There are the following protocol FSMs:

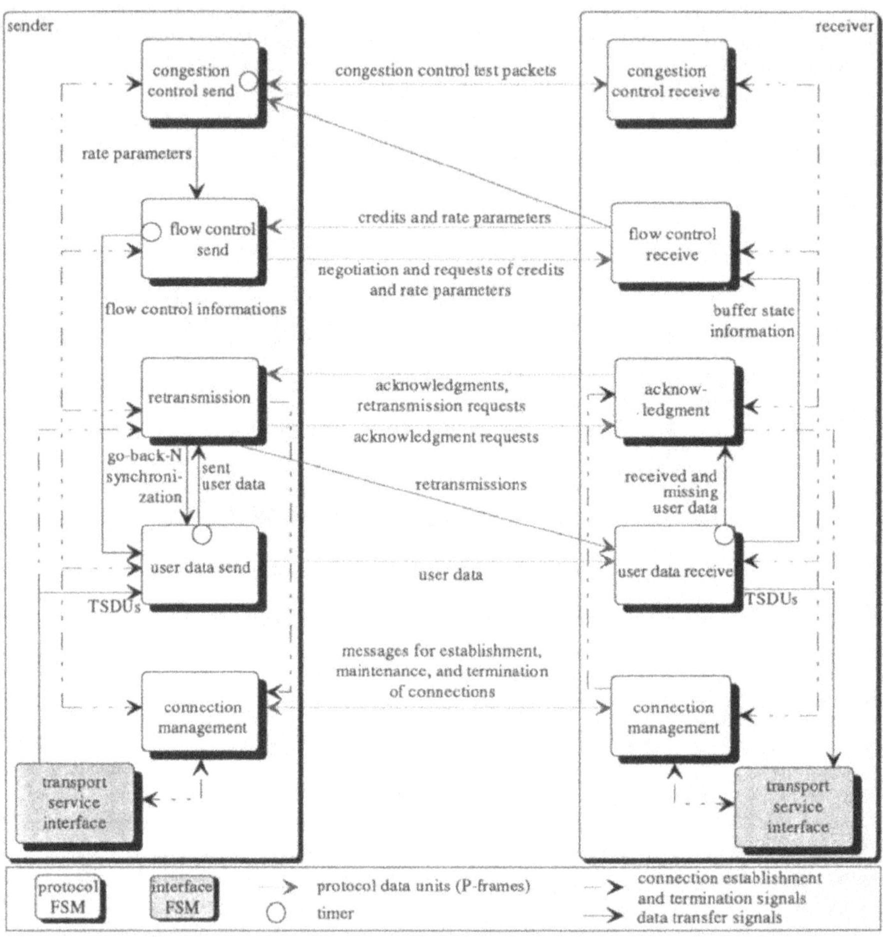

Figure 2: Communication Among FSMs

The ***connection management*** FSM is responsible for establishment, management, and termination of a full duplex connection. Therefore, it has to activate and initialize the other FSMs of the same entity for each new connection and has also to deactivate these FSMs after connection termination. The peer connection management FSMs exchange P-control-frames for connection establishment, initialize the connection context, and forward quality-of-service require-

ments to the other FSMs of the entity. The connection management FSMs also guarantee the unique use of connection identifiers and perform signalling tasks, e.g. in B-ISDN networks.
User data transfer can be divided into a send path and a receive path. User data transfer FSMs send and receive user data. The main tasks of both user data transfer FSMs (user data send and user data receive FSM) are formatting and analyzing P-data-frames. Additional functions are segmentation and reassembly. Therefore, each user data transfer FSM consists of two subordinate FSMs. On sending side, the segmentation FSM segments TSDUs into P-data-frames. The data send FSM formats the P-data-frame headers to describe the user data (e.g., sequence numbers). On receive side, the data receive FSM operates on these informations to detect duplicates or errors and checks the packet lifetime. The reassembly FSM reassembles P-data-frames into TSDUs. Another FSM in intermediate systems performs functions such as ***routeing & relaying***, route management, and monitoring.
Because data transfer mostly performs the critical path in communication protocols, ***retransmission and acknowledgment*** are handled by two special FSMs. The FSM on the send side (retransmission FSM) requests and receives acknowledgments from the peer's FSM on the receive side (acknowledgment FSM). The retransmission FSM analyzes the acknowledgment P-frames, initiates retransmissions dependent on the acknowledgment information and the selected strategy, and releases the buffers of acknowledged data packets. Retransmissions are sent to the peer's user data receive FSM. The retransmission FSM receives the control over sent user data packets from the user data send FSM periodically with an interval lower than one round trip time (rtt). The user data receive FSM signals informations about received data to the acknowledgment FSM periodically. The acknowledgment FSM sends negative selective and positive cumulative acknowledgments to the peer's retransmission FSM. Communication between the FSMs participating in user data transfer is reduced to a minimum. In the case of error-free communication, the FSMs in one entity communicate periodically instead of exchanging messages after each packet. In error cases, the informations are sent directly to avoid unnecessary delays.
Two ***flow control*** FSMs exist in each entity, one FSM for each data flow direction. They negotiate flow control parameters with the remote peer and use these parameters to control the data flow. The flow control receive FSM transmits rate parameters or credits according to the current traffic and available buffer space, which is signaled by the data receive FSM.
The ***congestion control*** FSMs control data flow by the interpretation of feedback signals from the network or the peer entity, e.g., congestion indications, round trip times, and inter-arrival times of test P-frames, which are exchanged between peer FSMs to detect congestions. The FSM may also react upon the feedback signals of intermediate systems.

2.2. Transport Services

2.2.1. Enhanced OSI Transport Service

At first, PATROCLOS provided a transport service similar to the transport service definition developed within the OSI95 project. The OSI95 transport service definition comprises a normal connection-mode, a fast connection-mode, an unacknowledged connectionless-mode, an acknowledged connectionless-mode, and a request/response connectionless-mode transport service. These services are grouped into three classes as shown in Table 1. Additional options and parameters are offered for the connection-mode service. The connection establishment may be performed by a 2-way- or a 3-way-handshake. Furthermore, an implicit connection establishment (fast connect) is provided.

Table 1: Enhanced OSI Transport Service

Service Type	Service Parameters	
Connection-oriented Service	establishment parameter	termination parameter
	2-way-handshake acknowledged 2-way-handshake unacknowledged 3-way-handshake acknowledged 3-way-handshake unacknowledged fast connect (implicit)	graceful normal (abrupt) timer-controlled
Connectionless Service	acknowledgment parameter	
	acknowledged unacknowledged	
Request/Response Service		

For connection termination the transport service user can select a graceful release, which makes sure that all in-transit data are delivered, in contrast to the normal (abrupt) termination, which is used in OSI-TP4. To provide a graceful connection release, the connection management FSM has to ensure that all data, which are delivered to the transport system before initiating connection termination, have been sent and acknowledged. The graceful release requires co-operation with other FSMs such as the retransmission FSM to be sure that all transmitted user data are correctly received and delivered at the receiver.

Another option is the timer-controlled termination. A connection is terminated, if there is no user data exchange for a certain inactivity interval. The data send and the data receive FSM send inactivity signals to the connection management FSM, if there is no user data exchange for the time period t_{send} or $t_{receive}$ respectively. Otherwise, they send activity signals. The transport service user has to specify a time value t_{idle}. If the connection management FSM receives $\lceil \frac{t_{idle}}{t_{send}} \rceil$ inactivity signals successively from the data send FSM or $\lceil \frac{t_{idle}}{t_{receive}} \rceil$ inactivity signals from the data receive FSM, the connection is terminated. The inactivity interval is bounded by the following formula:

$$t_{idle} < inactivity\_interval < \max(\lceil \frac{t_{idle}}{t_{send}} \rceil \cdot t_{send} + t_{send}, \lceil \frac{t_{idle}}{t_{receive}} \rceil \cdot t_{receive} + t_{receive})$$

2.2.2. F-CSS Transport Service

Because of the modular structure of PATROCLOS the required enhancements to offer other transport services are very low. PATROCLOS offers a number of different protocol functions. Especially, mechanisms appropriate for high-speed network environments with a large bandwidth-delay-product have been integrated. Because of the ability of PATROCLOS to be configured, we integrated the PATROCLOS protocol architecture and the F-CSS approach, which provides a framework for communication subsystems to offer a flexible transport service dependent on application and network requirements. F-CSS [2] provides an application interface specifically designed to support service flexibility. It relieves the application from dealing with a large number of available protocol functions by providing a small number of pre-defined service classes. The application interface provides the application with the possibility of formulating its individual requirements in terms of service parameters and protocol functions.

The F-CSS session model allows to comprise different types of communications. A session is an association between two or more communicating applications and may consist of several application data streams, each of which is being processed by a separate protocol machine. F-CSS configures different protocol machines, each being tailored to the requirements in one data flow direction. For configuration and management purposes, F-CSS provides several tools. A session configuration component receives session-setup requests from the local or the remote application and delivers a message to a protocol configurator to start the configuration of a suitable protocol machine for each individual data stream requested in the session-setup request. To allow a configuration based on protocol functions, a protocol description of PATROCLOS in a special description language (F-PDL), which has been developed within the F-CSS project, is required. The output of the F-CSS tools is directly passed to the connection management FSM of PATROCLOS, which initializes all other FSMs and delivers the selected protocol functions and parameters. Figure 3 shows the integration of PATROCLOS and the F-CSS components.

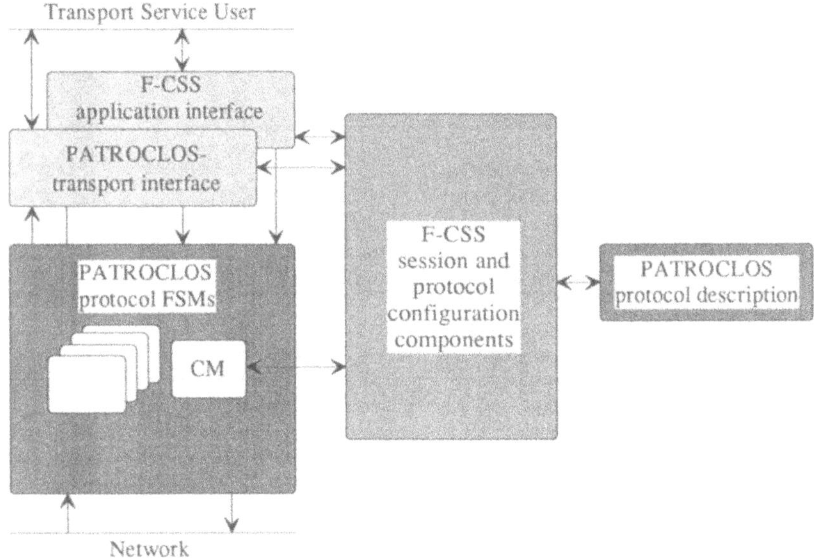

Figure 3: Integration of PATROCLOS and F-CSS configuration components

2.3. Protocol Functions and Mechanisms

In addition to several types of connection establishment and termination, the protocol offers several options for data transfer. These options may be specified by the transport service user or may be calculated by the F-CSS configuration tools. Table 2 gives an overview of the offered protocol functions.

User data may be delivered to the transport service user as TPDUs (P-frames), i.e. all received TPDUs are delivered in the same order they arrived, or as TSDUs. For delivery on a TSDU basis a sequenced or an unsequenced delivery can be selected.

For unreliable data transfer, retransmission can be switched off. For reliable data transfer, selective or go-back-N strategy can be selected. While selective retransmission has benefits in networks with a large bandwidth-delay product, it requires significantly more buffer memory at the receiver. For go-back-N strategy, however, the data send and the retransmission FSM are not able to work as independent as for selective retransmission strategy. When the retransmission FSM detects that a go-back-N retransmission is required, it first stops the data send FSM by sending a message to it. The response of the data send FSM contains a sequence number, which indicates the user data already sent. The retransmission FSM retransmits the user data already sent sends a signal to the data FSM to indicate that all retransmissions have been done and that the normal data transfer can proceed.

Table 2: Protocol functions and parameters

Protocol Functions	Protocol Function Parameters
Reassembly	delivery of TSDUs (reassembly) delivery of TPDUs (no reassembly)
Sequencing	no sequenced delivery of TSDUs to the transport service user
Retransmission	no selective repeat Go-Back-N
Flow control	no Stop-And-Go window-based rate control Stop-And-Go and rate control window-based and rate control
Congestion control	no 2P algorithm AACC algorithm

Window-based, rate-based, and stop-and-go mechanisms are provided for flow control. The flow control send FSM controls data flow by the interpretation of the peer's feedback signals. It signals periodically the maximum sequence number to be sent (limited by window-based flow control) to the data send FSM. For rate-based flow control, it starts a rate timer, calculates the maximum sequence number of user data to be sent for the next time period, and signals this value to the send control FSM, which controls that no user data with an illegal sequence number are sent.

Based on test P-frame delays and inter-arrival times, congestions are detected and the rate can be decreased by sending signals to the flow control send FSM. Currently, we use the adaptive admission control (AACC) algorithm [16] and the packet pair algorithm (2P) [17]. Both algorithms have slightly been modified. The original 2P algorithm sends packet pairs with a certain inter packet gap. Dependent on the gap between the corresponding acknowledgments, the inter packet gaps are decreased or increased. PATROCLOS does not use user data packets for this algorithm but special test packets, which take the same path as the user data packets. Furthermore, PATROCLOS does not consider the acknowledgment gaps, but the gaps of the test packets at the receiver. Therefore, the congestion control receive FSM has to inform the congestion send FSM about the detected inter arrival times of the test packets. That means that in PATROCLOS possible congestions in the reverse direction do not influence congestion detection in the other direction. Another difference is that the congestion analysis results in a modified rate in contrast to the original algorithm, which modifies the inter packet gap. The AACC algorithm is based on a calculation of a virtual delay between sender and receiver. The virtual delay is the difference between the sending time at the sender and the receiving time at the re-

ceiver. Dependent on increasing or decreasing virtual delays the inter packet gaps are increased or decreased. Again, in PATROCLOS we calculate and modify rate parameters. Therefore, the flow control receive FSM sends the flow control parameters not only to the flow control send FSM but also to the congestion send FSM. This is done with a single P-control-frame, which is addressed to both FSMs. A demultiplexer component has to deliver the P-control-frame to both FSMs.
Because of the high degree of modularity, the FSMs can be configured highly independent of each other. There are only a few protocol functions, which have influence to the behavior of other FSMs. Most of the functions, which can be configured have influence on one or two FSMs.

2.4. Other Parallel Transport Protocols Approaches

In addition to XTP, there are other transport protocols supporting parallel implementations. SNR [7] is a transport protocol to support high-speed communication in datagram and connection oriented networks. SNR has been developed at AT&T Bell Laboratories for parallel implementation on a M68030 based multiprocessor architecture. Control and user data transfer have been separated by exchanging control information periodically and independently from user data exchange. The protocol processors on send side and on receive side have to share a common memory. This fact allows an implementation with two memory busses, where control and user data packets are flowing across the same memory bus. Another difference to PATROCLOS is that retransmission processing and regular user data transfer are performed by the same process.
The MultiStream Protocol (MSP) [18] is also decomposed into several units, which are called protocol machines. MSP is based on concepts introduced by the HOPS architecture (HOPS: horizontally oriented protocol structure) [4]. The transport service user can switch off or on protocol functions for acknowledgment, retransmission, and delivery to the transport service user by selection of one of eight so-called streams. The offered protocol functions can not be selected independently of each other and user data transfer is not fully separated from control functions, e.g., acknowledgments are exchanged by piggy-backing.
TP++ [19] is an approach of high-speed protocol designed at Bellcore. Special data formats shall enable independent user data and control processing. Furthermore, most of the protocol functions can process incoming packets in an arbitrary order.
HTPNET [9] is another protocol, which has been strongly influenced by SNR. HTPNET has been developed at the University of New South Wales (Australia) and consists of four finite state machines, which are the result of separating a protocol into send/receive and data/control part. Error control, flow control, and rate control can be switched on or off. The combinations result in eight protocol classes.

3. IMPLEMENTATION

The main advantage of the PATROCLOS protocol architecture with its modular FSMs is its support for parallel implementation. This section describes the PATROCLOS implementation for an ATM network on a hybrid multiprocessor architecture, which uses transputers as universal processors and several specialized hardware units. Another implementation architecture for PATROCLOS consists of VLSI components only [20]. In contrast to other approaches (e.g., [21]), the transport protocol is processed outboard to prevent the host from transport protocol processing. Another reason is that with outboard processing no control data such as acknowledgments must cross the host bus.

3.1. Mapping of FSMs to Parallel Processes

The interface and protocol FSMs of PATROCLOS have been implemented as processes. For each FSM there is only one process, which can handle several connections. Processes of the same entity communicate with each other via asynchronous message passing across interconnecting inter-process channels. Timer processing is supported by special timer processes, which do not need to run on the same processor as the protocol process, and, therefore, may run in parallel. Protocol and timer processes co-operate by message passing, too.
A parallel C language with additional elements for message passing and process management has been used for implementation. Figure 4 shows the data receive process as an example for the general process implementation structure. Other processes have the same process structure.

```
process data_receive(){
while (TRUE){
  read_message_from_queue();  /* receive message
*/
  switch (message_type){
   case FRAME_FROM_NETWORK:   /* processing step 1
*/
     analyze_frame();
     deliver_frame();
     break;
   case TIME_OUT:             /* processing step 2
*/
     send_info_to_ack_and_flow_control();
     restart_timer;
     break;
    ...
}}}
```

Figure 4: Data Receive Process Implementation

Different processing steps have to be performed dependent on the incoming message type, which is received by the function `read_message_from_queue`. Processing step 1 includes all operations for an incoming packet from the network. The packet is analyzed (e.g. sequence control, lifetime control, address analysis etc.) and delivered to the reassembly process, if it is error-free. Because the acknowledgment process must have some knowledge about correctly received or missing packets, the data receive process has to send informations to that process. Usually, the data receive process submits these informations periodically. To control the information exchange, the data receive process starts a timer by sending a message to a timer process. The timer process sends a message back to the data receive process to signal a timeout. In processing step 2, the data receive process reacts upon the timeout message from the timer process by sending informations about received packets to the acknowledgment process. In the same processing step buffer state informations are sent to the flow control process. Finally, the data receive process restarts the timer. In addition to temporal parallelism (pipelining), a high degree of spatial parallelism by parallel user data and control processing is achieved. Furthermore, timer processing is an appropriate candidate to be performed in parallel.
As mentioned above the protocol architecture supports a variety of different protocol functions. The selected protocol functions and parameters are delivered during the initialization phase of a connection by the connection management FSM to all other processes. Each process extracts the relevant informations and stores them in the local context. The process code is constructed in a way, that a process can perform all possible combinations of mechanisms and parameters. The required operations are selected by if-then-else elements or by function point-

ers, which are also stored in the local context. Because only a few protocol functions influence the behavior of a FSM, only a few if-then-else statements or function pointers in the implementation are required. This strategy has the advantage that we can share the same code for several connections, which require different protocol functions and parameters. The amount of code and the local memory space is considerably reduced.

3.2. Hybrid Multiprocessor Architecture

Protocol functions on P-frame and TSDU level are mostly implemented in software, while hardware is required for cell level functions. The combination of universal processors and dedicated hardware results in a hybrid multiprocessor architecture. The complete implementation architecture has to be connected via a DMA unit to a host system. Figure 5 shows the implementation architecture and the processor module for the user data receive FSM.

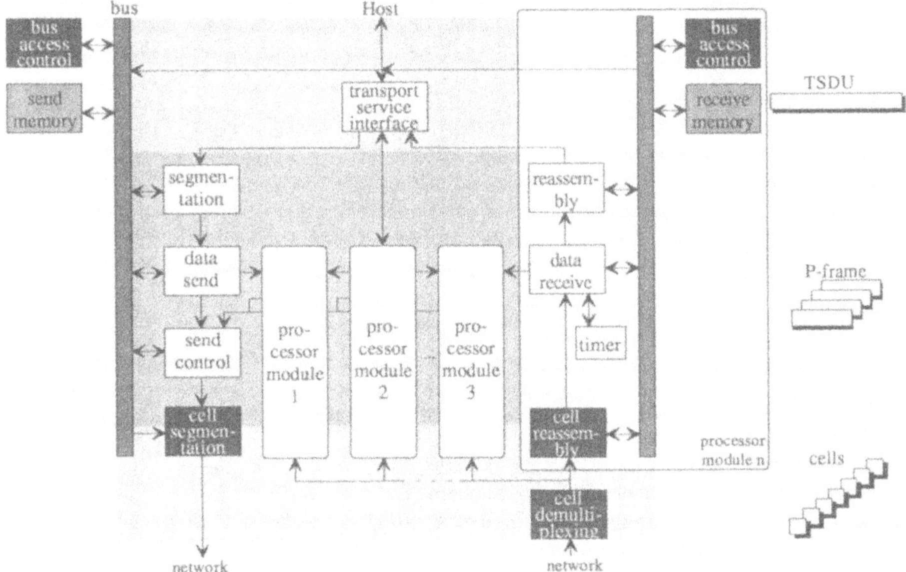

Figure 5: Hybrid Implementation Architecture

Each process derived from the protocol FSMs of the PATROCLOS protocol architecture may be mapped to a processor module of a suitable multiprocessor configuration. Furthermore, the processes resulting from implementation of the F-CSS components can run on one or more separate processor modules. A processor module consists of one or more universal processors (in our case transputers) for FSM processing, a cell reassembly unit to reassemble ATM cells into P-frames, a shared memory, and a bus access control unit to control the concurrent access to the shared memory by the processors and the reassembly units (cf. Figure 5).

Transputers are interconnected by hardware channels for inter-processor communication, which is performed in parallel to other computations on transputers. Memory management is distributed among several processors and requires special data structures, which have been designed in order to minimize copying of data and to permit an efficient memory management and access by several processors without any inconsistencies and conflicts [22].

The bus access control units and the reassembly processing unit on cell level (according AAL 3/4) have been implemented by dedicated hardware components to support the high network bandwidth [22]. Other hardware components are a cell segmentation unit on send side and a demultiplexing component on receive side to perform demultiplexing below the cell reassembly level and to distribute very efficiently incoming cells to the reassembly hardware units of the processor modules. In contrast to other approaches (e.g., [23]), which use linked lists for reassembly, incoming cells are reassembled such that consecutive cells are stored in consecutive memory areas.

The demultiplexer is the interface between the network access unit and the processor modules. A similar component to support efficient demultiplexing has been proposed in [24]. The demultiplexer [22] has to distribute incoming P-frames to the appropriate processor module, but distribution is performed on the level of ATM cells. The corresponding processor module of a P-frame is identified by the FSM address, which addresses a single FSM and is an extension of the transport entity address. The address extension is a simple bit array. Because each bit represents a certain FSM, the address format allows a simple mapping of an FSM address to the processor modules. The FSM address is always located in the P-frame header of a PATROCLOS protocol data unit (P-frame), and therefore in the first cell of a P-frame (BOM cell, begin of message). All other cells of a P-frame do not contain the FSM address. ATM cells of a single P-frame are identified by a unique value of VCI (virtual channel identifier), VPI (virtual path identifier), and MID (multiplex identifier). The FSM address is determined dependent of the cell type and delivered to the select generator. The FSM address for BOM and SSM cells can be extracted directly from the cell body, COM and EOM cells require a table look-up in a look-up table, which is usually realized by a content addressable memory (CAM).

4. PERFORMANCE ANALYSIS

The demultiplexer implementation provides a high degree of parallelism: writing cells into the FIFO buffer, determination of the FSM address, and reading the cells from the FIFO buffer are performed in parallel. Therefore, the demultiplexer needs only 55 clock ticks to process a 53 byte ATM cell assuming a 40 MHz clock. This results in a throughput of more than 700 000 cells/s (300 Mbit/s). The throughput of the demultiplexer has to be higher than the throughput of the reassembly units, which can reassemble up to 470 000 cells/s in our implementation. Generally, the throughput values of the hardware components are limited by the used FIFO and shared memory technology.

The following analysis discusses the achievable software performance of the receive part, which requires the most processing time of all receive processes and is the bottleneck of the PATROCLOS implementation. To measure the processing times for receive processing (cf. Figure 6), we used a hybrid monitor system. We implemented the protocol with a parallel C language, but we did not yet optimize the code. For the performance measurements, we selected rate-based and window-based flow control, selective retransmission, and inactive congestion control. Furthermore, we did not perform segmentation and reassembly in the measurement scenario. However, the lack of congestion control and segmentation/reassembly does not influence the performance as shown below. For performance measurement we implemented additional test environment processes for the transport user and a loopback driver.

4.1. Receive Throughput

To measure the performance for receiving, all processes of the send part, which are inactive during receive processing, are mapped to a single processor. Each of the receive processes has an own processor. Because P-control-frames are sent and received periodically and, therefore, less frequently than P-data-frames, processing of received P-data-frames will be the critical path. Figure 7 shows a more detailed analysis of the data receive process, which is the bottleneck of receive processing although running on a separate processor. In addition to analyze the received data packets, the data receive process informs the acknowledgment process periodically about correctly received P-data-frames and the flow control receive process about the buffer state. The data receive process sends messages to the acknowledgment process and the flow control receive process (processing step 2, cf. Figure 4) for every five P-data-frames (packets). Receiving a data packet is indicated by processing step 1. Each processing step is performed after receiving the corresponding message.

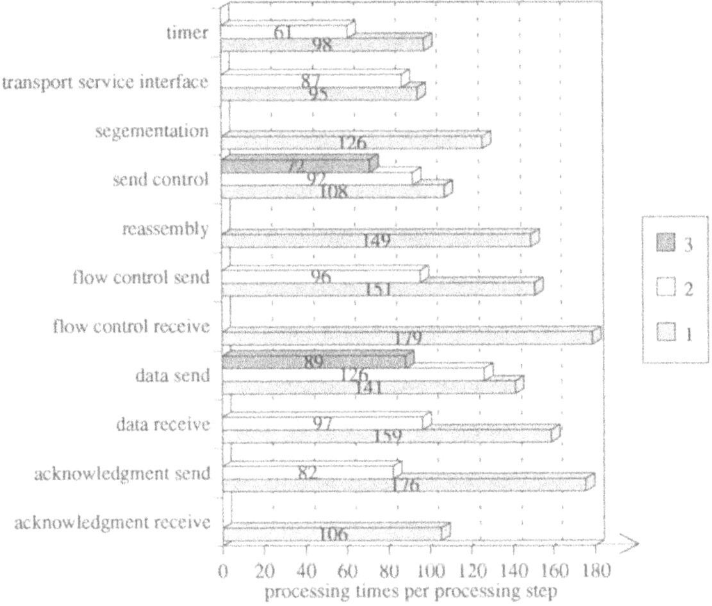

Figure 6: Receive Processing Times

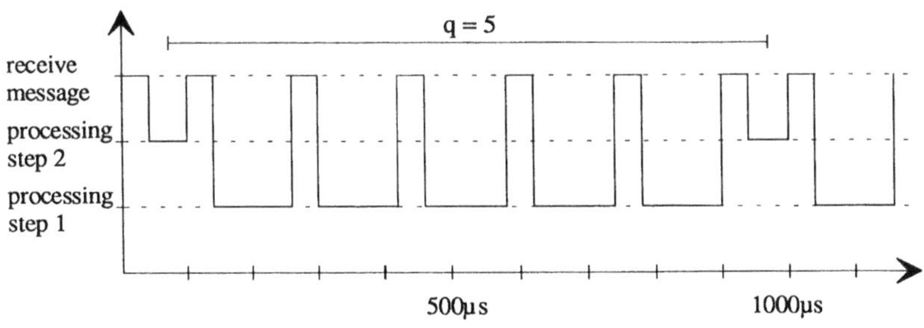

Figure 7: Analysis of the Data Receive Process

With q as the number of packets received per acknowledgment information, and $t_{dr,x}$ as the time of the data receive process for processing step x, the throughput P_r (receive performance) of the data receive process and, therefore, for receive processing is calculated by

$$P_r = \frac{1}{t_{dr,1} + q \cdot t_{dr,2}}$$

Increasing the value of q results in a higher performance. The throughput exceeds 5000 packets/s for q > 5 and 6000 packets/s for q > 10, but it is limited to 6300 packets/s by the time of the data receive process to analyze a P-data-frame (159 µs). The data receive process will also be the bottleneck, if packets have to be reassembled, because the reassembly process reassembles packets faster than the data receive process analyzes them. Congestion control has no influence on P-data-frame processing and does therefore not influence receive performance. The throughput values of the receive path have been calculated based on the processing times of the receive processes and have been verified by real measurements. The measured and the calculated results differ only by 2 %. Asynchronous inter-process communications allow very exact performance predictions.

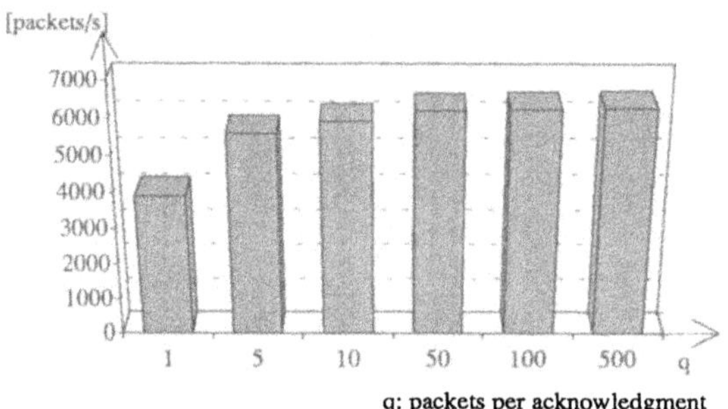

Figure 8: Throughput of receive path

4.2. Send Throughput

Similar to the analysis of the receive path, we evaluated the performance of the send path. Two parameters influence the performance of the data send process, which is the bottleneck process of the send path. The data send process has to signal periodically informations to the retransmission process, which describe the user data already sent (processing step 3). The time of this period must be lower than the round trip time, which includes propagation delay as well as processing of user data and acknowledgments at the sender and the receiver. Therefore, rtt will exceed 1 ms. Furthermore, the data send process receives a credit information message every q packets from the flow control receive process (processing step 2). Processing the P-data-frames received from the segmentation process is performed in processing step 1. Similar to the data receive process we can derive a formula for the throughput P_s (send performance) with $t_{ds,y}$ for the time of the data send process for processing step y:

$$P_s = \frac{1 - \frac{t_{ds,3}}{rtt}}{(q \cdot t_{ds,2} + t_{ds,1})}$$

The throughput depends on the parameters q and rtt as shown in Figure 9. For q > 5 we achieve more than 6000 packets/s. The throughput is limited to 8000 packets/s by the time of 126 µs required for processing step 1.

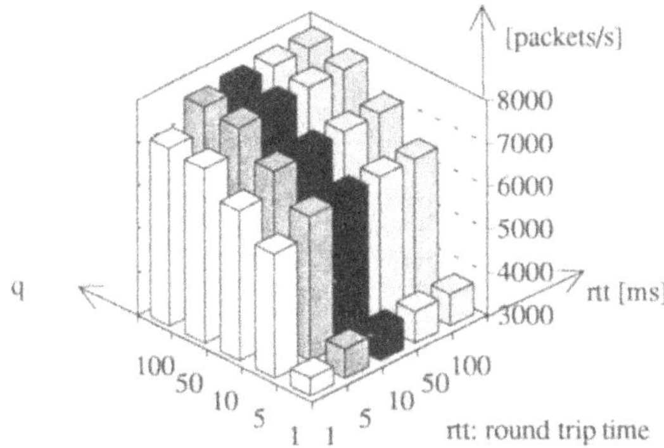

Figure 9: Throughput of Send Path

4.3. Speed-up

To compare the performance of the parallel implementation with an implementation on a single processor, we calculated the single processor performance by subtracting the overhead caused by interprocess communication from the parallel processing times. As illustrated in Figures 11 and 10, the parameters rtt and q have different effects on the performance. Decreasing rtt and q results in a lower performance because of the increasing overhead for control processing. The effects of control processing are stronger for an implementation on a single processor. In a parallel implementation, control processing has less impact on throughput performance, because control processing is performed by special processors. Therefore, speed-up increases with a high control processing overhead. This result is a major benefit of the presented parallel implementation. Usually, the speed-up of the discussed implementation will be 2-2.5 (receive path, cf. Figure 11) and 2.5-3.5 (send path, cf. Figure 10) dependent on the parameters q and rtt [22]. The cases, where the speed-up values are even higher, are unusual cases with an extremely frequent control information exchange.

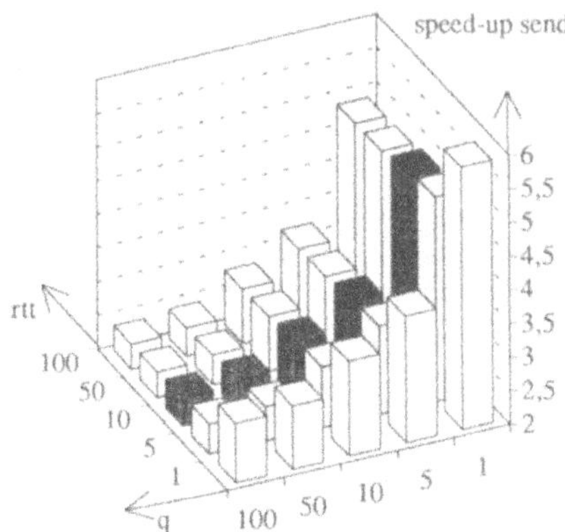

Figure 10: Send Speed-up

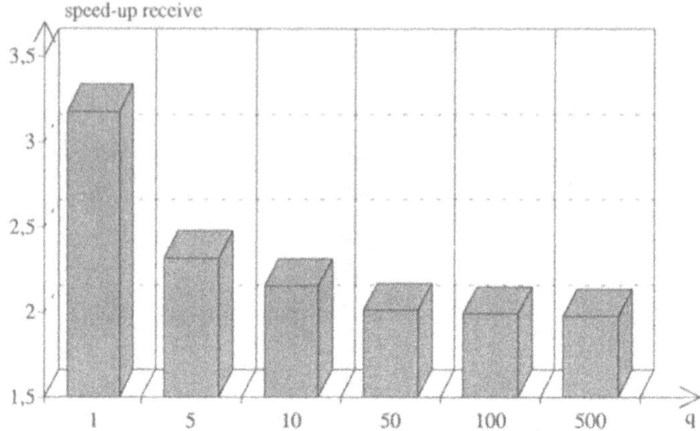

Figure 11: Receive Speed-up

4.4. Discussion

Compared to other transport protocol implementations on transputers, the PATROCLOS implementation achieves a very good performance. A TCP/IP implementation [5] running on the same platform as used for performance evaluation of PATROCLOS achieves less than 3000 packets/s and lower speed-up values. The bottleneck processes of that TCP/IP implementation require approximately the double processing time as the PATROCLOS bottleneck processes. One reason for this difference is that a lot of control functions such as acknowledgment processing or flow control, which are within the critical path of conventional protocols, have been moved to special FSMs of the PATROCLOS architecture and can be performed in parallel.
Our experiences cope with other researchers that it is difficult to separate a single packet into pieces and process these pieces in parallel efficiently [25]. However, the separation of control functions from user data processing functions is a very successful approach, because it improves the critical path of user data processing significantly and allows a more efficient parallel processing. In addition to control/data parallelism, pipelining is also a concept of parallelism that is applicable to transport protocol processing, especially if time-consuming functions as segmentation and reassembly are part of the critical path.

5. CONCLUSIONS

The general goal of PATROCLOS has been an integrated design of protocol and implementation architecture issues. Performance evaluations show significant improvements compared to similar implementations of other protocols. The good performance results, which we achieved without any code optimizations, indicate that a protocol design appropriate for efficient parallel implementation is very useful.
In addition to parallel processing, the presented implementation architecture has the advantage that the memory bottleneck, which usually occurs in network interfaces and is a very serious problem, can be reduced to a minimum, because we provide separate memory busses for incoming and outgoing user data. Because in contrast to other network interfaces, control information packets do not flow across the user data bus, the full bandwidth of the busses can be

used for user data transfer. Furthermore, implementing a transport subsystem with a transport service interface on outboard processors has the benefit that host interrupts and DMA transfers between network interface memory and host memory depend on the frequency of TSDUs. Generally, the results indicate that multiprocessor systems with several independent memory busses may be applicable candidates for future high-speed networks.
Although we developed a special protocol, which is well-suited for the parallel implementation architecture, some of the presented concepts, e.g., parallel processing of acknowledgments and user data, can also be applied to standard protocols such as TCP/IP, if we are able to extract control information from the data flow in the same efficient way as the demultiplexer of the PATROCLOS implementation. However, TCP/IP requires much more complex header parsing and composition functions, which are not easy to implement. Because of this fact, the drawbacks of TCP/IP in high-speed networks with large bandwidth-delay products [3], and the lack of the required flexibility to support a large variety of different applications, we developed the PATROCLOS transport subsystem. The PATROCLOS FSMs provide a lot of different protocol functions and can be configured dependent on application and network requirements. The modular protocol architecture simplifies the implementation of different kinds of transport service interfaces.
Recent work integrated advanced scheduling algorithms into the implementation to support quality-of-service guarantees and multicast extensions to be able to provide reliable and unreliable multicast services.

REFERENCES

[1] Danthine, A.; Baguette, Y.; Leduc, G.; Leonard, L.: *The OSI 95 Connection-mode Transport Service: The Enhanced QoS*, in: Danthine, A.; Spaniol, O.: High Performance Networking, IV, IFIP, North-Holland, 1993, pp. 397-412

[2] Zitterbart, M.; Stiller, B.; Tantawy, A.: *A Model for Flexible High-Performance Communication Subsystems*, Journal of Selected Areas in Communication, Vol. 11, No. 4, May 1993, pp. 507-518

[3] LaPorta, T.F.; Schwartz, M.: *Architectures, Features, and Implementation of High-Speed Protocols*, IEEE Network Magazine, Vol. 5, No. 3, May 1991, pp. 14-22

[4] Haas, Z.: *A Protocol Structure for High-Speed Communication over Broadband-ISDN*, IEEE Network Magazine, Vol. 5, No. 1, January 1991, pp. 64-70

[5] Rütsche, E.; Kaiserswerth, M.: *TCP/IP on the Parallel Protocol Engine*, in: Danthine, A.; Spaniol, O. (eds.): High Performance Networking, IV, IFIP, North-Holland, 1993, pp. 119-134

[6] Björkman, M.; Gunningberg, P.: *Locking Effects in Multiprocessor Implementations of Protocols*, ACM Sigcomm'93 Conference Proceedings, San Francisco, CA, September13-17, 1993, pp. 74-83

[7] Sabnani, K.; Netravali, A.; Roome, W.: *Design and Implementation of a High Speed Transport Protocol*, IEEE Transactions on Communications, Vol. 38, No. 11, November 1990, pp. 2010-2024

[8] Strayer, W.T.; Dempsey, B.J.; Weaver, A.C.: *XTP: The Xpress Transfer Protocol*, Addison-Wesley Publishing Company, 1992

[9] Chan, T.S.; Gorton, I.: *HTPNET: A New Transport Protocol for High-Speed Networks*, SCS&E Report 9402, School of Computer Science and Engineering, University of New South Wales, 1994

[10] Braun, T.; Zitterbart, M.: *Parallel Transport System Design*, in: Danthine, A.; Spaniol, O. (eds.): High Performance Networking, IV, IFIP, North-Holland, 1993, pp. 397-412

[11] Braun, T.; Zitterbart, M.: *A Parallel Implementation of XTP on Transputers*, 16th Annual IEEE Conference on Local Computer Networks, October 14-17, 1991, Minneapolis, USA, pp. 172-179

[12] Clark, D., Tennenhouse, D.: *Architectural Considerations for a New Generation of Protocols*, ACM SIGCOMM ´90, Philadelphia, U.S.A., September 24-27, 1990, pp. 200-208

[13] Sterbenz, P.G.; Parulkar, G.M.: *Design of a Gigabit Host-Network Interface*, Journal of High Speed Networks, Vol. 2, 1993, pp. 27-62

[14] Holzmann, G.J.: *Design and Validation of Protocols: A Tutorial*, Computer Networks and ISDN Systems, Vol. 25, 1993, S. 981-1017

[15] Watson, R. W.: *The Delta-t Transport Protocol: Features and Experiences*, in: Rudin, H. and Williamson R. (eds.): Protocols for High-Speed Networks, North-Holland, 1989, pp. 3-17

[16] Haas, Z.: *Adaptive Admission Congestion Control*, Computer Communication Review, Vol. 21, No. 5, October 1991, pp. 58-76

[17] Keshav, S.; Agrawala, A.; Singh, S: *Design and Analysis of a Flow Algorithm for a Network of Rate Allocating Servers*, in: Johnson, M.J. (ed.): Protocols for High-Speed Networks, II, North-Holland, 1991, pp. 55-72

[18] LaPorta, T. F.; Schwartz, M.: *The MultiStream Protocol: A Highly Flexible High-Speed Transport Protocol*, Journal of Selected Areas in Communication, Vol. 11, No. 4, Mai 1993, pp. 519-530

[19] Feldmeier, D.C.: *An Overview of the TP++ Transport Protocol Project*, in: Tantawy A. (ed.): High Performance Networks: Frontiers and Experience, Kluwer Academic Publishers, 1994

[20] Schiller, J.; Braun, T.: *VLSI-Implementation Architecture for Parallel Transport Protocols*, IEEE Workshop on VLSI in Communications, Stanford Sierra Camp, Lake Tahoe, CA, September 15-17, 1993

[21] Steenkiste, P.A.; Zill, B.D.; Kung, H.T.; Schlick, S.J.; Hughes, J.; Kowalski, B.; Mullaney, J.: *A Host Interface Architecture for High-Speed Networks*, Danthine, A.; Spaniol, O. (eds.): High Performance Networking, IV, IFIP, North-Holland, 1993, pp. 31-46

[22] Braun, T.: *A Parallel Transport Subsystem for Cell-Based High-Speed Networks*, Ph.D. Thesis (in German), University of Karlsruhe, Germany, VDI-Verlag, Düsseldorf, 1993

[23] Brendan, C.; Traw, S.; Smith, J.: *Hardware/Software Organization of a High-Performance Host Interface*, IEEE Journal of Selected Areas in Communications, Vol. 11, No. 2, february 1993, pp. 240-253

[24] Thekkath, C. A.; Nguyen, T.D.; Moy, E.; Lazowska, E.D.: *Implementing Network Protocols at User Level,* ACM Sigcomm'93 Conference Proceedings, San Francisco, CA, September13-17, 1993, pp. 64-73

[25] Partridge, C.: *Gigabit Networking*, Addison Wesley Publishing Company, 1994

14

A Reduced Operation Protocol Engine (ROPE) for a multiple-layer bypass architecture

Y.H. Thia (*)[1] and C.M. Woodside (**)

Newbridge Networks, Inc., Ottawa, Canada (*) and
Dept. of Systems and Computer Engineering, Carleton University, Ottawa, Canada (**)

Abstract — The Reduced Operation Protocol Engine (ROPE) presented here offloads critical functions of a multiple-layer protocol stack, based on the "bypass concept" of a fast path for data transfer. The motivation for identifying this separate processing path is that it involves only a small subset of the complete protocol, which can then be implemented in hardware. Multiple-layer bypass also eliminates some inter-layer operations such as queue and buffer management, context switching and movement of data across layers, all of which are a significant overhead. ROPE is intended to support high-speed bulk data transfer. The paper describes the design of a ROPE chip for the OSI Session and Transport layer protocols, using VHDL. The design is practical in terms of chip complexity and area, using current gate array technology, and simulation shows that it can support a data rate approaching 1 gigabit per second, in a connection attached to an end-system.

Keyword codes: C.2.2, B.4.1
Keywords: Network Protocols, Data Communications Devices

1 Introduction

The advent of Fibre Optic technology, which offers high bandwidth and low bit error rates, has shifted the performance bottleneck from the communications channel to the communications processing in the end-points of the system [26]. Other trends such as improved quality-of-service guarantees will reinforce this effect. The heavy processing load is due to a combination of operating system overhead, protocol complexity, and per-octet processing on the data stream. To alleviate the end-system bottleneck one may consider new protocols [10], improved software implementation of existing protocols [5, 35], parallel processing techniques [14, 21, 38], special protocol structures [15, 30] and hardware assist [22] by offloading all or part of the protocol functions to an adaptor. This paper takes the latter approach.

The key problems associated with offboard processing include:

- ☐ Partitioning the functionality between the host and the adaptor is difficult and may easily lead to a complex additional protocol between the two parts, which may cancel out or offset the potential gain from offloading. For example, the buffer management task [36] may be offloaded, but this leaves the problem of control for accessing it within the full protocol logic.

[1] This research was done while Dr. Thia was at Carleton University

- Non-protocol-specific processing is a large part of the total load, as shown in [35]. Examples include interrupt handling, context switching and data copying at layer boundaries in deeply layered protocol stacks.
- The choice of hardware for the adaptor depends on the complexity of the functions it supports. In [2, 22] where the transport protocol layer is offloaded or in [7] where the full protocol stack can be offloaded, general purpose microprocessors are used. Probably because of the complexity of existing protocols, VLSI [24] implementation above the data link layer has been disappointing so far. In [8], dedicated VLSI chips are used to support TCP checksums. Also, some newer lightweight transport protocols are specially designed for VLSI implementation [1, 3].
- There is a tradeoff between performance, flexibility and cost. If the key functions of the frequently executed portion of the protocol remain relatively stable, there can be significant advantage in providing hardware support for these functions leaving the other tasks in the host software for flexibility.
- As host processing speed continues to outpace memory bandwidth and as the network bandwidth approaches the processor memory bandwidth, it is important to keep data movement on the workstations down to the minimum [4, 9, 28].

This paper presents a feasibility study for a new approach to hardware assistance. It combines the relatively simple operations needed for data transfer across multiple layers and provides a hardware “fast path” for them, which will be efficient for bulk data transfer. It is based on the “protocol bypass concept” [37] which is a generalization of Jacobson’s "Header Prediction" algorithm [20] for TCP/IP. Bypass solves the problems identified above, which may limit the use of offboard processing, by implementing an entire service through all layers for certain cases. This simplifies the interface between the host and the adaptor chip and minimizes their interaction, which is supported by an access test, some DMA processing and a simple command protocol. The chip design based on bypassing is called ROPE, for Reduced Operation Protocol Engine. The contribution of this paper is to define the host/chip interface and the chip operation, and to report on a VHDL-based feasibility study of the chip design. It appears to be feasible to support an end-system single-connection data rate approaching 1 Gbps.

The next section introduces the bypass concept, its architecture and implementation. Section 3 analyzes the key protocol processing overheads and discusses the requirements for a bypass VLSI implementation. Sections 4, 5 and 6 describe a design study of a ROPE chip using the industry standard hardware description language, VHDL, with conclusions in Section 7.

2 The Bypass Concept

A bypass adds an additional path for certain operations, with minimal changes to the original software. Conformance to the protocol is maintained by doing all the other operations through the normal "heavyweight" path. A bypass path can be provided for send, for receive, or for both together, and is compatible with other end-systems implemented without a bypass.

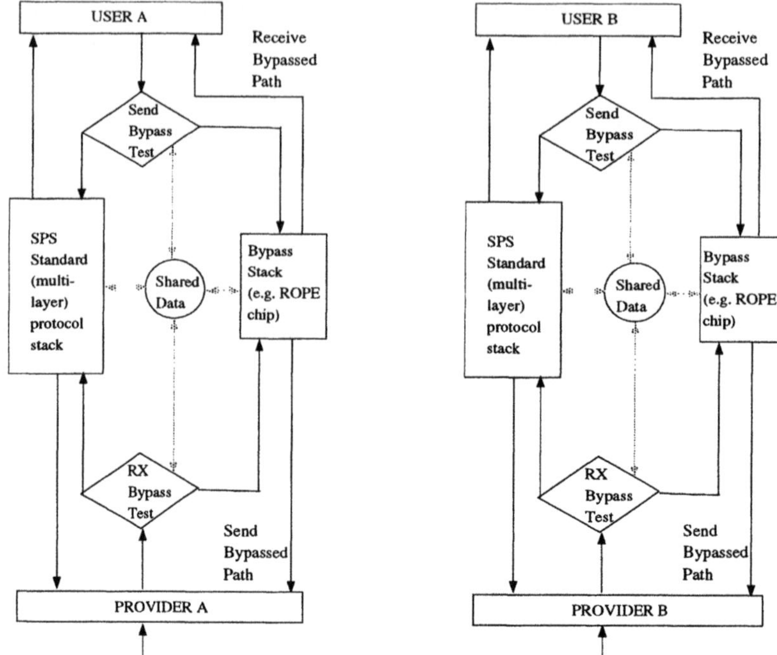

Figure 1 Bypass Architecture

2.1 Bypass Architecture

Figure 1 illustrates the architecture of a bypass implementation for any standard protocol. The standard protocol stack (SPS) is the processing path taken by all PDUs during a connection without the bypass. The SPS may refer to a single layer or to multiple adjoining layers of a layered protocol stack. The bypass has 4 key components:

- *Send Bypass* Test;
- *Receive Bypass* Test;
- *Bypass Stack*;
- *Shared data* for access by the two tests, the Bypass stack and the SPS.

The send bypass test identifies outgoing packets that are data packets in the data transfer phase. The receive bypass test matches the incoming PDU headers with a template that identifies the predicted bypassable headers. The bypass stack performs all the relevant protocol processing in the data transfer phase. The shared data are used to maintain state consistency between the SPS and the bypass stack, including window flow control parameters and connection identifiers. Whenever there is a change in the processing path between the SPS and the bypass stack, checks are performed to ensure that there are no outstanding packets in the current path, i.e. "no in-transit PDUs", before the change is made. A more detailed discussion on this is presented in another paper [33].

2.2 Efficient logic for the bypass test

The "no-in-transit PDU" test can often be avoided. At the beginning of data transfer on a new connection, it is automatically satisfied. It holds as long as no packet fails a bypass test, and it is sufficient to maintain a flag to indicate this. Once a packet fails, and goes to the SPS, then a full "no-in-transit PDU" test must be performed for each packet until the test succeeds, after which control can go back to the flag. Token management and synchronization points of the session layer [19, 18] which are mapped by equivalent application and presentation service primitives can be used as synchronization points within the bypass architecture, as it switches between the SPS and the bypass stack. Since they are only inserted periodically in bulk data transfer and can be controlled by the application process, these overheads are not excessive.

2.3 Multiple-layer bypass

A bypass for multiple layers instead of just one gives additional gains by avoiding:

- ☐ Overhead of encoding and decoding the interface control information passed between layers;
- ☐ Executing the full general protocol logic for the layers to decide how to manipulate the data;
- ☐ Queueing of data at layer boundaries.

The advantage is increased further in cases where some layers, like the network and application layers, have been further subdivided into sublayers.

A multiple-layer bypass path is a concatenation of processing procedures performed by the adjacent layers when they are simultaneously in the data transfer phase. Meanwhile, the separate layers in the SPS path handle the other phases.

In summary, the separation of the bypass path offers the following advantages:

- ☐ The processing path of data PDUs can be optimized;
- ☐ The number of possible PDU formats in the bypass path is reduced to data transfer PDUs;
- ☐ The finite state machine of the protocol is now reduced to only the "OPEN" state, for as long as processing remains in the bypass path. The state of the system does not change during the entire data transfer phase and the protocol processing is reduced to ensuring reliable transfer of data across the communications network.

3 Design Considerations for a Hardware Bypass

3.1 Factors affecting system performance

This section summarizes the major factors affecting throughput performance in a deeply layered stack on an end system. It follows the description by Heatley and Stokesberry [16].

Protocol procedures can be characterized as per-octet, per-packet or per-group-of-packets operations. The per-octet operations take an average time A per octet (e.g., checksum), and the per-packet procedures take time B per packet (e.g. address decoding and multiplexing). Per-group-of-packets operations include for example transmission of acknowledgments, whose frequency is implementation-dependent, and their timing will be aggregated and included in parameter B.

The throughput bound imposed by protocol processing alone, λ_{max} in octets per second, is then given by the equation:

$$\lambda_{max}(x) = \frac{x}{A.x + B.\lceil x/M \rceil} \tag{3.1.1}$$

where x is the size of the user message in octets and M is the maximum PDU size. In bulk data transfer, as x becomes large,

$$\lim_{x \to \infty} \lambda_{max}(x) = \frac{1}{A + B/M} \tag{3.1.2}$$

The protocol processing load on an end system is typically shared between the host and the network adaptor. As the raw data bit rate supported by optical networks approaches the main memory bandwidth of the end system, the cost of moving data and of per-octet processing limits the effective throughput presented to the application process, especially for bulk data transfer. The data portion of a PDU may be physically moved for the following reasons:

- Copying between the adaptor buffer and the host system memory;
- Crossing protection domains (address spaces) — e.g. at the user/kernel boundary. This problem becomes more pronounced in microkernels which treats a protocol task as a server process outside the kernel domain [11];
- Per-octet processing like presentation conversion and checksum routines.

Hardware implementation is particularly efficient for per-octet operations.

3.2 Requirements of a bypass VLSI implementation

The problems associated with separate or offboard processing, which were discussed in the introduction, are addressed by the VLSI design as follows:

- □ A clean separation of functionality requiring only a simple protocol to communicate between the host and adaptor is desired, and is provided by a bypass. Its particular set of functions are complete in themselves and have a focussed interface with the host software at the packet entry point. There is relatively infrequent switching between the SPS and the bypass stack;
- □ Reduced non-protocol-specific processing overhead. For example the processing of acknowledgment packets is dominated by interrupt handling, typically a few hundred instructions, rather than by the protocol processing itself. Our approach removes acknowledgment handling altogether from the host. Also, the bypass system can be extended to incorporate multiple-layer stacks and remove overhead that way;
- □ VLSI implementation complexity: only the data transfer functions are implemented.
- □ Tradeoff between performance, flexibility and cost. The host software processes non-data-transfer packets which are typically small but require more flexible and more complex processing. They are also only per-packet, so they have less impact overall. They can be efficiently handled by the host given the projected increase in host processing speed [17]. The operations implemented in VLSI are understood to have less need of flexibility.

Layer	*Procedure*	***Bypass Chip***	*Host*	*Per-Octet (A)*	*Per-Packet (B)*	*Per-Group-Of-Packets Aggregated to Per-Packet for bulk data transfer (B)*	*Remarks*
Presentation	Encoding	X		X			
	Encryption	X		X			
	Compression	X		X			
	Context Alteration		X			X	
Session	Synchronization Management		X			X	
	Token management		X			X	
Transport (Class 4)	Checksum (Optional)	X		X			
	Timer Management	X			X	X	Depends on Implementation
	Generation of ACK packets (Flow Control)	X				X	
	Resequencing	X			X		
All 3 layers	Header Construction	X			X		
	Header Decode	X			X		
	Buffer Management	X			X		Minimized (Simple scheme)
	Context Switching	X			X	X	Moved away from host OS
	Data Copying	With multiple-layer bypass, data Copying within layers is elminated.			X		Use of dual-ported memory and DMA..

Table 1 Bypassable versus Non-bypassable functions

- Remove operations that are inefficient on the host. It is important to look at both the processing requirements and data movement of the Application PDUs. The critical resource is often the host bus/memory path, and caching is used to increase its efficiency. However long traverses through the data for per-octet operations like checksum and encoding interfere with efficient caching, so it is particularly appropriate to offload them.

Table 1 identifies procedures which are strong candidates for implementation in the bypass chip, and those which are better handled by the host, during the data transfer phase. Besides these, the send bypass test is done on the host and the receive bypass test is done on the Network Interface Adaptor.

4 VLSI implementation of a Reduced Operation Protocol Engine (ROPE) chip using VHDL

The VHSIC Hardware Description Language (VHDL) [6, 27] was used to model and synthesize the chip. VHDL is an industry standard language which can be used to represent all levels of abstraction, from logic gates to the system level. By utilizing VHDL as a specification tool, it is possible to begin simulation and debugging of complex systems

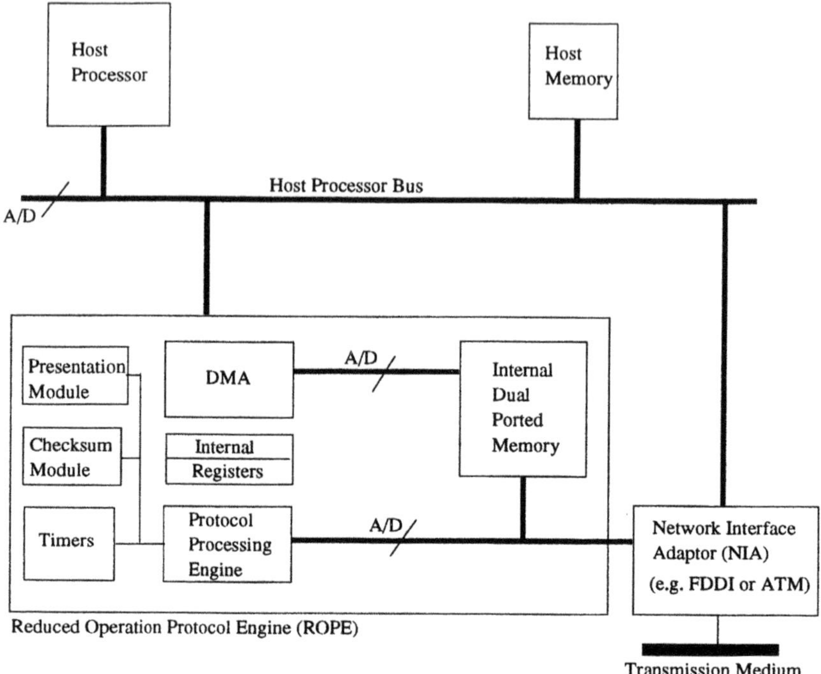

Figure 2 Block Diagram of VLSI bypass system

before details regarding the implementation are fully specified. It also offers the potential of an automatic path from the protocol specification to VLSI implementation, in which any modifications to the specification can be easily propagated to the gate level design.

4.2 Architectural Description

Figure 2 shows the block diagram of the system. The host processor and NIA components provide logical interfaces for simulation of behaviour, but they insert no timing delays. For modeling purposes they were described as being infinitely fast, either as a source or as a sink. This places the maximum stress on the ROPE chip.The architectural considerations involved in the chip design can be summarized as follows:

- Movement of data across the host bus interface are minimized by using an on-chip DMA for fast block data transfer to/from the host system memory.
- On-chip dual-ported memory is used, rather than the host memory, to avoid critical constraints on bus access latency and throughput.
- The control registers of the bypass chip are I/O mapped to the host processor. This enables the host processor to configure the bypass chip directly.

The presentation module shown in the Figure was allowed for in the data structures but was not fully designed.

4.3 First Design: Design Steps

Figure 3 shows the steps followed in this study. There were three stages, a behavioural model, a structural or RTL model, and a gate level design. These gave us two kinds of feasibility check, that the logic we specified will execute the protocol within the environment we envisage, and that the design is technically feasible, for instance in a reasonable chip area.

A VHDL behavioral model for the system was initially written and tested. It includes the bypass chip, the host processor with a simplified bus/memory architecture, and a vestigial high-speed network interface adapter which acts simply as an infinite source/sink for data packets. The first design implemented the Basic Combined Subset (BCS) of the session layer and the Transport protocol class 2 (TP2) protocols. Only the logic for the data transfer phase is included in this model, as it is during this phase that the bypass chip is active. A host processor submodel generates a simple test sequence that initiates bypass connections and a series of bypassable data packets. Non-bypassable packets are also inserted to simulate for example the arrival of a connection release PDU. The test sequence allowed us to verify the following:

- Window flow control logic.
- Generation of the header field including the sequence number.
- Generation of acknowledgment packets on receive.
- Synchronization between the SPS and the bypass stack.

Next, the behavioural model of the bypass chip was manually converted to a structural (RTL) model for synthesis, leaving the other components as behavioral constructs. A behavioral description has no implied architecture in its representation, while an RTL level description has a definite architecture and clocking scheme, and characterizes the system in terms of registers, switches (multiplexors), and operations. An initial assignment of operations into clock cycles is also made. These descriptions, like behavioral descriptions, are technology-independent (silicon library independent). In a VHDL RTL level description, not all of the functionality of VHDL can be used because some of the language features do not map into the RTL model. For example a WAIT FOR statement has no meaning in hardware, but is useful in behavioral simulation. Features that can be easily mapped to hardware include IF THEN ELSE statements and signal assignments. The same test sequence was again generated by the host processor, and its results were compared with the original model to ensure behavioral consistency.

The structural model was then passed through the SYNOPSYS synthesis tool [31] for gate level generation with the 0.8 μm BiCMOS macro library from Texas Instruments [32]. The timing information generated by this process was back-annotated to the structural model to obtain the simulation throughput results presented in section 5. This was sufficient for our present purposes, to estimate the space complexity and timing of the chip, and the final step of generating a chip layout for fabrication and fault analysis was not performed. As the model was too large for the SYNOPSYS package we were using, it was divided into 3 submodels. The dual-ported SRAM was not synthesized as its gate counts and performance characteristics can be easily extracted from data books.

The second design, with additional functionality, is described in the next section.

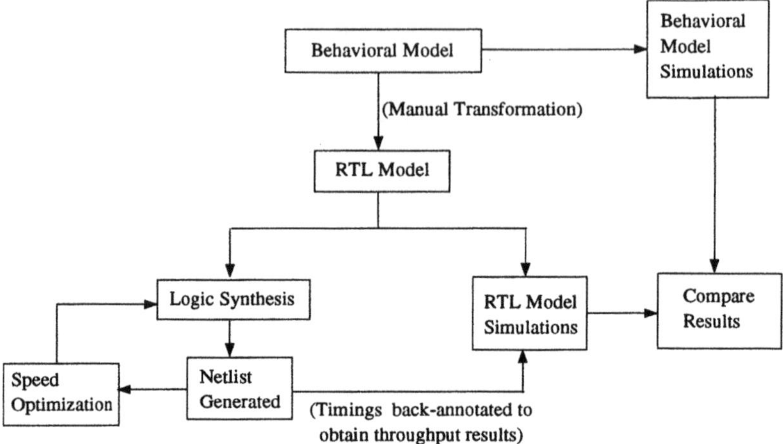

Figure 3 Design flow diagram

4.4 Behavioural description

The sequence of operation in the bypass system is summarized as follows:

1) Four high level procedures are made available to the host processor to control the bypass chip, namely: BYPASS_START, BYPASS_DMA, BYPASS_SYNC and BYPASS_RESTART. On receiving the first bypassable PDU, the BYPASS_START procedure is called. This procedure sets up a bypassable connection by sending information like its initial window flow control parameters and the DST_REF field to the bypass chip. These are stored in a process control block for the particular connection in fixed on-chip memory locations and are also accessible by the host (I/O mapped). The maximum number of connections allowed for simultaneous bypassing will be equal to the number of these control blocks allocated in the bypass chip (5 in this study — See figure 4).
2) For subsequent bypassable packets, the host processor initiates the BYPASS_DMA procedure which checks for free buffer space in the bypass chip and programs the DMA by sending the starting address pointer where the PDU is located, and its total length. The destination address is supplied by the bypass chip. Arbitration for the host processor bus between the host and DMA is provided by the DMAreq and DMAack lines. DMA transfers the PDU into the internal dual-ported SRAM (Static RAM). Buffers are pre-allocated in fixed sizes and are accessed by a simple round robin scheme using a set of buffer pointers.
3) The protocol engine polls the status field of a packet to check if there are any data to be processed. If the status is FILLED, protocol processing of the bypass stack can proceed. The dual-ported structure allows protocol processing to proceed concurrently with any DMA transfer to/from the host processor. A precomputed header template is used to construct the header field very quickly.

4) Processed in-sequence packets within the transmit window are passed to the network interface adapter, which in this study acted as an instantaneous sink.
5) Whenever the host processor encounters a switch in the processing path, i.e. from the bypass stack to the SPS, it will issue a BYPASS_SYNC procedure. This procedure will flush the bypass chip of any "in-transit PDU" for that particular connection and return any updated information, for example window control parameters, from the bypass chip to the host in order to maintain state consistency between the two paths. This may occur when it receives, for example, a connection release primitive or a session control primitive of the session layer (see section 6) during the data transfer phase.
6) Whenever the host processor wishes to re-enter the bypass path after a switch in the processing path, the host will issue a BYPASS_RESTART procedure which will pass only those data that were updated in the standard protocol stack, like window flow control parameters, back to the bypass chip. Parameters like the DST-REF field which is not changed for the duration of the connection need not be updated.

4.5 Second Design, including major procedures for Transport Class 4 (Implemented)

This section describes extensions to the first design, which only supports Session BCS and TP2 functionality, to include some common TP4 functionality. Procedures for checksum, retransmission on timeout and resequencing were implemented. Extensions to the Session layer functionality and procedures for presentation layer conversion were not implemented, but are also discussed in section 6.

4.5.1 OSI Checksum The transport protocol class 4 checksum algorithm [12] was implemented. It is often difficult to perform it on the fly at the sender end as the two-byte checksum field is placed in the variable part of the header. However, with the bypass system, the header structure and the position of the checksum field are known in advance. Also, calculation of the checksum can now be simplified further by precomputing the partial checksum of the header fields.

4.5.2 Timers

During the data transfer phase TP4 uses a Retransmission timer (T1), a Window timer (W) and an Inactivity timer (I). Only the retransmission timer with one interval per connection and the window timer were implemented here.

Timer management in software is an expensive process due to software interrupt handling overhead and update processing of the timer queue [34], but can be easily handled by VLSI implementation. On-chip timers are very efficient and can be executed concurrently with other protocol processes. The only overhead is in starting and stopping the timers. Once it is started, a timer is an autonomous process until an interrupt signal is activated. A separate area of on-chip memory is reserved to store the state information of the timer list.

4.5.3 Retransmission and Resequencing

At the receiver end, out-of-sequence PDUs outside the flow-control window will be discarded. Otherwise, a PDU is buffered for resequencing. Duplicate TPDUs can be detected

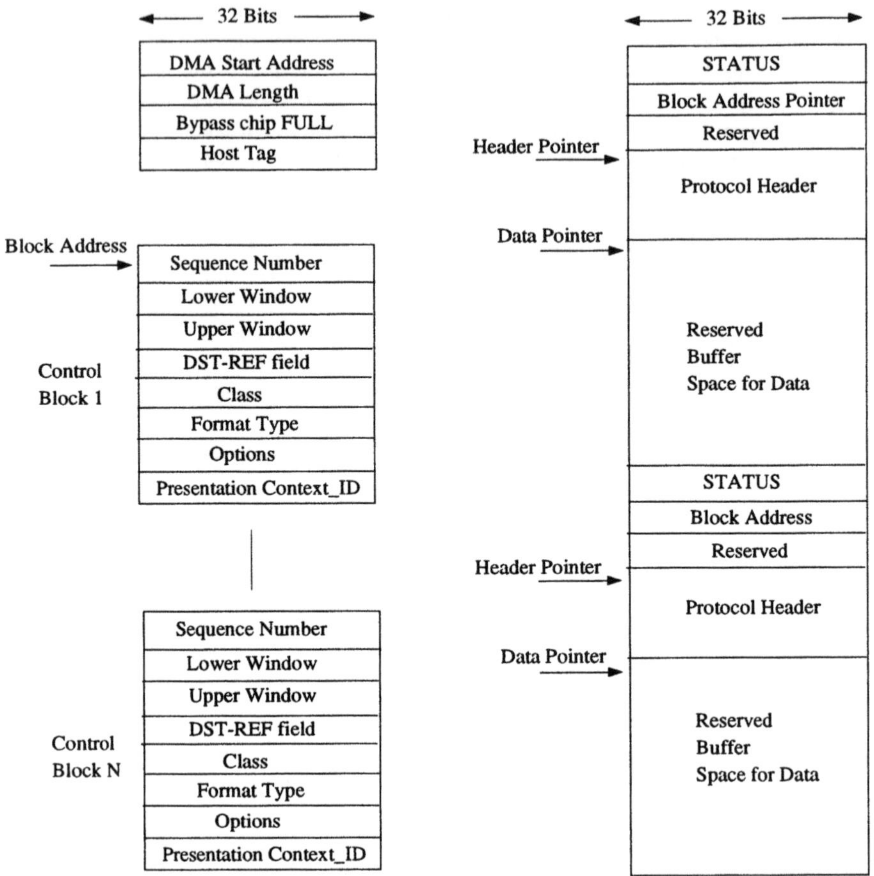

Host Tag : This tag is set on receipt of a host command, e.g BYPASS_START, BYPASS_DMA, BYPASS_SYNC or BYPASS_RESTART.

STATUS : Indicates the status of the buffer, e.g. EMPTY, FILLING, FILLED or CLOSED.

Figure 4 Organization of internal bypass chip memory

easily because the sequence number matches that of a previously received TPDU. At the sender end, if timer T1 expires, the transport entity can retransmit either the first TPDU, or all TPDUs (Go-back-N) waiting for acknowledgment. In this design the Go-back-N retransmission strategy was used. For a large window, the on-chip buffer may not be sufficient to hold the unacknowledged data packets for retransmission or to buffer data packets for resequencing, and slower external memory would be needed.

Procedures	*Combinational Area (Equivalent NAND2 gates)*	*Non Combinational Area (Equivalent NAND2 gates)*	*Total Area (Equivalent NAND2 gates)*	*Throughput performance with 1 Kbyte packet length (Mbps)*
Session (BCS)/ Transport Class 2	*6318*	*2929*	*9247*	*2,362.8*
Dual Ported SRAM (4 Kbyte)	*N/A*	*N/A*	*Approximately 41,984*	*N/A*
Session (BCS)/ Transport Class 4 with the addition of the Checksum Procedure with timer circuitries	*8334*	*4227*	*12561*	*313.3*

Table 2 Throughput Performance and gate count of bypass VLSI chip

5 Results

The timing information obtained from the netlist was back-annotated to the structural model to obtain throughput performance results. The operating parameters and assumptions made in this study are:

- A chip clock rate of 66 MHz.
- 1 Kbyte data packets
- Window size of 64. An acknowledgment packet is sent for every 20 packets received.
- The host bus/memory subsystem and network interface adapter were assumed to be infinite sinks/source of data packets.
- One thousand packets were processed for each iteration.
- 4 Kbyte of internal dual ported SRAM.

Table 2 shows the throughput performance of the ROPE chip. The throughput value includes the time taken to move the data packet out from the internal memory of the bypass chip to the network interface adapter, but not the data copy operation from the host memory. Hence the throughput result includes just one copy operation of the data packet. The total gate count for the bypass chip with Session (BCS), TP2 and 4 Kbyte internal dual ported SRAM is 51,231 equivalent NAND2 gates. With the additional TP4 procedures like checksum and timer circuitries, the total gate count increased to 54,545 gates. Texas Instruments offers a gate array package with 112,000 usable gates (TGB1150). This leaves enough chip area for extra on-chip memory or hardwired presentation procedures. With no per-octet operations, i.e. just protocol processing of packet headers (TP2), the bypass stack could effectively achieve a throughput of 2.362 Gbps. This is consistent with results presented by Clark [5] which claims that TCP processing can achieve up to 800 Mbps (without checksum) on current RISC-based processors. With OSI checksum, the throughput performance drops to 313.3 Mbps. This is still more than an order of magnitude higher than an equivalent 19.2 Mbps optimized hand assembled software implementation of the OSI checksum algorithm on the VAX 8800, reported by Sklower [29].

6 Considerations for the Session and Presentation layers (not implemented)

The next steps would logically be to enhance the Session [18, 19] processing to the BSS (Basic Synchronized Subset) or BAS (Basic Activity Subset) level, or to add Presentation processing

In the BSS, synchronization points occur but the session services do not actually save session SDUs and do not themselves perform the recovery operations. These activities are the responsibility of the application layer or the application program. The session layer merely decrements the serial number back to the synchronization point and the user must apply it to determine where to begin recovery procedures. Hence these activities are best handled on the host processor and are not very suitable for bypassing. The benefit they would confer (if bypassed) would be to reduce the frequency of switching paths.

Presentation processing can definitely be bypassed. During the data transfer phase, it consists only of the Presentation data encoding/decoding functions. Substantial performance gains could result if the presentation conversions are simple and are used consistently, although the inflexibility of a hardware version is an evident weakness. One possible application of ROPE with hardwired presentation conversion is in video servers with the proposed encoding standards such as MPEG [13].

7 Summary

It can be concluded from this study that it is feasible to implement the bypass stack (at least for the transport and session layers) in VLSI and that the performance would be at least an order of magnitude higher than software protocol processing. The bypass system offloads the critical protocol functions and the associated non-protocol-specific functions onto a "Reduced Operation Protocol Engine" (ROPE). The gate count for the bypass chip can easily fit into a commercially available gate array Integrated Circuit. Per-octet operations are particularly efficient when performed on the chip. The host processor is relieved of a significant proportion of protocol processing and can concentrate on the application processing. The speed of communication processing in the host system can now match the transmission bandwidth of high-speed networks, e.g. ATM technology, thereby increasing the application-to-application throughput performance. (In an ATM system we assume that the segmentation and reassembly or SAR operation would also be in hardware, since it is done frequently.) The host processor is also relieved of acknowledgment processing. An existing implementation of the OSI stack can be adapted for bypassing with only a small modification of the original software, thus providing an easy migration path for current systems.

The scope of functions included in a bypass may be narrowly defined, or more extended. A bypass does not include fast connection setup but also does not interfere with it. There is no segmentation/reassembly within the bypass path, but we do not see this as a major restriction, as research suggests that fragmentation of PDUs should be restricted only to the lower layers and should occur only once in the protocol stack [23]. The Segmentation and Reassembly sublayer of the ATM adaptation layer is a good place for such functions [25].

In the first design, the bypass chip with a 66 MHz clock can support a throughput rate of 2.3 Gbps (Session BCS and TP2) for 1 Kbyte TPDU packets using current $0.8\mu m$ BiCMOS technology. In the second design, extended to include the TP4 checksum, retransmission on timeout and resequencing procedures, the throughput decreased to 313.3 Mbps. Further advances in speed can be obtained in proportion to technology improvements, making the approach viable for some considerable time to come.

Acknowledgments

Bell-Northern Research provided the VHDL tools used in this study, and Dr. Simon Curry, Hemi Thakar, Dr. Parviz Yousefpour, Bernard Doray, Mike Majid and Mustapha Bourahla helped with their use. This research was supported by the Ontario government program of Centers of Excellence, through the Telecom Software Methods Project of TRIO, the Telecommunications Research Institute of Ontario.

References

[1] Balraj T.S. and Yemini Y., "Putting the Transport Layer on VLSI - the PROMPT protocol chip". IFIP, Stockholm, May 13-15, 1992.

[2] Beach B., "UltraNet: An Architecture for Gigabit Networking," in Proc. 15th Conference on Local Computer Networks, Minnesota Oct, 1990.

[3] Chesson G., "XTP/PE Design Considerations," in Proc. IFIP Workshop Protocols for High-Speed Networks, Zurich, May 9-11, pp. 27-33, 1989.

[4] Clark D. and Tennenhouse D., "Architectural Considerations for a New Generation of Protocols," in ACM SIGCOMM 1990.

[5] Clark D., Jacobson V., Romkey J., and Salwen H., "An analysis of TCP processing overhead," in IEEE Commum. Mag., vol. 27, pp. 23-29, June 1989.

[6] Coelho D.R., "The VHDL Handbook," Kluwer Academic Publishers, 1989.

[7] Cooper E.C, Steenkiste P.A., Sansom R.D. and Zill B.D., "Protocol Implementation on the Nectar Communication Processor," in ACM SIGCOMM'90, 1990.

[8] Dalton C., Watson G., Banks D., Calamvokis C., Edwards A. and Lumley J., "Afterburner," in IEEE Network July 1993.

[9] Davie B.S., "Architecture and Implementation of a High-Speed host Interface," in IEEE Journal on selected areas in communications, Vol. 11, No. 2, February 1993.

[10] Doeringer W.A., Dykeman D., Kaiserwerth M., Meister B., Rudin H., and Williamson R., "A survey of Light-Weight Transport protocols for High Speed Networks," in IEEE Trans. on Comm., vol. 38, No. 11, pp 2025-2039, Nov. 1990.

[11] Druschel P., Peterson L.L., "Fbufs: A High-Bandwidth Cross-Domain Transfer Facility", in the Proceedings of the 14th ACM Symposium on Operating Systems Principles, Dec 1993.

[12] Fletcher J.G., "An Arithmetic Checksum for Serial Transmissions" in IEEE Transactions on Communications, Vol. Com-30, No. 1, Jan 1982.

[13] Gall D.L., "MPEG," in Communications of the ACM, Apr. 1991.

[14] Giarrizzo D., Kaiserswerth M., Wicki T. and Williamson R., "High-Speed Parallel Protocol Implementation," in Proc. IFIP Workshop Protocols for High-Speed Networks, Zurich, May 9-11, 1989.

[15] Haas Z., "A Communication Architecture for High-Speed Networking," in IEEE INFOCOM, pp. 433–441, June 1990.

[16] Heatley S., Stokesberry D., "Analysis of Transport Measurements Over a Local Area Network," IEEE Communications Magazine, Jun 1989.

[17] Hennessy J.L and D.A. Patterson, "Computer Architecture: A quantitative approach," Palo Alto, California, Morgan Kaufmann Publishers 1991.

[18] Information processing systems — OSI Basic connection oriented session protocol specification, standard ISO-8327, 1987.

[19] Information processing systems — OSI Basic connection oriented session service definition, standard ISO-8326, 1987.

[20] Jacobson V., "4BSD TCP Header Prediction," in ACM SIGCOMM, Comp. Commun. Review, vol. 20, no. 2, pp. 13–16, Apr. 1990.

[21] Jain N., Schwartz M. and Bashkow T.R., "Transport Protocol Processing at Gbps rates," ACM SIGCOMM '90 Symp, Philadelphia, pp. 188-199, Sep. 24-27, 1990.

[22] Kanakia H. and Cheriton D., "The VMP network adapter board (NAB): High-performance network communication for multiprocessors," ACM SIGCOMM '88 Symp., Stanford, CA, pp. 175-187, Aug. 16-19, 1988.

[23] Kent C.A., Mogul J.C., "Fragmentation Considered Harmful," ACM SIGCOMM '87 Workshop Comp Commun Review, vol. 17, no. 5, Special Issue, Aug 1987.

[24] Khrishnakumar A.S., Sabnani K., "VLSI Implementations of Communication Protocols — A Survey," IEEE Journal on Selected Areas in Communications, vol. 7, no. 7, pp. 1082–1090, Sep 1989.

[25] Lyles J.B., Swinehart D.C., "The Emerging Gigabit Environment and the Role of Local ATM," in IEEE Communications Magazine, Apr 1992.

[26] Partridge C., "Gigabit Networking," in Addison Wesley Professional Computing Series, 1993.

[27] Perry D.L., "VHDL," McGraw-Hill Inc., 1991.

[28] Ramakrishnan K.K., "Performance Considerations in Desigining Network Interfaces". IEEE Journal on Selected Areas in Communications, Vol. 11, No. 2, Feb 1993. .

[29] Sklower K., "Improving the Efficiency of the OSI Checksum Calculation," in ACM SIGCOMM Computer Communications Review, Vol. 19, No. 5, pp. 32-43, Oct. 1989.

[30] Sterbenz J.P.G. and Parulkar G.M., "AXON: A High Speed Communications Architecture for Distributed Applications," in IEEE INFOCOM, Jun 1990.

[31] SYNOPSYS Design Analyzer™ Reference Manual, Version 2.0, May 1991.

[32] TGB 1000 Series 0.8um BiCMOS Gate Arrays Macro Library Summary, Application Specific Integrated Circuits, Texas Instruments, Dec. 1991.

[33] Thia Y.H., Woodside C.M., "High-Speed Protocol Bypass Algorithm with Window Flow Control", in Proc. of the 3rd IFIP International Workshop on Protocols for High-Speed Networks, Stockholm, May 13-15, 1992.

[34] Varghese G. and Lauck T., "Hashed and Hierarchical Timing Wheels: Data structures for the efficient implementation of a Timer facility," in Proc. of the 11th ACM Symp. on Operating System Principles, Nov. 1987.

[35] Watson R.W. and Mamrak S.M., "Gaining efficiency insmith transport services by appropriate design and implementation choices," in ACM trans. on Computer Systems, vol. 5, no. 2, pp. 97–120, May 1987.

[36] Woodside C.M. and Montealegre J.R., "The effect of buffering strategies on protocol execution performance," in IEEE Trans. Commun., vol. COM-37, pp.545-554, June 1989.

[37] Woodside C.M., Ravindran K. and Franks R.G., "The protocol bypass Concept for High Speed OSI Data transfer," in Proc. IFIP Workshop on Protocols for High-Speed Networks, Zurich, May 9-11, 1990.

[38] Zitterbart M., "High-Speed Protocol Implementations based on Multiprocessor-Architecture," in Proc. IFIP Workshop Protocols for High-Speed Networks, Zurich, May 9-11, 1989.

PART SIX

Implementation and Performance

Deadlock situations in TCP over ATM

Kjersti Moldeklev[a] and Per Gunningberg[b]

[a]Norwegian Telecom Research, P.O. Box 83, N-2007 Kjeller, Norway

[b]Swedish Institute of Computer Science, P.O. Box 1263, S-16428 Kista, Sweden*

Abstract

The implementation of protocols, such as TCP/IP, and their integration into the operating system environment is very decisive for protocol performance. Putting TCP on high-speed networks, e.g. ATM, with large maximum transmission units causes the TCP maximum segment size to be relatively large. What Nagle's algorithm consider a "small" segment is not small anymore which affects the TCP throughput. We report on TCP/IP throughput performance measurements for various sizes of send and receive socket buffers, using two Sun IPX machines running SunOS 4.1.1 connected to FORE System's ATM network. For some common combinations of socket buffer sizes we observe a dramatic performance drop to less than 3% of normal throughput. The drop is caused by a deadlock situation in the TCP connection which is resolved by the 200 ms spaced timer generated TCP acknowledgment. The factors which in combination force the TCP connection into deadlocks are a large maximum segment size, asymmetry of the socket buffer sizes, use of Nagle's algorithm, the delayed acknowledgment strategy, the sequence of actions on acknowledgment reception, and at last the socket layer optimization for efficient memory management. We explain what causes the deadlock situations, discuss some alternatives to avoid or prevent the situations and present measurement results from proposed changes in the implementation.

Keyword Codes: C.2.2; C.2.5; C.4
Keywords: Computer-Communication Networks, Network Protocols; Local Networks; Performance of Systems

1. INTRODUCTION

The TCP/IP protocols are often the first protocol suite to be put on high-speed networks such as ATM (Asynchronous Transfer Mode). This in spite of the fact that TCP/IP originally was not designed to match the characteristics of high-performance networks. Several extensions of the TCP protocol [1] [2] have been suggested to make it perform better over these networks and for connections with a high bandwidth-delay product [3]. New high bit-rate demanding applications, such as multimedia conference systems, and the characteristics of high-speed networks, have triggered a lot of research on new transport protocols [4].

* Email: kjersti.moldeklev@tf.tele.no, per@sics.se

However, it is well known that the *implementation* of protocols, and their *integration* into the operating system environment may dominate all improvements of protocol mechanisms. Implementation optimizations for TCP/IP on low-speed networks (with smaller data units) may be totally wrong for high-performance networks. In this paper we show that some implementation optimizations for the "ethernet era" in fact degrade TCP/IP performance on high-speed networks with large data units.

We have measured the throughput performance of TCP over ATM between two Sun IPX machines running SunOS 4.1.1, for various sizes of send and receive socket buffers. Normally we measured around 20 Mbit/s sustained throughput, but for some combinations of socket buffer sizes we observed a dramatic drop to between 0.65 to 0.16 Mbit/s. This is less than 3% of the normal throughput. It happened for common combinations of socket buffer sizes, such as a send socket buffer of 16 kbytes and a receive socket buffer of 32 kbytes.

The ATM network transfers small data units, called cells, which are 53 bytes long with a 48 byte payload. For computer communication the recommended end-to-end ATM service is through an ATM Adaptation Layer (AAL), either AAL3/4 or AAL5. The adaptation layer aggregates cells into much larger data units which are more efficiently handled by higher layers and better match the application data units. The TCP/IP protocols use the size of these AAL data units - the ATM network MTU (Maximum Transmission Unit) - to compute the Maximum Segment Size (MSS) [5] [6]. In our measurements the AAL3/4 payload is 9244 bytes. The normal TCP/IP header is 40 bytes which means that the MSS used by TCP is 9204 bytes.

The dramatic drop in performance is caused by a deadlock situation in the TCP connection which is broken up by the 200 ms timer generated TCP acknowledgment. It causes TCP to behave as a stop-and-go protocol with one or two data segments sent every 200 ms. The deadlock occurs when the amount of data sent is not enough to trigger a TCP window update packet at the receiver, and at the same time there is not enough space in the send buffer to create a segment of size MSS bytes. Nagle's algorithm prohibits the sending of non-MSS segments if there are unacknowledged bytes. Since TCP piggybacks acknowledgments onto window updates, the connection is deadlocked until the receiver sends a timer generated acknowledgment.

The deadlock problem also exists for small MSS's and low-speed connections, but it is not so likely. Actually, it will not happen for socket send buffers which are larger than three MSS segments. Furthermore, for small MSS's the discrepancy between the normal and degraded throughput is not that big, which makes it difficult to detect or uninteresting to investigate into. We know of one paper reporting on performance degradation due to the interaction between the delayed acknowledgment strategy and Nagle's algorithm, namely Crowcroft et al [7]. They report on a boundary effect of remote procedure call (RPC) response time over TCP and ethernet.

The most straightforward way to *prevent* many of the deadlock situations is to switch off Nagle's algorithm. As will be presented, there is hardly any performance penalty having it switched off. A straightforward *avoidance* solution is to ensure that the send socket buffer is equal or greater than the receive socket buffer or than three MSS's. These and other alternatives which require small changes in the TCP implementation will be discussed.

This paper explains what causes the deadlocks and discusses some alternatives to solve the underlying problems. As many of today's TCP implementations incorporate many of the same SunOS optimization mechanisms, we feel that this should be of interest to a broader audience than the SunOS users. The rest of this paper is outlined as follows. Section two summarizes the socket layer, the TCP protocol, and operating system and implementation issues of importance to understand the protocol behavior. Section three goes into a detailed description of the cause of throughput degradation. Section four discusses possible solutions. Section five contains conclusions. The reader with experience on TCP implementations in Unix BSD environments may want to skip the next section.

2. TCP IN A BSD-BASED UNIX ENVIRONMENT

In most Unix systems, the transport and lower layer protocols are implemented as part of the kernel. There are several reasons for doing this, see for example [8] [9]. The user data to be transmitted is located in user space. On a write system call, user data in the write call is copied from application address space to kernel address space so that TCP and other protocols can do the further processing of the data. Similarly, on reception the requested amount of user data is copied from kernel address space to application space.

TCP is a bidirectional protocol. It establishes a connection and each peer informs the other about its current window size. The window size refers to the number of bytes rather than the number of packets. The sender sends segments which could be as small as a single byte and as large as 64 kbytes. However, TCP sets an MSS per connection. It is normally set to the network MTU minus the size of the TCP/IP header [5]. The MTU is 1500 bytes for ethernet and 9244 bytes for our FORE ATM SBA-100 driver version 2.0. The normal TCP/IP header is 40 bytes which means that the MSS for the measurements in this paper is 9204 bytes.

2.1. Network memory management in the socket layer

In BSD based systems, as for instance SunOS 4.1.1, the socket layer acts as the interface between the application in user space and the protocols within the kernel. The socket layer offers an application programming interface, i.e. system calls such as write and read. A system provided identifier is used in the system calls in order to identify different connections with application processes. Associated with this identifier are two socket data buffers, one for data to the kernel (write) and one for data to the application (read). Each socket buffer consists of

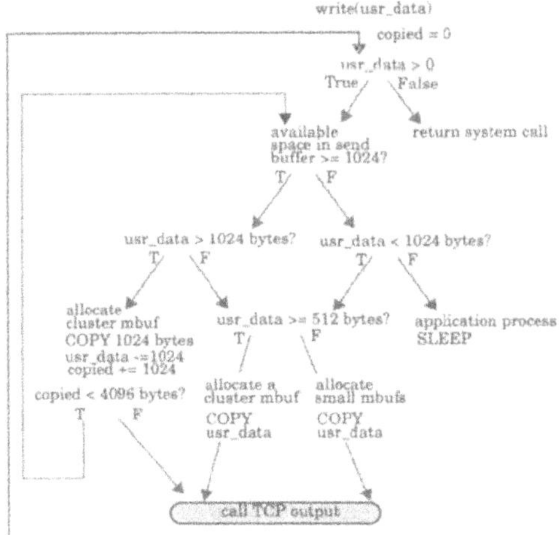

Figure 1. Transmit socket layer data copy strategy

an ordered chain of mbufs [10]. An mbuf is a data structure used by all kernel protocols in SunOS 4.x and by many other BSD systems as well. Data to and from the application is copied to and from these mbuf chains. Associated with the socket identifier is a variable which holds the number of used mbufs in each direction and a variable for the current amount of bytes in these chains of mbufs. The user can set a maximum allowed number of bytes in these chains by using the SO_RCVBUF and SO_SNDBUF socket options.

The application process is put to "sleep" if the socket layer is unable to copy all the application data in the write system call into the buffer. For further progress it has to wait until buffer space is released.

There are two types of mbufs, the "small" mbufs which hold 112 bytes of data and "cluster" mbufs which can take 1 kbyte of data [10]. Whenever possible, the system tries to use cluster mbufs when copying data from user address space to the socket mbuf chain roughly as illustrated in Figure 1. Use of cluster mbufs means that the system can avoid copy operations within the kernel by using a pointer and a reference count instead. Note in Figure 1 that the TCP protocol output routine is called either after all data in the write system call has been copied or after 4096 bytes of data have been copied, whatever occurs first. The motive for copying 4096 bytes into the socket send buffer is to exploit parallelism between the kernel protocol execution and the network interface packet transmission. However, this is only true for networks with a maximum transmission unit (MTU) smaller than 4096 bytes.

2.2. TCP acknowledgment strategy and flow control

TCP's end-to-end flow control is through a sliding window mechanism where the receiver announces its free buffer space to the transmitter. Therefore, when data is copied to the application the receiver checks if a window update packet should be returned. The algorithm for sending a window update [10] is roughly depicted in Figure 2. An update is sent if the window can slide more than either a) 35% of the receive buffer size or b) two MSS segments.

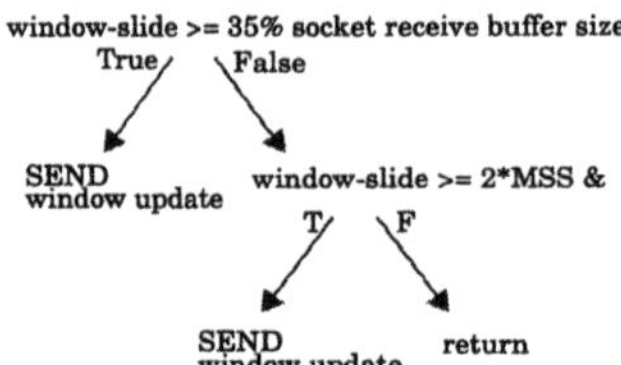

Figure 2. TCP window update algorithm

Both the socket send and receive buffers limit the amount of data that can be outstanding between two communicating TCP peers. The available space (maximum - current amount of data bytes) of the socket receive buffer is used to set the announced TCP window size to ensure that the sender will not send more data than can be received. The send socket buffer is used as the repository for TCP segments in case of retransmissions. Since data bytes remain in the send socket buffer until they are acknowledged, the available space for copying in new data into the socket send buffer is further limited.

The receiving peer sends acknowledgments to the sending peer. In case of lost segments, a time-out function at the sender will generate a retransmission of the unacknowledged bytes. Acknowledgments are delayed until they can be piggybacked onto either data segments in the

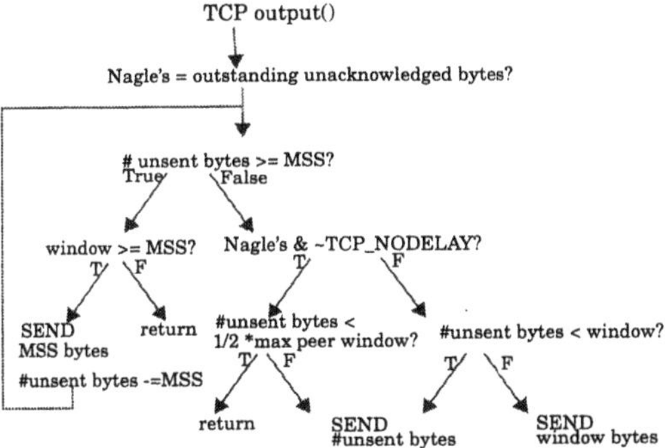

Figure 3. TCP segmentation of the send socket buffer

reverse direction or on window update packets [11]. (A compile option can set TCP to acknowledge each incoming segment.) In addition, explicit acknowledgments are cyclically generated every 200 ms. These 200 ms spaced timer generated acknowledgments are independent of the point of time of connection set-up or the last (not necessarily final) segment reception on this connection.

2.3. Nagle's algorithm

Nagle's algorithm [12] was introduced as a solution to the "small-packet problem". It was observed that TCP sent many small segments which resulted in unnecessary header and processing overhead. Nagle's algorithm inhibits sending TCP segments which are smaller than the assigned MSS if any previously transmitted data on the connection remains unacknowledged, see Figure 3. Nagle's algorithm can be switched off by the TCP_NODELAY option. Switching off Nagle's algorithm is necessary for applications which send small amount of data with no replies, such as a stream of mouse events which has no data in the reverse direction. Still, Nagle's algorithm is recommended on both telnet and ftp connections, and the default TCP configuration uses Nagle's algorithm.

According to Figure 3, a "small" segment is less than MSS bytes. A reflection is that in high-speed networks with large MTUs, Nagle's "small" segments are not actually small anymore. For our ATM network a "small" segment is less than 9204 bytes.

2.4. Invocation of TCP routines

On transmit the TCP output routine is initiated by the write system call. In Unix, a process calling the kernel is never preempted by another process while executing a system call [10]. The kernel call must explicitly give up the processor by a sleep call or run to completion of the system call. System calls appear *synchronously* to the application, i.e. the application process is blocked until the system call returns. The time until return includes any sleep calls while in the kernel. This means that a write system call with large user data sizes is blocked until all data are processed by both the socket and the protocol layers. As can be seen in Figure 1 the process does a sleep when there is not enough space in the socket buffer for all the application data.

System call execution might however be interrupted by the bottom half of the kernel, by hardware interrupts. They occur *asynchronously* and unrelated to the current system call processing.

The receiving application does a read system call which is blocked until there is something to read from the socket layer. On a frame arrival the network interface will receive the frame and generate a hardware interrupt. The hardware interrupt routine runs the device driver, copies data to mbufs, and thereafter initiates a software interrupt. This interrupt handling routine calls the higher-layer protocol e.g. IP which calls the TCP input routine. After TCP has processed data and put it into the receive socket buffer a wake-up call is executed which puts the application process back on the scheduling queue.

An incoming segment with window update and/or acknowledgment information may trigger new data segments to be sent in the other direction. This depends on the current number of bytes in the socket send buffer, Nagle's algorithm and the size of the announced window as previously described. The initiation of segment transfer(s) on acknowledgment reception is one of the points which causes the deadlocks. It may be anticipated that the send algorithm should strive to form as large segments as possible in order to reduce overhead. The obvious thing to do would be to copy as much as possible into the buffer before TCP output is called, especially since an acknowledgment will release buffer space. This is not the case. On the contrary, the action is to first transmit available bytes and then to ask for a refill of the buffer. The argument for doing it in this order is that the application process must be woken up and put into the scheduler queue in order to copy more data. This could cause an unacceptable delay. As a consequence TCP may send small segments. As will be shown later, this order has a more devastating consequence when Nagle's algorithm decides not to send a non-MSS segment.

3. OBSERVED TCP THROUGHPUT DEADLOCKS

All performance measurements in this paper are run by letting TCP transfer 16 Mbytes memory-to-memory between two Sun IPXs using the FORE ASX-111 ATM switch and FORE ATM SBA-100/175 cards (140 Mbit/s physical transmission) with the 2.0 device driver. The user data size of the write/read system call was 8192 bytes. The ATM network interface MTU is 9244 bytes, making TCP compute its MSS to 9204 bytes. Each reported throughput measure is an average of 25 runs. Table 1 presents the throughput for different sizes of the socket send and receive buffers.

In the following we will describe throughput degradations depicted as grey shaded entries in Table 1. S is the size of the send socket buffer. R is the size of the receive socket buffer. As can be seen from the table, dramatic drops in throughput occur when the receive buffer space is equal or less than the sender space. The slow-start [13] behavior is not an issue in these TCP performance measurements, since both the sender and receiver reside on the same IP subnetwork. The degradations in Table 1 are instead due to either the inherent delayed acknowledgment strategy, a combination of the delayed acknowledgment strategy and use of Nagle's algorithm, the sender-side silly-window avoidance rule, or cell loss in the ATM receive interface.

3.1. Classification of the throughput anomalies

In this section we will classify the shaded areas in Table 1. The degradation caused by cell loss or the silly-window avoidance effect will not be discussed in detail. The cell loss happens for large TCP windows which results in buffer overflow at the receiving ATM interface. A cell loss will cause a TCP segment retransmission. The sender-side silly-window syndrome avoidance [10] may occur when the send socket buffer size is more than MSS bytes larger than the receive socket buffer size. We will focus on the other classes.

Table 1 TCP Mbit/s throughput on an ATM network for combinations of socket buffer sizes and a user buffer size of 8192 bytes

S \ R	4k	8k	16k	24k	32k	40k	48k	52k*
4k	11.58	11.78	0.16	0.16	0.16	0.16	0.16	0.16
8k	13.31	15.02	0.16	0.16	0.16	0.16	0.16	0.16
16k	13.47	16.39	18.67	0.34	0.34	0.33	0.49	0.49
24k	13.68	16.46	19.67	19.07	19.98	19.69	1.85	6.63
32k	13.67	16.52	16.88	19.93	21.16	20.91	20.11	20.31
40k	13.07	16.33	17.02	20.02	21.12	21.01	20.58	20.66
48k	13.74	16.19	17.06	19.95	21.20	20.98	6.06	5.39
52k*	13.59	16.08	16.83	20.02	21.28	21.08	6.86	6.78

*52428 bytes, max SunOS socket buffer size

Predictable from the acknowledgment strategy

Combination of socket copy rule and Nagle's algorithm

Combination of timer acknowledgment and Nagle's algorithm

Cell loss -> packet loss

Sender-side silly-window syndrom avoidence

For the grey entries in Table 1 the throughput drop is caused by the same phenomenon, the sender cannot transmit enough data to trigger a window update and a piggybacked acknowledgment from the receiver. We have a deadlock situation where the sender refrains from sending more data, and the receiver refrains from returning an acknowledgment. This deadlock can only be resolved by the cyclically timer generated acknowledgment. Thereafter, the sender starts to send again, but after a while the connection will be back in the same deadlocked situation. Hence, the connection has a stop-and-go behavior of which data is prompted by the 200 ms cyclically generated acknowledgment. The *deadlock* throughput is decided by the cycle for timer generated acknowledgments and the amount of data transmitted until the next deadlock. (The difference in deadlock throughput is only due to the size and the number of the segments transmitted in-between connection deadlocks.)

The combinations of send, S, and receive, R, socket buffer sizes which cause these deadlock situations are marked with two shades of grey in Figure 4. For combinations in the darker area, the connection goes directly into the deadlock situation. For the lighter grey area it may take some time before it happens. It takes longer time for the S=24k R=48k/52k entries in Table 1 compared to the other entries. Due to the measure method we therefore get a higher average throughput than the deadlock throughput for these entries. When the connection gets into the deadlock situation, the throughput is 0.67 Mbit/s.

The throughput deadlocks can be partitioned into three classes depending on the reason for causing the deadlock:

- Deadlocks predictable from the window update rules
- Deadlocks caused by the socket layer data copying rules and Nagle's algorithm
- Deadlocks caused by the timer generated acknowledgment and Nagle's algorithm

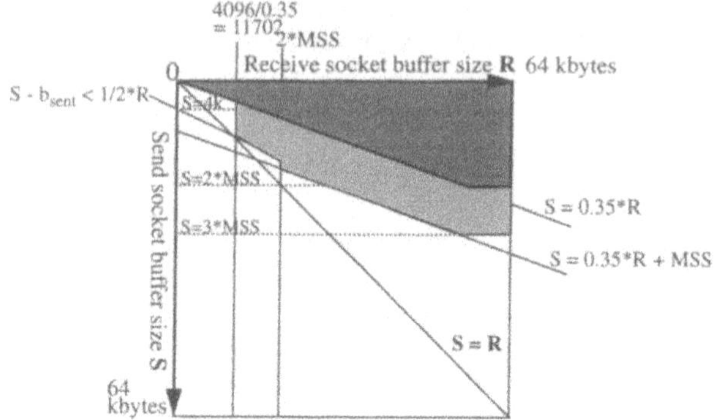

Figure 4. Anomalous socket buffer size combinations for MSS=9204 bytes

3.1.1. Deadlocks predictable from the window update rules

In this section we will discuss the dark grey area in Figure 4. In this class, the send socket buffer size is less than 35% of the receive socket buffer size, and also less than twice the MSS of the ATM network. Knowing the window update algorithm and acknowledgment strategy of TCP, these results are predictable. Even if the whole send socket buffer is sent, it is not enough to trigger a window update onto which to piggyback an acknowledgment.

The following inequalities hold between the send socket buffer and the receive socket buffer marked with the dark grey area in Figure 4:

$$(S < 35\%R) \,\&\, (S < 2*MSS) \tag{1}$$

For example, consider the entry S=8k R=24k in Table 1, which yields 0.16 Mbit/s. The socket layer copies 4 kbytes into the socket send buffer before it calls TCP. At the receiving side, these 4 kbytes are less than 35% of 24k and less than 2*MSS. Therefore, no window update will be returned. The sender can, and will copy another 4 kbytes into the send socket buffer, but due to Nagle's algorithm the sender refrains from sending these 4 kbytes until an acknowledgment has been received. There is now no more space for copying in additional data. The connection is deadlocked and the returned acknowledgment will be timer generated. When this acknowledgment is received by the sender, the sender first transmits the remaining bytes in the send socket buffer before it starts copying more data from the application to the socket buffer. Thus, the connection is stop-and-go with 4 kbytes sent every 200 ms which gives a throughput of 0.16 Mbit/s. Actually, the connection will deadlock independent of Nagle's algorithm. Even if the whole 8 kbyte send socket buffer is sent, it will still be less than 35% of the receive buffer.

Now consider the entries S=16k R=48k/52k which yield 0.49 Mbit/s. Here 3084 bytes + 9204 bytes are immediately sent as two segments when the timer generated acknowledgment releases buffer space. Otherwise the behavior is as described above.

3.1.2. Deadlocks caused by the data copying rules and Nagle's algorithm

In this class S is larger than 35%R or 2*MSS. The deadlock situations are therefore not predictable according to the TCP window update strategy. Anyhow, the TCP connection sooner or later phases into a behavior which relies on timer generated acknowledgments to resolve deadlocks.

The light grey area in Figure 4 is bounded by the dark grey area and the following inequalities between the send socket buffer, S, and the receive socket buffer, R:

$$(S < 35\%R+MSS) \ \& \ (S < 3*MSS) \tag{2}$$

For small S and R there are some boundary effects, which will be discussed later. The upper limit 3*MSS is caused by the implementation decision to first send available bytes in the socket send buffer and thereafter copy from user space. If it were done the other way round, the upper limit would instead be 2*MSS .

For example, consider the entry S=8k R=16k which yields 0.16 Mbit/s. S is big enough (50% of R) to trigger the 35% window update rule. This is what happens; The first write system call of 8 kbytes results in only 4 kbytes being copied into the send socket buffer before TCP is called, see Figure 1. Figure 5 (a) illustrates this by showing for the S=8k R=16k entry, the sender side data segment transmission and acknowledgment reception, that is, the number of outstanding unacknowledged bytes. TCP transmits a segment with 4096 bytes, and the last 4096 bytes of the write call are copied into the send socket buffer. At this point in time there are 4k unacknowledged bytes, and 4k new unsent bytes in the send buffer. Due to Nagle's algorithm these new 4 kbytes can not be transmitted since they are less than MSS. At the receiver the window can slide only 25% (4k/16k) so there is no window update to piggyback an acknowledgment onto. At this stage, the connection is deadlocked and TCP acts as a stop-and-go protocol with a window of 4096 bytes, and acknowledgments generated every 200 ms. The achieved throughput is 20 kbyte/s or 0.16 Mbit/s. After a small initial phase the entry S=16k R=40k also goes directly into deadlock and transmits 8 kbytes in-between deadlocks.

Common to these two examples is that the connection immediately gets into deadlock. A deadlock situation occurs when b_{sent} bytes are sent which are not enough to advance the window but are enough to block the sender to create a new MSS segment. Connection deadlock happens when

$$(b_{sent} < 2*MSS) \ \& \ (b_{sent} < 35\%R) \ \& \ ((S_{byte} - b_{sent}) < MSS) \tag{3}$$

S_{byte} is the number of data bytes in the socket send buffer. The inequality is illustrated in Figure 5 (b).

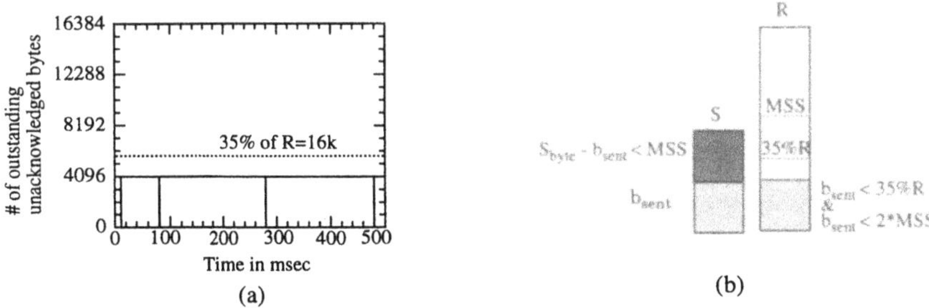

Figure 5. (a) Traces of b_{sent}, S=8k R=16k
(b) Snapshot of bytes in send and receive socket buffers at deadlock

The sender may send several segments in-between deadlocks, thus the throughput may vary. The absolute upper throughput limit after deadlock is though (35%R+MSS)/200ms.

3.1.3. Deadlock caused by the timer generated acknowledgment and Nagle's algorithm

For entries like S=16k R= 24k/32k it seems that there should be no problem because after 4 kbytes are transmitted, a full MSS of 9204 bytes can be constructed, which should prompt a window update. But assume now instead that at some point in time there will be, say, an 8 kbyte segment sent. This 8 kbyte segment is not big enough to advance the window and the available space in send socket buffer is not big enough to create a full MSS. But how could there be an 8k segment sent? In short, the timer generated acknowledgment may arrive just after that 8k has been copied into an empty send buffer. When this situation is reached, 8k is sent every 200 ms. Observe that this is caused by the fact that the implementation on reception of an acknowledgment first sends what is available in the send buffer and thereafter copies more bytes into the buffer. Appendix A gives a detailed presentation of the 0.34 Mbit/s throughput result with a 16k send buffer and a 32k receive buffer.

Figure 6 (a) depicts how a S=16k R=32k connection gets into deadlock after about 350 ms. Assume the sender has transmitted b_{sent} bytes. A timer generated acknowledgment which acknowledges b_{ack} of these b_{sent} bytes such that $(b_{sent} - b_{ack})$ satisfies (3), that is

$$((b_{sent} - b_{ack}) < 35\%R) \ \& \ ((b_{sent} - b_{ack}) < 2*MSS) \ \& \ ((S_{byte} - (b_{sent} - b_{ack})) < MSS) \qquad (4)$$

is the reason for the deadlock. Figure 6 (b) presents a snapshot of the socket send buffer in this situation. Due to the segment flow on the TCP connection, the probability of a timer generated acknowledgment to actually acknowledge b_{ack} bytes as above is very high; The connection deadlocks within 600 ms. See Appendix A for details.

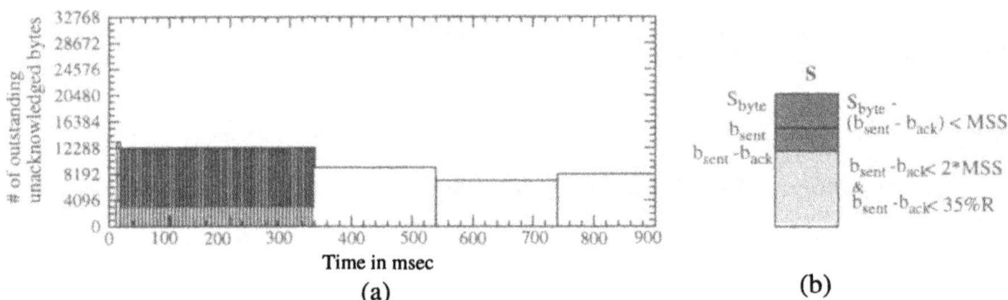

Figure 6. (a) Trace of a S=16k R=32k connection (b) Snapshot of the socket send buffer

3.1.4. Boundary effects

With the SunOS 4.1.x socket layer optimization of calling the protocol for at least every 4 kbytes results in the first segment being maximum 4 kbytes, independent of the user data in the write call as long as this is larger than 4 kbytes. This will get the connection directly into deadlock if R is larger than 4096/35% = 11702 bytes. This gives a vertical line at 4096/0.35 = 11702 bytes in Figure 4. If R < 11702 the deadlock depends on the user data size. A smaller user data size than 4k will move this boundary to the left.

Another boundary effect is as follows. When b_{sent} bytes are sent, there is potential for another S_{byte} - b_{sent} bytes to be transmitted. Depending on b_{sent} and S_{byte}, this segment may be less than MSS, and with Nagle's algorithm in use, according to Figure 3 such a small segment is sent only if

$$(S_{byte} - b_{sent} > 1/2*R) \qquad (5)$$

Inequality (5) is due to TCP sending a segment if the size of the segment is at least half the maximum advertized receive window. This was done to cope with an initial problem of the sender avoidance of the silly-window syndrome when communicating with hosts with tiny buffers, e.g. 512 bytes [10]. Thus, if S_{byte} - b_{sent} is larger than half the maximum advertized window, R, for R less than 2*MSS a small segment is transmitted independent of Nagle's algorithm. In Figure 4 (5) is drawn with b_{sent} equal to 4 kbytes.

3.1.5. Deadlocks with small MSS

The deadlocks above occur also on other networks. The larger the network interface MTU, the larger the hazardous send and receive socket size combination areas. The hazardous socket size combinations for ethernet is (without the boundary effects) depicted in Figure 7. Due to the smaller MTU the number of combinations to avoid is much smaller.

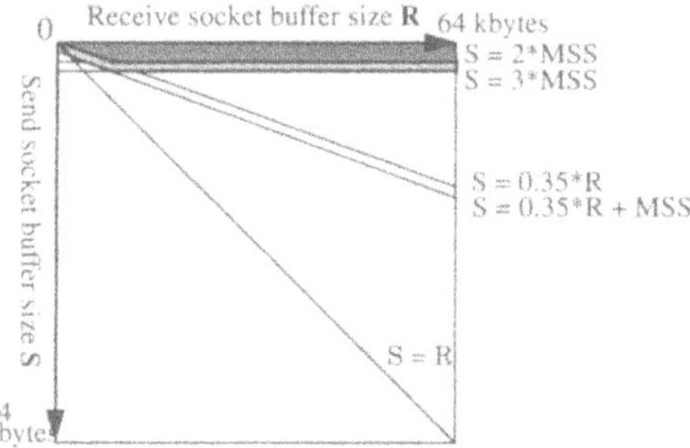

Figure 7. Ethernet anomalous socket buffer size combinations, MSS = 1460 bytes

4. DEFEATING THE DEADLOCKS

This section discusses how to avoid or prevent the deadlock situations. The obvious avoidance solution is to keep away from the dangerous S and R combinations, i.e. to ensure that S >= 3*MSS or S >= R. Another straightforward prevention solution is to turn off Nagle's algorithm. Other alternatives are to change Nagle's algorithm, the size of MSS, and to remove the 4k limit on the number of bytes copied to the socket buffer before TCP is called.

4.1. Preventing deadlocks by turning Nagle's algorithm off

Setting the TCP_NODELAY option removes the lightly shaded low throughput entries without a significant performance penalty for other socket size combinations, see Table 2 for pairwise comparison of the upper (Nagle's on) and lower (Nagle's off) rows of each entry. As expected, the throughput performance in the darkly shaded area is still doomed to be low. In some of the darkly shaded entries the throughput increase is due to the fact that the sender

Table 2 TCP Mbit/s throughput over ATM
(a) with Nagle's algorithm (b) without Nagle's algorithm

S \ R	4k	8k	16k	24k	32k	40k	48k	52k*
4k	11.58 / 11.77	11.78 / 11.78	0.16 / 0.16	0.16 / 0.16	0.16 / 0.16	0.16 / 0.16	0.16 / 0.16	0.16 / 0.16
8k	13.31 / 13.31	15.02 / 15.13	0.16 / 15.30	0.16 / 0.33	0.16 / 0.33	0.16 / 0.33	0.16 / 0.33	0.16 / 0.33
16k	13.47 / 13.47	16.39 / 16.23	18.67 / 18.76	0.34 / 18.98	0.34 / 19.31	0.33 / 19.10	0.49 / 0.66	0.49 / 0.66
24k	13.68 / 13.68	16.46 / 16.13	19.67 / 19.77	19.07 / 20.59	19.98 / 20.29	19.69 / 20.33	1.85 / 19.28	6.63 / 19.26
32k	13.67 / 13.68	16.52 / 15.88	16.88 / 16.42	19.93 / 19.74	21.16 / 20.98	20.91 / 20.92	20.11 / 19.86	20.31 / 19.84
40k	13.07 / 13.44	16.33 / 16.27	17.02 / 16.91	20.02 / 19.87	21.12 / 20.94	21.01 / 20.89	20.58 / 20.45	20.66 / 20.70
48k	13.74 / 13.07	16.19 / 16.10	17.06 / 16.47	19.95 / 19.53	21.20 / 20.92	20.98 / 20.80	6.06 / 17.53	5.39 / 17.17
52k*	13.59 / 13.75	16.08 / 15.65	16.83 / 16.76	20.02 / 19.78	21.28 / 21.21	21.08 / 20.73	6.86 / 14.84	6.78 / 14.39

*52428 bytes, max SunOS socket buffer size

aa.aa / bb.bb: (a) Nagle's algorithm on (b) Nagle's algorithm off

Predictable from the acknowledgment strategy

Combination of socket copy rule and Nagle's algorithm

Combination of timer acknowledgment and Nagle's algorithm

transmits the whole send buffer size in-between deadlocks. The throughput degradation due to packet loss is reduced, since the receiver copes better with a more even distribution of (smaller) packets in time.

4.2. Incorporating smaller changes in TCP

To cope with the deadlock due to the timer generated acknowledgment and Nagle's algorithm requires changes to current implementations. We identify two possible changes which we think have little impact on TCP and existing implementations:

- A change in Nagle's algorithm in order to allow small packets to be transmitted if there are more than MSS outstanding unacknowledged bytes. This is a small adjustment which implies expanding Nagle's statement with a conditional test on the *number* of unacknowledged bytes.
- Set the MSS to a power-of-two multiple of the 1024 byte cluster page size. Protocol and network memory management optimizations are thereby performed on corresponding chunk sizes.

In the following we present performance results after the changes above.

4.2.1. A change in Nagle's algorithm

By changing Nagle's algorithm to allow small segments we will significantly reduce the probability of entering deadlock situations. As can be seen from the upper line of each of the entries in Table 3, this change to Nagle's algorithm significantly improves the throughput performance of the S=16k R=24k/32k and S=24k R=48k/52k entries. These are the cases of which the timer generated acknowledgment gets the connection into an anomalous behavior.

Table 3 TCP throughput over ATM in Mbit/s,
(a) with a change to Nagle's algorithm (b) TCP MSS of 8192 bytes

S \ R	4k	8k	16k	24k	32k	40k	48k	52k*
4k	11.94 12.21	11.88 11.72	0.16 0.16	0.16 0.16	0.16 0.16	0.16 0.16	0.16 0.16	0.16 0.16
8k	13.08 13.47	15.24 15.01	0.16 0.16	0.16 0.16	0.16 0.16	0.16 0.16	0.16 0.16	0.16 0.16
16k	13.14 13.74	16.35 16.72	18.93 18.95	13.01 17.80	11.16 17.83	0.49 0.49	0.49 0.49	0.49 0.49
24k	13.17 13.76	16.35 16.50	19.01 19.08	19.61 20.80	20.24 20.55	20.27 20.46	20.30 19.14	20.32 19.06
32k	13.16 13.75	16.21 16.46	17.07 20.56	19.92 21.10	21.21 21.14	20.91 21.02	20.92 20.59	20.69 20.62
40k	13.19 13.76	16.30 16.02	16.78 20.19	19.92 20.86	21.19 21.22	20.96 21.07	20.67 20.87	20.57 21.02
48k	13.21 13.80	16.57 16.86	17.41 21.12	20.56 21.39	21.53 21.29	20.61 21.08	15.96 16.43	16.48 15.23
52k*	13.07 13.74	16.76 16.83	17.77 21.17	20.66 21.47	21.37 21.24	20.97 21.04	12.89 19.94	12.12 18.97

*52428 bytes, max SunOS socket buffer size

aa.aa Nagle's change
bb.bb TCP MSS 8 kbytes

Predictable from the acknowledgment strategy

Combination of socket copy rule and Nagle's algorithm

The lack of improvement of the S=8k R=16k and S=16k R=40k entries is due to the socket layer 4k copy rule. The average throughput reported in the S=16k R=24k/32k entries is improved since it now takes longer time to get into the deadlock.

4.2.2. A 8192 byte TCP MSS

A TCP MSS of 8192 bytes can be achieved by either TCP rounding the computed MSS to its nearest lower integer multiple of 1024 bytes, or by setting the ATM MTU to 8232 bytes. In the first case the TCP code must be changed. The second case requires a change in the ATM network driver. Setting the ATM MTU to 8232 bytes makes TCP compute its MSS as 40 bytes less, that is 8192 bytes.

The bottom line of each of the entries in Table 3 present the throughput results with the TCP MSS of 8192 bytes. The S=16k R=48k/52k entries are now classified as medium grey, since S now is big enough to hold two MSS segments. The throughput of these entries and the S=8k R=16k and S=16k R=40k entries remain low because it is the socket layer copy optimization which causes this low throughput. The S=16k R=24k/32k and S=24k R=48k/52k entries have improved significantly. Note that as a side effect an MSS of 8192 bytes improves the throughput for the sender-side silly-window-avoidance entries, and the packet loss entries.

4.3. No explicit 4k limit in the socket copy rules

A change to the socket layer removes the explicit limit of 4 kbytes copied to the socket buffer before the protocol is called. The inevitable limit is now only the available space in the socket buffer. This change makes the possibility of deadlock more dependent on the user data size, but does not prevent the deadlocks. With a user data size of 8 kbytes as in the previous measurements this change avoids only the S=8k R=16k deadlock. On the contrary, with a user data size of 4 kbytes the connection would deadlock as described earlier.

5. CONCLUSIONS

In this paper we have pointed out common socket buffer size settings which force TCP into deadlock situations over high-speed networks with large MTUs. The number of socket size combinations increases with an increasing MTU relative to the send and receive socket buffer sizes. The window update rules and Nagle's definition of a "small" segment are of importance. On high-speed networks with large transmission units, the unit which Nagle's algorithm considers small is not small anymore.

The predictable deadlocks are due to the TCP window update rules, the other deadlock situations are due to:

- A bad interaction between Nagle's algorithm and the delayed acknowledgment strategy
- The socket layer copy rule (apparently optimized for ethernet)
- The action sequence of the TCP implementation on acknowledgment reception
- The explicit timer generated acknowledgment which operates independent from the rest of the protocol

We presented ways of getting round the throughput deadlocks. The conclusions from these are

- A send socket buffer equal or larger than three MSS's or equal or larger than the receive socket buffer avoids all deadlocks.
- Turning off Nagle's algorithm prevents non-predictable deadlocks.
- A change to Nagle's algorithm to allow small segments to be transmitted if there are at least MSS unacknowledged bytes, greatly reduced the probability of connection deadlock.
- Setting the MSS to a power-of-two multiple of 1 kbyte causes segment sizes to be more in line with the internal network memory management rules and common socket buffer sizes. To avoid some of the deadlocks an appropriate TCP [1] MSS size is 8k.

REFERENCES

1. Postel, J. Transmission Control Protocol, protocol specification. RFC 793, September 1981.
2. Comer, D. Internetworking with TCP/IP, principles, protocols, and architecture. Englewood Cliffs, NJ, Prentice-Hall. ISBN 0-13-470188-7. 1988.
3. V. Jacobsen, B. Braden, and D.Borman "TCP extensions for high-performance". RFC 1323, May 1992.
4. Partridge, C. Gigabit networking. Addison-Wesley. ISBN 0-201-56333-9. 1993.
5. Postel, J. The TCP maximum segment size and related topics. RFC 879, November 1983.
6. Braden, R (ed). Requirements for Internet hosts - communication layers. RFC 1122, October 1989.
7. Crowcroft, J, Wakeman, I, Wang, Z, and Sirovica, D. Is layering harmful? IEEE Network, 6 (1), 20-24, January 1992.
8. Clark, D D. Modularity and efficiency in protocol implementation. RFC 817, July 1982.
9. Mogul, J C, Rashid R F, and Accetta, MJ. The packet filter: an efficient mechanism for user-level network code. Proc. of ACM SOSP, 1987, 39 - 51.

10. Leffler, S J et al. 4.3 BSD Unix operating system. Reading, Mass., Addison-Wesley. ISBN 0-201-06196-1. 1989.
11. Clark, D D. Window and acknowledgment strategy in TCP. RFC 813, July 1988.
12. Nagle, J. Congestion control in TCP/IP internetworks. RFC 896, January 1984.
13. Jacobsen, V. 1988. Congestion avoidence and control. Proc. of ACM SIGCOMM'88, 314-329, Palo Alto, USA, 16-19 August 1988.

APPENDIX

A Throughput with a 16k send and a 32k receive socket buffer

In the following we will describe in detail how a throughput deadlock arises with a socket send buffer of 16 and a socket receive buffer of 32 kbytes. The diagram in Table A.1 is used to illustrate TCP internal actions and state and the packet flow on the network.

The sender is a loop of write(8k) calls, while the receiver is a loop of read(8k) calls. The data segments are represented with $DATA_X$ where X is the number of user bytes in the packets, and acknowledgments with ACK_Y where Y is the number of bytes the packets acknowledge.

The data transfer phase starts with a write(8k). The socket layer copies 4 kbytes into the socket buffer before it calls the TCP protocol through tcp_output(). A packet with 4 kbyte user data is transmitted. After the return from the tcp_output() routine, the socket layer continues by copying more bytes from the current and next write(8k) call until the socket buffer is full. At this stage, the call produces a segment of length 9204 (MSS) bytes. 13300 bytes have now been transmitted, but are not acknowledged. The last 3084 bytes in the send socket buffer are not transmitted since they are less than MSS (line 7).

After the segments have arrived at the receiver, the application reads them in two chunks. After the second read(), the window can slide 13300/32768 = 40.6%, and a window update with an acknowledgment is returned. The acknowledgment releases 13300 bytes in the send socket buffer and acknowledges all outstanding bytes. When the window update is received, TCP first sends the remaining 3084 data bytes in the send socket buffer before the application is scheduled to run. Thereafter, another segment of MSS bytes will be transmitted. Due to the generated segment size pattern on the sender side, an acknowledgment will be returned immediately after having received a 9204 byte segment, since the window will slide more than 35%.

The sequence of sending a 3084 byte and a 9204 byte segment followed by an acknowledgment is repeated until the sender receives a timer generated acknowledgment, TO. Such an acknowledgment is generated on line 21, and it arrives at the sender on line 25 and acknowledges less than MSS bytes.

The timer generated acknowledgment on line 21 releases only 3084 bytes in the socket send buffer when it is received on line 25. Thus, in the socket send buffer, 9204 bytes still remain unacknowledged. The 3084 bytes ready to be transmitted on line 25 are not sent due to Nagle's algorithm. The reception of the acknowledgment wakes up the application. Another 4 kbytes can be copied into the send buffer which now contains 16 kbytes. There are 7180 (16384-9204) bytes left in the socket send buffer ready to be transmitted. Due to Nagle's algorithm, no segment is transmitted because there are outstanding unacknowledged bytes.

On line 28 a new 200 ms spaced timer generated acknowledgment is returned. It acknowledges 9204 bytes and at this time the sender does not have any outstanding unacknowledged bytes. Therefore, the remaining 7180 bytes in the send socket buffer are

Table A.1 ATM 16k send and 32k receive socket buffer, user data size of 8 kbytes

	TRANSMIT 8192 bytes in each system call	Socket copy before TCP call	From TCP to proc in sleep	Bytes in socket send buffer	Number of unacked bytes	PACKETS on the wire	Bytes in socket receive buffer	From TCP to proc in sleep + TO	RECEIVE maximum 8192 bytes in each system call
1	write(8k)	4096		4096	4096	>> DATA$_{4096}$ >>	4096		
2		4096		8192	4096		4096		
3	write(8k)	4096		12288	4096		4096		
4		4096		16384	13300	>> DATA$_{9204}$ >>	13300	wkup	
5	write(8k)	sleep			13300		5180		read(8192)
6			wkup	3084	0	<< ACK$_{13300}$ <<	0		read(5108)sleep
7				3084	3084	>> DATA$_{3084}$ >>	3084	wkup	
8		4096		7180	3084		0		read(3084)sleep
9		4096		11276	3084		0		
10	write(8k)	4096		15372	12288	>> DATA$_{9204}$ >>	9204	wkup	
11		sleep		15372	12288		1012		read(8192)
12			wkup	3084	0	<< ACK$_{12288}$ <<	0		read(1012)
13				3084	3084	>> DATA$_{3084}$ >>	3084	wkup	
14		4096		7180	3084		0		read(3084)
15	write(8k)	4096		11276	3084		0		
16		4096		15372	12288	>> DATA$_{9204}$ >>	9204		
17	write(8k)	sleep		15372	12288		1012		read(8192))
18			wkup	3084	0	<< ACK$_{12288}$ <<	0		read(1012)
19				3084	3084	>> DATA$_{3084}$ >>	3084		
20				3084	3084		0		read(3084)sleep
21		4096		7180	3084	ACK$_{3084}$ <<	0	TO	
22		4096		11276	3084		0		
23	write(8k)	4096		15372	12288	>> DATA$_{9204}$ >>	9204	wkup	
24		sleep		15372	12288		9204		read(8192)
25			wkup	12288	9204	<< (ACK$_{3084}$)	0		read(1012)sleep
26		4096		16384	9204		0		
27	write(8k)	sleep		16384	9204		0		
28			wkup	7180	0	<< ACK$_{9204}$ <<	0	TO	
29				7180	7180	>> DATA$_{7180}$ >>	7180	wkup	
30		4096		11276	7180		0		read(7180)sleep
31		4096		15372	7180		0		
32	write(8k)	sleep		15372	7180		0		
33			wkup	8192	0	<< ACK$_{7180}$ <<	0	TO	
34				8192	8192	>> DATA$_{8192}$ >>	8192	wkup	
35		4096		12288	8192		0		read(8192)sleep
						.			
38		4096		16384	8192		0		
39	write(8k)	sleep		16384	8192		0		
40			wkup	8192	0	<< ACK$_{8192}$ <<	0	TO	
41				8192	8192	>> DATA$_{8192}$ >>	8192	wkup	
42		4096		12288	8192		0		read(8192)sleep
43		4096		16384	8192		0		
44	write(8k)	sleep		16384	8192		0		
45			wkup	8192	0	<< ACK$_{8192}$ <<	0	TO	
46				8192	8192	>> DATA$_{8192}$ >>	8192	wkup	
47		4096		12288	8192		0		read(8192)sleep
48		4096		16384	8192		0		
49			wkup	8192	0	<< ACK$_{8192}$ <<	0	TO	
50				8192	8192	DATA$_{8192}$	4096	wkup	
51				8192	8192		0		read(8192)sleep
52				0	0	ACK$_{8192}$	0	TO	

transmitted in one segment. At the receiver this segment (line 29) does not trigger a window update. At the sender side, there are now only 7180 bytes in the send socket buffer. Hence, there is space for another 9204 (16384 - 7180) bytes, that is MSS bytes! The acknowledgment on line 28 wakes up the application and twice are 4 kbytes copied into the send socket buffer. A new segment is not transmitted, since its length is less than MSS. The write call on line 32 immediately does a sleep because there is not enough free space in the socket buffer to copy an additional 1024 bytes. This is the case even if there is space for an additional 1012 bytes which would create a full MSS, see Figure 1. At this point in time, there are 15372 bytes in the socket buffer of which 7180 have been transmitted.

The next action is the timer generated acknowledgment on line 33. It acknowledges 7180 bytes. On reception of this acknowledgment at the sender, 8192 bytes are transmitted. Again, no progress is made since the sender can not send an MSS segment. The receiver does not return an acknowledgment as the window can slide only 25%. We have reached the point at which 8192 bytes will be sent every 200 ms. This behavior is repeated (lines 38 thru 52) and gives a throughput of about 40 kbyte/s, or 0.32 Mbit/s.

Figure A.1 presents traces of the data segment transmission and acknowledgment reception on a S=16k R=32k connection. (a) presents the segment transfer until the connection deadlocks, (b) is a magnified vertical slice of the initial segment flow of (a), and (c) presents the timer generated acknowledgment. Taking a closer look on Figure A.1 (b) it is evident that the connection will deadlock with a probability of 1 within 600 ms. The sender transmits the 3084 byte segments approximately 6 ms apart. About 1.7 ms after the transmission of the 3084 byte segment, the 9204 byte segment is transmitted. Within each 200 ms cycle, the time interval the receiver has only 3084 unacknowledged bytes is shifted by approximately 2 ms relative to the time of the timer generated acknowledgment. That is, within the first few 3-5 timer generated acknowledgments the connection is deadlocked.

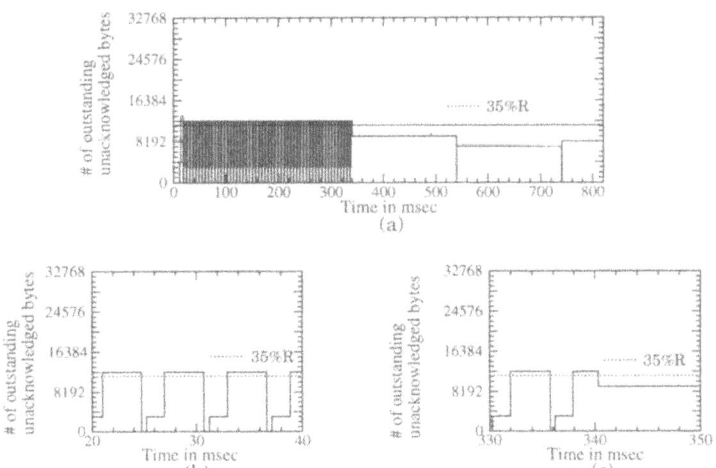

Figure A.1 Trace of data segments and acknowledgments, S=16k R=32k

16

A Guaranteed-Rate Channel Allocation Scheme and Its Application to Delivery-on-Demand of Continuous Media Data[1]

T. Kameda[a], J. Ting[b] and D. Fracchia[c]

[a]School of Computing Science, Simon Fraser University, Burnaby, B.C., Canada V5A 1S6

[b]Bell Northern Research, P.O.Box 3511, Station C, Ottawa, Ontario, Canada K1Y 4H7

[c]School of Computing Science, Simon Fraser University, Burnaby, B.C., Canada V5A 1S6

Abstract

We propose a new reservation-based bandwidth allocation method to satisfy the requirements of real-time data. It reserves bandwidth in the form of uniformly distributed time slots on a link. It is conceptually simple and should be easy to implement. Delay-insensitive or loss-insensitive data can be transmitted concurrently, using the unreserved bandwidth.

As an example of its application, we discuss the on-demand delivery of pre-recorded, continuous media data. Given the file size, the playback rate, and the buffer space available at the client site, we compute the range for data transmission rate that prevents client buffer underflow and overflow. Based on this range, our channel allocation scheme can reserve bandwidth along the connection from the server to the client. We first discuss in detail the case where data are not compressed, so that all playback-frames of data consist of the same number of bytes, and then report similar results on the compressed data case, in which different playback-frames may consist of different numbers of bytes.

Keyword Codes: C.2.3; H.3.5; H.5.1
Keywords: Network Operations; Online Information Services; Multimedia Information Systems

1. INTRODUCTION

1.1 ATM and Real-Time Applications

With the recent advances in digital communication technology, it has become possible to transmit many different types of information in digital form in an integrated network, sharing the network resources. The *Asynchronous Transfer Mode* (ATM) of *Broadband Integrated Systems Digital Network* (B-ISDN) [11, 19] is intended to take advantage of the reliability and fidelity of modern digital communication facilities, in order to provide much faster packet switching than has been available in the past. ATM has often been

[1]This work was partially supported by grants from the Natural Sciences and Engineering Research Council of Canada and the Advanced Systems Institute of British Columbia.

hailed as a panacea for all communication needs. However, it's been realized that, by its very nature, it is not straightforward for ATM to deliver the kind of quality-of-service required by many real-time applications (e.g., [3, 7, 9]). A great deal of effort has recently been made to address this problem (e.g., [7, 10, 13, 20]). We propose a new bandwidth allocation scheme, which represents our contribution towards solving this problem.

Although our allocation scheme has a wide range of applications, we focus on the delivery-on-demand of pre-recorded, continuous media data. It is envisioned that consumers will be able to select video programs in the comfort of their home, instead of renting video tapes from a video rental store. The simplest method for delivering a movie would be to make the client wait until the entire movie had been transmitted to the buffer of the client's machine before viewing the movie. This approach is not satisfactory if the client does not want to wait that long, or there isn't enough storage in the client's machine to store the entire movie. In such situations, we need to start viewing the movie before the data transfer is complete.

In this paper, we discuss in some detail:

1. a new bandwidth allocation scheme for a communication link that meets strict real-time requirements without packet loss, and
2. its application to the on-demand delivery of pre-recorded, continuous media data (compressed at variable rate) for real-time playback, when the client has a finite storage capacity.

1.2 Previous Work

In addition to those contained in the papers we cited above, a number of other solutions have been proposed for the bandwidth allocation for real-time data transfer in delay-sensitive packet-switching networks [4, 5, 14, 15, 16], and storage and retrieval of multimedia information [2, 6]. However, even in the absence of failures and errors, meeting stringent real-time deadlines with packet-switching is difficult, and most of the solutions cannot offer 100% guarantees.

In most real-time applications, end-to-end delay and jitter are probably the most important quality-of-service parameters, while minimizing the amount of buffer required inside the network is perhaps of secondary importance. However, depending on the application, the relative importance of these parameters varies. For one-way, real-time video or voice transmission, for example, the absolute amount of delay is, in most cases, less important than the jitter. In this section, we first review some representative bandwidth allocation and scheduling schemes to put our contribution in perspective. Based on a recent paper by Li, et al. [10], we compare several schemes with respect to three criteria: (i) maximum end-to-end queueing delay (excluding the propagation delay), (ii) maximum jitter, and (iii) the worst case buffer requirement at a network node. In Table 1 below, the following parameters are used.

- B – Link bandwidth (bits/sec)
- r_m – smallest connection rate (the transmission rate of the narrowest bandwidth assigned to a connection)
- h – route length (in number of hops)

	Max queueing delay	Jitter	Buffer required per node
Stop-and-Go[7]	$2h(l/r_m)$	$h(l/r_m)$	$2B(c/r_m)$
VirtualClock[20]	$h(l/r_m)$	$h(l/r_m)$	$B(c/r_m)$
Immediate forwarding[10]	hT_f	T_f	BT_f
2-frame choice[10]	$2hT_f$	hT_f	$2BT_f$
Binary Preallocation	$h(c/B)$	0	c

Table 1: Comparison of various schemes
(All but the last row is from [10].)

- c – cell size (in ATM, $c = 53 \cdot 8 = 428$ bits)
- l – cell payload size (in ATM, $l = 48 \cdot 8 = 384$ bits)
- T_f – frame time

ATM service is based on the transmission of small, fixed-size packets called *cells*. In this paper, we use the terms, cell and packet, interchangeably.

We refer the reader to the discussion of the first four rows of the above table to reference [10]. It is commented in [10] that *Stop-and-Go queueing* has a limitation in that its frame time (or scheduling interval) must be fixed at c/r_m, which in turn determines the maximum queueing delay for all connections. If a voice connection is to be made at 64Kb/sec through an ATM network, for example, $l/r_m = (48 \cdot 8/64 \cdot 10^3)sec = 6msec$, so that the maximum queueing delay per node is $12msec$. Delays of this magnitude may be intolerable for other applications with higher-speed connection rates.

The basic idea of the *Immediate forwarding* scheme is that since the maximum queueing delay, for example, is proportional to the frame time T_f, T_f should be kept small, independently of the desired minimum connection rate, r_m. If r_m satisfies $kT_f = c/r_m$, then the minimum-rate connection can use every kth frame to transmit its cells.

The last row of the above table summarizes the properties of *Binary Preallocation* that we propose in Section 3 of this paper. They are derived based on the assumption that all the local clocks of the nodes tick at exactly the same rate but they are not synchronized with each other [8, 12], so that the cell slots of an input link are not necessarily aligned with those on the corresponding output link. If they were perfectly aligned, taking the cell processing time at the node into account, then both the max queueing delay and buffer required per node would be 0 (the queueing delay does not include the cell processing time). Another important feature of our scheme is that it imposes no lower bound, r_m, on the available connection rate, so that an arbitrarily low connection rate can be provided without causing any wasted bandwidth.

Gemmell [6] discusses the retrieval from disk and playback of continuous media data. His model (which doesn't include a network) is developed specifically for the real-time requirements of digital audio playback. He shows how to describe these requirements in terms of the consumption rate and the data-retrieval rate. Lower bounds are derived on buffer space for certain common retrieval scenarios. Gemmell's work motivated us to investigate a more general case, where there is a network between the source and the playback unit. When pre-recorded, compressed video data are transmitted over a network and played back at the destination, a variable number of cells are consumed per playback-frame. To analyze such a system, we use a model in which the playback unit has a variable playback rate within a certain range.

The rest of the paper is organized as follows. In the next section, we present our model and define some useful terms. In Section 3, we propose a new bandwidth allocation scheme, called *Binary Preallocation*, mentioned above. The remaining sections discuss an application of *Binary Preallocation* to the on-demand delivery of pre-recorded, continuous media data. In Section 4, we first consider the buffer overflow and underflow problems for uncompressed data, and derive tight lower bounds on the required buffer size. Based on these bounds, we then derive the feasible range for the data transfer rate. Section 5 will investigate those problems for compressed data.

2. THE MODEL

We model a network by a directed graph, in which the directed edges represent unidirectional communication links. A server is represented by a node of outdegree 1 and indegree 0, and a client is represented by a node of outdegree 0 and indegree 1. All other nodes participate in transmitting packets (cells).

To model ATM, we assume that all cells sent over the network have the same size, and the bandwidth of a communication link is expressed in terms of the number of cells it can transmit per second. The unit of playback is called a *playback-frame*, and it consists of a sequence of cells. We further assume that the capacity of a link ℓ is of the form $R_0/2^{h(\ell)}$ (cells/sec), where R_0 (cells/sec) is a constant for the network but the non-negative, dimensionless integer $h(\ell)$ depends on the particular link ℓ.

Each request by a client is characterized by the following three parameters.

1. F: the file size (in the number of cells it consists of).
2. B: the buffer space (in the number of cells that it can store) available for this request at the client's site.
3. R_c: the *consumption* (or *playback*) *rate*, i.e., the rate of playback in cells/sec.

A service for a multimedia on-demand application can be provided in three phases: the *connection set-up, data transmission* and *clear-up* phases. We now elaborate on the connection set-up phase. Upon receiving a request from a client, a network node calculates the feasible transfer rate range and then forwards the request to the appropriate server. The network then uses a routing algorithm to find a route between the server and the client, which is called a *global logical channel* or *connection.* Every link on this connection has a *local logical channel* or, simply, a *channel*, corresponding to this global logical channel. To ensure that there are enough resources in every node (processing capacity and buffer space) and link (bandwidth) along a global logical channel, a routing algorithm performs an *admission test* at each node along the path it is trying to find. If the admission test succeeds, the required resources at the node are temporarily reserved for the request. Finally, if a feasible connection is found, the network accepts the request and proceeds to the data transmission phase. Otherwise, the request is rejected, and the temporarily reserved resources are released.

If the delivery rate of playback-frames from the network is lower than the consumption rate, R_c, playback-frames must first be prefetched and buffered, before playback can start.

We assume that, while playback is in progress, one playback-frame in the buffer is used exclusively for playback and no part of it should be modified. If the buffer does not hold the next playback-frame when the playback of the current playback-frame is complete, playback is disrupted. We call this problem the *buffer underflow problem.* On the other hand, if the playback-frame delivery rate is higher than R_c, unless enough buffer space is provided, the arriving cells may overflow the buffer before they are played back. We call this the *buffer overflow problem.* Gemmell [6] shows that starting playback as soon as possible minimizes the required buffer space and vice versa.

In the rest of this paper, a frame will mean a playback-frame (i.e., not the time frame we discussed in the previous section).

3. BANDWIDTH ALLOCATION

Given that we want to use ATM for all applications, the problem we face is how to support synchronous applications within the framework of ATM.

We divide the time on each link into *cell slots*, whose size is the time needed to output a single cell over that link. Suppose link ℓ can transmit $R(\ell) = 2^{-h(\ell)}R_0$ cells/sec. Number the cell slots of link ℓ as 0,1,2, ... We define the subchannel consisting of the even-numbered cell slots, i.e., 0, 2, 4, ..., and call it C_0^ℓ. So, for $i = 0, 1, 2, \ldots$, the ith cell slot belonging to C_0 (we omit the superscript ℓ, when it is clear from the context) is given by

$$C_0(i) = 2i.$$

Similarly, C_1 consists of the odd-numbered cell slots, i.e.,

$$C_1(i) = 2i + 1.$$

In general, the subscript α in C_α is a binary string whose length is denoted by k, i.e., $\alpha = d_1 d_2 \cdots d_k$. When $k = 0$, $\alpha = \Lambda$ (the null string). Formally, C_α can be defined recursively as follows: The logical channel consisting of the total bandwidth of link ℓ, i.e., C_Λ^ℓ, is defined by

$$C_\Lambda(i) = i.$$

Given $C_{d_1 d_2 \cdots d_{k-1}}$,

$$C_{d_1 d_2 \cdots d_{k-1} 0}(i) = \text{ the } i\text{-th even cell slot in } C_{d_1 d_2 \cdots d_{k-1}} = C_{d_1 d_2 \cdots d_{k-1}}(2i), \tag{1}$$

and

$$C_{d_1 d_2 \cdots d_{k-1} 1}(i) = \text{ the } i\text{-th odd cell slot in } C_{d_1 d_2 \cdots d_{k-1}} = C_{d_1 d_2 \cdots d_{k-1}}(2i + 1). \tag{2}$$

From the above definition, we can derive the following formula.

$$C_{d_1 d_2 \cdots d_k}(i) = 2^k \cdot i + d_1 + d_2 \cdot 2 + \cdots + d_k \cdot 2^{k-1}. \tag{3}$$

Binary Preallocation Algorithm (BPA)

Let R (cells/sec) be the requested bandwidth on link ℓ.

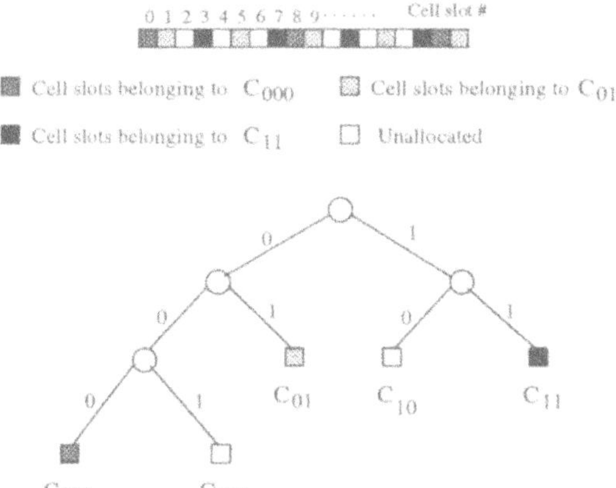

Figure 1: *Binary Preallocation* and its tree representation.

1. Find an unallocated (potential) channel, $C_{d_1 d_2 \cdots d_{j-1}}$ with the minimum transmission speed such that $R(\ell)/2^{j-1} \geq R$. Let $k = j$.
2. **while** $R(\ell)/2^k \geq R$ **do**
   divide the channel into two subchannels (each with the transmission speed of $R(\ell)/2^k$ cells/sec) according to the formulas (1) and (2) and pick one of them. Let $k := k+1$.
   end; □

BPA allocates the channel found in Step 1 to the request, if the while loop in Step 2 is executed 0 times; otherwise, it allocates the channel picked in Step 2. Fig. 1 depicts the result of applying BPA to three requests for bandwidths of $R(\ell)/8$, $R(\ell)/4$ and $R(\ell)/4$, respectively.

Comments:

1. Let ℓ and ℓ' be two links meeting at a node with bandwidths $2^{-h(\ell)}R_0$ and $2^{-h(\ell')}R_0$, respectively, where $h(\ell') \geq h(\ell)$. Then for an arbitrarily chosen channel for ℓ', BPA always finds a channel of ℓ with the same speed, provided there is enough unallocated bandwidth left on ℓ. Thus, we can establish a global logical channel out of local logical channels of the same bandwidth.
2. For a high-speed network, the switching (at the intermediate nodes) of cells belonging to the same connection would be implemented by hardware. The subscript α of the allocated channel C_α can be used as the control input to this switching hardware.
3. BPA has some resemblance to the "buddy system" of main memory allocation used in some operating systems. Therefore, the book-keeping of allocated and unallocated logical channels can be done in a similar way. For example, if two logical channels $C_{d_1 d_2 \cdots d_{k-1} 0}$ and $C_{d_1 d_2 \cdots d_{k-1} 1}$ are released, then they should be combined into one logical channel, $C_{d_1 d_2 \cdots d_{k-1}}$, for future allocation. Also, the unallocated logical

channels can be kept track of by placing them in different lists depending on their bandwidth.

Advantages:

1. Little buffer space is required (see Table 1) at each network node along the connection.
2. Scheduling at each network node is very simple. (See Comment 2 above.)
3. The cells on different connections do not interfere with each other, so that no congestion will result.
4. The unused bandwidth is always in the form of cell slots which are uniformly distributed over time. Thus, periodic cell slots can be allocated to future requests.
5. Unlike the other schemes discussed in Section 1, a connection with an arbitrarily low rate can be established without causing any wasted bandwidth.
6. Two types of synchronized media, e.g., voice and video, could be transmitted using two separate *virtual channels* in the same *virtual path* in an ATM network. The voice and video cells to be synchronized with each other arrive at the client within a fixed time bound.

Disadvantages:

1. Unallocated bandwidth may become fragmented. Thus, no single unallocated channel may be able to satisfy a new request, even though there is still enough unallocated bandwidth. Unlike memory allocation, however, fragmented bandwidth need not lead to waste. Two or more unallocated channels whose total bandwidth is not less than the required bandwidth can be allocated to satisfy a request.
2. Dedicating a fixed amount of bandwidth is wasteful for bursty real-time data transmission. This problem can be mitigated to a certain degree by the following approach, which we call *delayed forwarding.* (See Fig. 2. In the figure, the channel designators, α and α', have the same length, so that C^l_α and $C^{l'}_{\alpha'}$ have the same speed.) When cells belonging to a connection arrive on an input link of a network node, the node holds each cell for a fixed interval of time before forwarding it on the output link for the connection. This will enable the node to discover the unused cell slots and use them for other traffic. This approach works, of course, provided this delay is acceptable for the application.
3. The requested bandwidth R may not be equal to $R(\ell)/2^k$ for any integer k, so that a part of the allocated bandwidth $(R(\ell)/2^k - R)$ is wasted. This problem can be overcome by one of the following two methods: (i) Instead of allocating $R(\ell)/2^k$ to the request, allocate a set of channels including one with bandwidth $R(\ell)/2^{k+1}$ and some others with smaller bandwidths such that the total bandwidth is very close to R. (ii) *Delayed forwarding* discussed above.

The cell slots belonging to unallocated channels should be used for other delay-insensitive or loss-insensitive data.

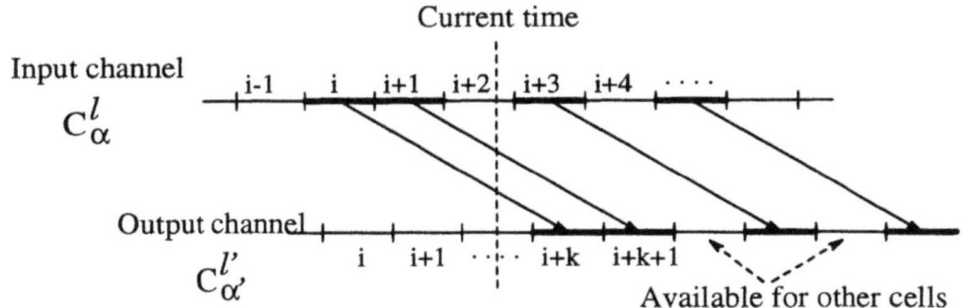

Figure 2: Delayed forwarding.

4. DELIVERY-ON-DEMAND OF UNCOMPRESSED DATA

Let f (cells) be the playback-frame size, F (cells) be the size of a multimedia file to be played back, R_c (cells/sec) be the consumption rate of the client's playback unit, and B (cells) be the buffer size available at the client site. In this section, given these parameters, we derive the minimum transfer rate, R_t^{min}, and maximum transfer rate, R_t^{max}, for uncompressed data. Any transfer rate R_t, satisfying $R_t^{min} \leq R_t \leq R_t^{max}$, can be used without causing buffer overflow or underflow. To avoid the discussion of trivial cases, we assume that $F > B$.

4.1 Range for Feasible Transfer Rate

Although the problem is to determine the range for feasible R_t for given B, F, and R_c, we first turn the problem around, and try to establish a lower bound on B for a given R_t. Since the interval time between two successive cells is the same on a given connection based on *Binary Preallocation* (see Section 3), to simplify analysis, we assume that the buffer fills with arriving cells as a linear function of time. (If the solution (i) suggested to solve Disadvantage 3 is used, this assumption does not hold and, therefore, the analysis must be slightly modified.)

Immediately after a frame is played back, the next frame should be available. This requires that $B \geq 2f$. We consider the two cases, $R_t < R_c$ and $R_t \geq R_c$, separately.

1. $R_t < R_c$:

To prevent buffer underflow, suppose we buffer as much data as possible before playback begins. Fig. 3 shows the number of cells in the buffer as a function of time. A played back frame is shown as being instantaneously consumed. Playback should start when there is only fR_t/R_c (cells) of free space left in the minimum-sized buffer. Thus, when the first frame has been played back, which takes f/R_c (sec), the buffer will be full except for one frame worth of free space, which has just been emptied. At this moment, the F–B remaining cells of the file are yet to be transmitted. The transmission of these cells must complete before the playback of the last frame of the file can start. We thus obtain

$$\frac{F-2f}{R_c} \geq \frac{F-B}{R_t}. \tag{4}$$

Let B_{min} denote the minimum possible value of B, which is a function of R_t/R_c. From

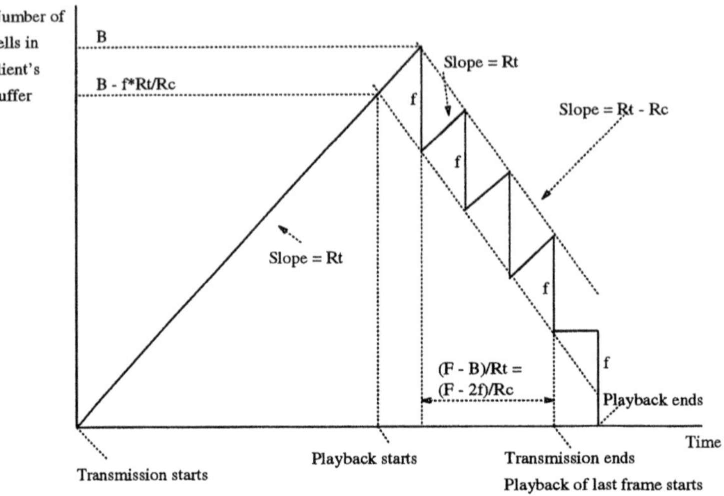

Figure 3: Buffered cells vs. time when $R_t < R_c$.

(4) we can derive

$$B_{min}/f = F/f - (F/f - 2)\frac{R_t}{R_c}. \tag{5}$$

2. $R_t \geq R_c$:

Since data arrive faster than they can be played back, we start playback as soon as a frame is available in the buffer. To analyze the buffer overflow problem, we consider two cases:

a) Transmission of the $F-f$ remaining cells of the file takes longer than the playback of the first frame, i.e., $f/R_c < (F-f)/R_t$. (Fig. 4.)

b) Transmission of the remaining cells completes before the playback of the first frame finishes, i.e., $f/R_c \geq (F-f)/R_t$.

In the case b), we clearly need $B \geq F$, which is the uninteresting case that we ruled out earlier. Therefore, in the rest of this section, we consider only case a), assuming that $f/R_c < (F-f)/R_t$. In Fig. 4, transmission begins at time A, while playback starts at time S. The length of the line segment QS, denoted by $\overline{QS}$, is f, i.e., the amount of the prefetched data at the time playback starts, and $\overline{MD}$ is the amount of data in the buffer when transmission completes at time D. $\overline{OC}$ is the amount of data in the buffer at the time frame l (the last frame that finishes its playback no later than the time transmission completes) has just been played back. If $\overline{MD} > \overline{OC}$, as shown in Fig. 4, the maximum buffer space requirement occurs at time D. However, if $\overline{MD} < \overline{OC}$, the buffer contains more data at time C than at time D. Based on the above observations, we can derive

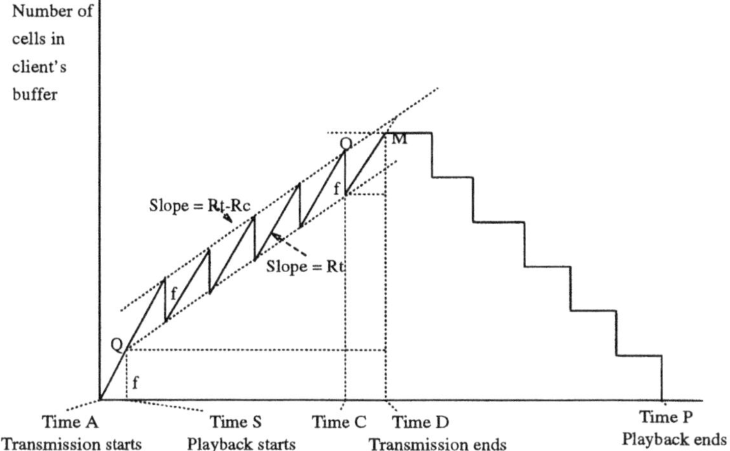

Figure 4: Buffered cells vs. time when $R_t > R_c$, $f/R_c < (F-f)/R_t$ and $f \leq F - f - \lfloor (F-f)R_c/fR_t \rfloor fR_t/R_c$.

the lower bound on B as $max\{\overline{MD}, \overline{OC}\}$, which can be broken down as in (6), which is more general than is necessary here: (We will need this general form in Section 5.)

$$
\begin{aligned}
B \geq\ & \text{space for the prefetched data before playback starts} \\
& +\text{space for data buffered between time } S \text{ and time } C \\
& -\text{space for data consumed between time } S \text{ and time } C \\
& \textit{if the amount of data in frame } l \textit{ is less than the amount of data buffered after} \\
& \textit{the playback of frame } l \textit{ finishes,} \\
& +\text{space for data buffered after the playback of frame } l \text{ finishes} \\
& \textit{else} \\
& + \text{space for frame } l.
\end{aligned} \tag{6}
$$

Note that $\lfloor ((F-f)/R_t)/(f/R_c) \rfloor$ is the number of frames which have been played back before transmission ends. From (6), we obtain

$$B \geq f + \left\lfloor \frac{F-f}{R_t} \cdot \frac{R_c}{f} \right\rfloor f \left(\frac{R_t}{R_c} - 1 \right) + max \left\{ f, F - f - \left\lfloor \frac{F-f}{R_t} \cdot \frac{R_c}{f} \right\rfloor \frac{fR_t}{R_c} \right\}, \tag{7}$$

which can be rewritten as

$$\frac{B}{f} \geq 1 + \left\lfloor \left(\frac{F}{f} - 1 \right) \frac{R_c}{R_t} \right\rfloor \left(\frac{R_t}{R_c} - 1 \right) + max \left\{ 1, \frac{F}{f} - 1 - \left\lfloor \left(\frac{F}{f} - 1 \right) \frac{R_c}{R_t} \right\rfloor \frac{R_t}{R_c} \right\}. \tag{8}$$

(8) involves only "normalized" parameters, B/f, F/f and R_t/R_c. Both (5) and (8) yield $B_{min} = 2f$ at $R_t = R_c$. The interpretation of this is, of course, that when $R_t = R_c$, we need only two frames worth of buffer, regardless of the file size F; f for buffering the arriving data and f for playback.

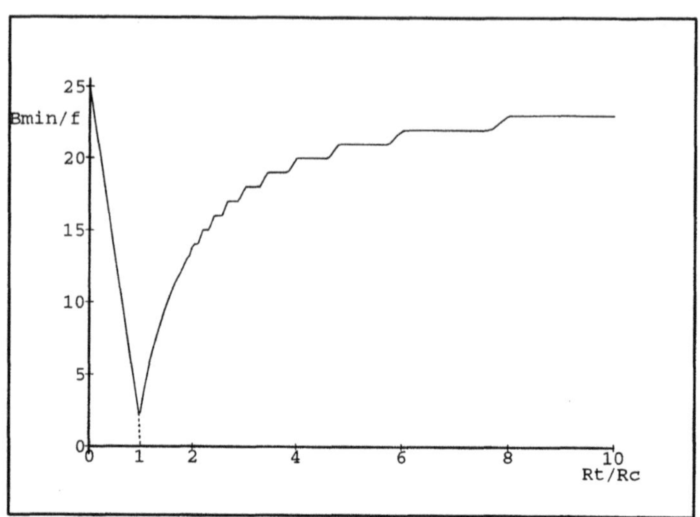

Figure 5: B_{min}/f vs. R_t/R_c. (*Maple* was used to draw this figure.)

Example 1: Based on (5) and (8), Fig. 5 illustrates the relationship between the (normalized) minimum required buffer size, B_{min}/f, and the (normalized) transfer rate, R_t/R_c, for $F/f = 25$. When $R_t/R_c \geq 1$, in some ranges the buffer space requirement remains constant while the ratio R_t/R_c increases. The flat segments correspond to the case illustrated in Fig. 4, i.e., where $f \leq F - f - \lfloor (F-f)R_c/fR_t \rfloor fR_t/R_c$, so that (8) yields $B_{min}/f = F/f + \lfloor (F/f-1)R_c/R_t \rfloor$. □

In practice, B/f (≥ 2) is a given constant, and we want to select a suitable transfer rate, R_t. In terms of a graph like Fig. 4, one can draw a horizontal line $B_{min}/f = b$, where bf is the available buffer size. In the cases of interest, this line intersects the curve at two points. The left and right intersections yield R_t^{min}/R_c and R_t^{max}/R_c, respectively. Any transfer rate, R_t such that $R_t^{min} \leq R_t \leq R_t^{max}$, can be used without causing buffer overflow or underflow.

4.2 Local Channel Allocation

In this subsection, we discuss how to select the transfer rate, R_t, within the range derived in the previous subsection. Recall that *Binary Preallocation* presented in Section 3 allocates a channel with bandwidth of the form $2^{-k}R(\ell)$ (cells/sec), in which k is an integer and $R(\ell)$ (cells/sec) is the total bandwidth of link ℓ. Therefore, if there is an integer k satisfying

$$R_t^{min} \leq 2^{-k}R(\ell) \leq R_t^{max}, \tag{9}$$

then a channel of bandwidth $2^{-k}R(\ell)$ can be allocated to this request, provided there is still enough unallocated bandwidth for the link. If there is one or more choices for k, any one can be chosen. However, if there is no integer k satisfying (9), then we should choose the smallest k such that $2^{-k}R(\ell) > R_t^{min}$, and set $R_t = 2^{-k}R(\ell)$. In this case, some slots in the allocated connection will go unused. (See the discussion about Disadvantage 3 at the end of the previous section.)

Note that when we select a transfer rate during the channel establishment phase, we need only reserve a *bandwidth* of size $2^{-k}R(\ell)$, instead of a particular *channel*, C_α, of bandwidth $2^{-k}R(\ell)$, until it's been determined that a connection of bandwidth $2^{-k}R(\ell)$ is feasible.

4.3 Global Channel Allocation

As stated in Section 2, in order to establish a connection (or a global logical channel), we need to use a routing algorithm and an admission test at each node visited. Clearly, a fast routing algorithm is very important. However, we shall not specifically discuss this topic in this paper, referring the reader, instead, to references on known routing algorithms [1, 17].

We only comment on the possible resource contentions in the channel establishment phase. The network may be routing several requests simultaneously, each reserving some resources at the visited nodes. We abort a request (or divert it to another route) if required resources are not available at a node, so that no deadlock is possible.

5. DELIVERY-ON-DEMAND OF COMPRESSED DATA

In the case of compressed data, different playback-frames may consist of different numbers of cells, but the frame consumption rate, R_{fc} (frames/sec), is still constant. We assume that for every file, the minimum and maximum frame sizes, f_{min} (cells) and f_{max} (cells), respectively, are known, where $f_{max} > f_{min}$, in addition to the other parameters, i.e., F, B, and N (the number of cells in the file).

Thus, the *maximum consumption rate* is given by

$$R_c^{max} = f_{max}R_{fc} \quad \text{(cells/sec)},$$

and the *minimum consumption rate* is given by

$$R_c^{min} = f_{min}R_{fc} \quad \text{(cells/sec)}.$$

To avoid the discussion of trivial cases, we assume that $F > B$ and $N > 2$.

5.1 Buffer Underflow and Overflow

If $R_t \leq R_c^{min}$, then, clearly, there is no danger of buffer overflow. Similarly, if $R_t \geq R_c^{max}$, then there is no danger of buffer underflow. So, we need to address the buffer underflow problem only when $R_t < R_c^{max}$, and the buffer overflow problem only when $R_t > R_c^{min}$. Given a particular file, let S be the set of its frames, whose sizes range from f_{min} to f_{max}. Consider now the family of all files whose set of frames coincides with S. These files differ from each other in the sequence in which different frames appear in it. It turns out that a particular sequence gives rise to a more severe buffer under flow or overflow problem than others. For example, if the frames appear in the decreasing order in their sizes, then we need more buffer space than otherwise, in order to prevent buffer underflow. Intuitively, this is because, once playback is started, the buffer is more quickly emptied. Consider two files, $\mathcal{F}_1$ and $\mathcal{F}_2$, which are the same except that, in $\mathcal{F}_1$, frame a of size f_a is followed immediately by frame b of size f_b ($< f_a$), whereas, in $\mathcal{F}_2$, the order is reversed. Suppose playback starts for both files at the same time. It is not difficult to see that, if

Value of R_t	Problem encountered	Worst case
$R_t < R_c^{max}$	Buffer underflow	largest frames first
$R_t > R_c^{min}$	Buffer overflow	smallest frames first

Table 2: Problems for compressed data

buffer underflow occurs for $\mathcal{F}_2$ at the time the playback of frame b has just completed, then buffer underflow occurs for $\mathcal{F}_1$ as well at the time the playback of frame a has just completed. In other words, $\mathcal{F}_1$ is no better than $\mathcal{F}_2$, as far as preventing buffer underflow is concerned.

Above, we considered only those files which share the set S. Since, in our model, the only information we have about a file is its parameters, F, N, f_{max}, and f_{min}, we don't know what S is. The worst S, as far as preventing buffer underflow is concerned, consists of as many frames of size f_{max} as possible, and these frames all appear in the beginning of the file. Let N_{max} be the maximum number of frames of size f_{max}. We have $N_{max} f_{max} + (N - N_{max}) f_{min} \leq F$, or $N_{max} = \lfloor (F - f_{min} N)/(f_{max} - f_{min}) \rfloor$. We can similarly show that the maximum number, N_{min}, of frames with size f_{min} can be expressed as $N_{min} = \lfloor (f_{max} N - F)/(f_{max} - f_{min}) \rfloor$. Table 2 summarizes the worst case for buffer space requirement.

5.2 Bounds on Transfer Rate

As in the case of uncompressed data, immediately after a frame is played back, the next frame should be available. This requires that $B \geq 2 f_{max}$.
We divide the range of R_t into three parts:

(1) Region A: $R_t \leq R_c^{min}$ (only buffer underflow is possible).

(2) Region B: $R_c^{min} < R_t < R_c^{max}$ (both buffer underflow and overflow are possible).

(3) Region C: $R_t \geq R_c^{max}$ (only buffer overflow is possible).

Below, we present the minimum buffer space required, B_{min}, as a function of R_t. For the derivation of those results, the reader is referred to [18].

(1) Region A ($R_t \leq R_c^{min}$):

$$B_{min} = F - \frac{N-2}{R_{fc}} R_t. \tag{10}$$

Let $R_t^{min} = R_{fc}(F - B)/(N-2)$. If $R_t^{min} > R_c^{min}$ (i.e., $(F-B)/(N-2) > f_{min}$), there is no feasible transfer rate in Region A. Otherwise, we can select any R_t such that $R_t^{min} \leq R_t \leq R_c^{min}$. (See Region A in Fig. 6 for an example of (10).)

(2) Region B ($R_c^{min} < R_t < R_c^{max}$): Let

$$D_{pref} = f_{max} N_{max} - (f_{max} N_{max} - f_{max}) \frac{R_t}{R_c^{max}} = f_{max} N_{max} - (N_{max} - 1) \frac{R_t}{R_{fc}}, \tag{11}$$

and $\tilde{R}_t = N_{min} R_c^{min} / (N_{min} - N_{max} + 2)$.

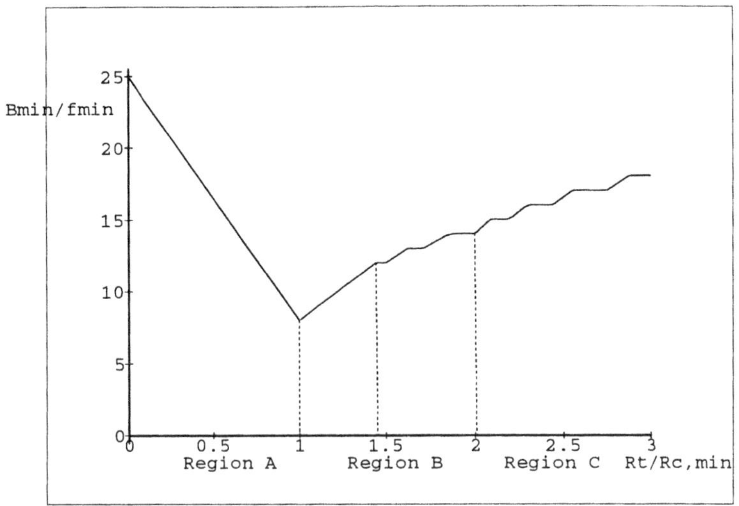

Figure 6: B_{min}/f_{min} vs. R_t/R_c^{min}. (*Maple* was used to draw this figure.)

a) When $N_{min} - N_{max} + 2 > 0$ and $max\left\{\tilde{R}_t, R_c^{min}\right\} \leq R_t \leq R_c^{max}$:

$$B_{min} = D_{pref} + \left\lfloor \frac{F - D_{pref}}{R_t} R_{fc} \right\rfloor \left(\frac{R_t}{R_{fc}} - f_{min} \right) + max\left\{ f_{min}, F - D_{pref} - \left\lfloor \frac{F - D_{pref}}{R_t} R_{fc} \right\rfloor \frac{R_t}{R_{fc}} \right\}. \quad (12)$$

b) When $N_{min} - N_{max} + 2 \leq 0$ and $R_c^{min} < R_t < R_c^{max}$, or when $N_{min} - N_{max} + 2 > 0$ and $R_c^{min} \leq R_t \leq \min\left\{R_c^{max}, \tilde{R}_t\right\}$:

$$B_{min} = D_{pref} + N_{min}\left(\frac{R_t}{R_{fc}} - f_{min} \right) + \frac{R_t}{R_{fc}}. \quad (13)$$

(3) Region C ($R_t \geq R_c^{max}$):

a) When $R_t \geq max\left\{\bar{R}_t, R_c^{max}\right\}$:

$$B_{min} = f_{max} + \left\lfloor \frac{F - f_{max}}{R_t} R_{fc} \right\rfloor \left(\frac{R_t}{R_{fc}} - f_{min} \right) + max\left\{ f_{min}, F - f_{max} - \left\lfloor \frac{F - f_{max}}{R_t} R_{fc} \right\rfloor \frac{R_t}{R_{fc}} \right\}. \quad (14)$$

b) When $R_c^{max} < R_t \leq \bar{R}_t$: (If $\bar{R}_t \leq R_c^{max}$, this case is impossible.)

$$B_{min} = f_{max} + N_{min}\left(f_{max} - f_{min}\right) + \left\lfloor \frac{F - f_{max}}{R_t} R_{fc} \right\rfloor \left(\frac{R_t}{R_{fc}}\right)$$
$$+max\left\{f_{max}, F - f_{max} + N_{min} f_{min} - \left\lfloor \frac{F - f_{max}}{R_t} R_{fc} \right\rfloor \frac{R_t}{R_{fc}}\right\}. \qquad (15)$$

Example 2: Fig. 6 illustrates the relationship between B_{min}/f_{min} and R_t/R_c^{min}. (Since R_c is variable for compressed data, we normalize R_t with respect to R_c^{min}.) The parameters used are $F = 250$ (cells), $N_{max} = 6$ and $N_{min} = 13$, $f_{max} = 20$ (cells), and $f_{min} = 10$ (cells). For these data, we have $\tilde{R}_t = 13/9$ and $\bar{R}_t = 23/13$. From Fig. 6, one can see that Region B ($R_c^{min} < R_t < R_t^{max}$) has two subregions, corresponding to case a) and case b), respectively. These two subregions meet when $R_t = \tilde{R}_t$. Since $\bar{R}_t < R_c^{max}$ for the specific values used in this example, only case a) is possible in Region C. If $\bar{R}_t > R_c^{max}$, there would be two subregions. We can also see that the minimum buffer space is required when $R_t = R_c^{min}$. For any fixed B_{min} larger than this minimum, a range of transfer rates, $[R_t^{min}, R_t^{max}]$, is feasible. □

6. CONCLUSION

We have presented a new channel allocation scheme for ATM-based networks, which has several important advantages over the currently known schemes. The most important advantage is that it will be able to satisfy the requirements of many real-time applications. We then discussed the delivery-on-demand of pre-recorded, continuous media data as a possible application of this scheme.

Although we assumed in our analysis that the buffer space available at the client site, B, was smaller than the file size, F, it is, of course, possible in practice that $B \geq F$. If so, then there is no restriction on the transfer rate, R_t. However, even in this case, our analysis is still useful in providing us with the minimum playback starting time, since minimizing the required buffer space also minimizes the playback starting time [6].

We are still in an early stage of investigation, and a lot of detail needs to be worked out. One such detail is the hardware implementation of the switches at network nodes. We are also currently investigating practical ways of allocating several subchannels of a link to a connection (see the discussion on Disadvantage 1 in Section 3).

References

[1] Bertsekas, D., and Gallager, R., *Data Networks,* Second Ed., Prentice-Hall, 1992.

[2] Clark J., "A Telecomputer," *Proc. SIGGRAPH '92*, Chicago, (July 1992), pp. 17-23.

[3] Decina, M., Open issues regarding the universal application of ATM for multiplexing and switching in B-ISDN, *Proc. ICC'91*, 1991, 1258-1264.

[4] Demers, A., "A high throughput bulk data transfer protocol," *Proc. SIGCOMM '89*, (Aug. 1989), pp. 154-167.

[5] Ferrari, D., "Client requirements for real-time communication services," *IEEE Comm. Magazine*, Vol. 28, No. 11 (Nov. 1990), pp. 65-72.

[6] Gemmell, J., "Principle of delay-sensitive multimedia data storage and retrieval," *ACM Trans. Information Systems*, Vol. 10, No. 1 (Jan. 1992), pp. 51-90.

[7] Golestani, S.J., Congestion-free communication in high-speed packet networks, *IEEE Trans. Comm.*, Vol. Comm-39, No. 12 (Dec. 1991), pp. 1802-1812.

[8] Kopetz, H. and Ochsenreiter, W., Clock synchronization in distributed real time, *IEEE Trans. Computers,* C-36, No. 8 (Aug. 1987), 933-940.

[9] Kurose, J., Open issues and challenges in providing quality of service guarantees in high-speed networks, *Computer Communication Review*, Vol. 23, No. 1 (Jan. 1993), 6-15.

[10] Li, C-S., Ofek, Y., Segall, A., and Shoraby, K., Pseudo-isochronous cell switching in ATM networks, *Proc. INFOCOM'94*, June 1994, pp.428-437.

[11] Minzner, S.E., "Broadband ISDN and asynchronous transfer mode (ATM)," *IEEE Comm. Magazine* Vol. 27 (1989), pp. 17-24.

[12] Ofek, Y., Generating a fault tolerant global clock using high-speed control signals for the MetaNet architecture, *IEEE Trans. Comm.* Vol 42, No. 5 (May 1994), pp. 2179-2188.

[13] Peha, J.M., and Tobagi, F.A., Evaluating scheduling algorithms for traffic with heterogeneous performance objective, *GLOBECOM'90*, 1990, pp. 21-27.

[14] Press, L., "The Internet and interactive television," *Comm. ACM,* Vol. 36, No. 12 (Dec. 1993), pp. 19-23.

[15] Ramanthan, S., and Rangan P.V., "Feedback techniques for intra-media continuity and inter-media synchronization in distributed multimedia systems," *Computer J.*, Vol. 36, No.1 (Jan. 1993), pp. 19-31.

[16] Rathgeb, E.P., "Comparison of policing mechanisms for ATM networks," *Proc. 3rd RACE Workshop*, (Oct. 1989), pp. 211-224.

[17] Stallings, W., *Data and Computer Communications,* Third Ed., Macmillan, 1991.

[18] Ting, J., Kameda, T., and Fracchia, D., Delivery-on-demand of continuous media data, *Tech. Rept.* LCCR/CSS 94-04, School of Computing Science, Simon Fraser University, 1994.

[19] Woodruff, G.M., and Kostpaiboon, R., "Multimedia traffic management principles for guaranteed ATM network performance," *IEEE J. Selected Areas in Comm.*, Vol. 8, No. 3 (Apr. 1990), pp. 437-446.

[20] Zhang, L., "Virtual Clock: A new traffic control algorithm for packet switching networks," *ACM Trans. Computer Systems*, Vol. 9, No. 2 (May 1991), pp 101-124.

17

A Hybrid Deposit Model for Low Overhead Communication in High Speed LANs

Randy B. Osborne
osborne@merl.com
Mitsubishi Electric Research Labs*

Abstract

This paper presents a new, "hybrid deposit" model for low overhead communication wherein the sender directly deposits messages into the destination user-level memory. The destination address is a function of both sender state and destination state. The motivation is to increase the sender's role in communication in order to simplify the destination's role and thus enable fast, low-cost communication interfaces. The model separates data delivery from synchronization so as to enable the optimization of simple data delivery while leaving more difficult synchronization to other mechanisms.

The paper specializes this hybrid deposit idea to ATM LANs and examines implementation methods. We first describe a software implementation with stock workstations and ATM interface cards. This implementation achieves a best case application-to-application latency of 24.5μsec. For consistent low latency, high bandwidth, and reduced burden on the host processor, hardware support is required. We describe an interface architecture called DART that provides basic functionality for hybrid deposit.

With hardware support, the hybrid deposit model looks promising for applications in parallel, distributed, and real-time computing.

Keyword Codes: C.2.1, C.2.2, B.4.1
Keywords: Network Architecture and Design; Network Protocols; Data Communication Devices

1 INTRODUCTION

The considerable recent work on workstation network interfaces for high speed LANs (e.g. [Dav93, BP93, TS93]) has focussed on achieving high bandwidth. This paper focuses, in contrast, on low overhead communication — by which we mean both low latency and low impact on the host processor — for high speed LANs. We present a low level protocol and network interface architecture for low overhead communication. The key idea of this paper is to use both sender information and destination information to demultiplex messages directly to where they are needed. We show that this combination has flexibility for a wide range of application requirements.

*201 Broadway, Cambridge, MA 02139

We have built a software implementation to evaluate these ideas. We get a best case latency for a 32 byte application to application data transfer of 24.5μsec on DECStation 5000s connected via an ATM network (*sans* switch; 30μsec with an ATM switch). This is about 10 times faster than the fastest conventional approach (using Fore Systems' AAL3/4 implementation) on the same hardware[Ke94]. With appropriate hardware support, we believe latencies of under 10μsec for a 155Mbps ATM LAN and under 3μsec for a 622Mbps ATM LAN are possible in the workstation LAN environment.[1]

Our interest in low overhead communication is motivated by applications in parallel, distributed, and real-time computing. Latency, one of the fundamental issues in parallel computing [AI87], limits the maximum parallelism that can be exploited in an application. In distributed computing, low latency can help improve the performance of remote procedure call (RPC) and thereby enhance the performance of the ubiquitous client-server model. Low latency can also be useful for real-time computing, but far more important is predictability: it is important to insulate real-time tasks on the host processor from unrelated asynchronous communication events. Finally, low latency communication increases the flexibility in structuring a system.

Our goal is to achieve minimal communication latencies between application processes across high speed LANs. We assume the compute nodes are multiuser with their own address spaces and memory (e.g. workstations) and we assume the network is multiuser, so there must be protection against accidental or malicious interaction of users. It is important that the communication allow random access data transfers for parallel computing.

Section 2 describes the problem and related work. Section 3 presents a "hybrid deposit" communication model that uses both sender and destination information, maintaining full multiuser protection. Section 4 specializes this hybrid deposit model to ATM LANs. Section 5 describes a software implementation of this model. Section 6 presents an interface architecture called DART. Finally, Section 7 concludes and discusses further work.

2 THE PROBLEM AND RELATED WORK

The main problem to achieving low overhead communication is how to handle asynchronous message arrivals at the destination to meet all four requirements of low latency, low impact, low cost, and multiuser protection simultaneously. (Sending is easier because it's synchronous — to a first approximation.) Many global address space parallel machines sacrifice protection to get a very cheap solution in which messages write directly into destination memory. Most workstations sacrifice latency and impact to get a low cost solution by multiplexing application and message processing on a single processor. Consequently, an asynchronous message arrival incurs overhead in *both* the communication and whatever application happens to be running at the time. High performance systems sacrifice cost and duplicate resources to bypass the host processor and operating system e.g. parallel machines such as the Intel Paragon and *T [NPA92] and real-time systems such as Spring OS [SR91]. These systems devote hardware for the worst case requirements.

An intermediate solution is to control asynchronous events by separating events by their

[1] For one transit of a fast ATM switch.

	Data delivery	Interrupts
Destination-based	$address_{msg} = f(dest)$	$intr_{msg} =$ always
Sender-based	$address_{msg} = f'(sender)$	$intr_{msg} = g(sender)$
Hybrid	$address_{msg} = f''(sender, dest)$	$intr_{msg} = g'(sender, dest)$

Table 1: Message Passing Options

need for the host processor. Events, such as data delivery, that don't require the host processor can be handled directly e.g. by depositing data directly into user memory. Events — chiefly synchronization — that do require the host processor can be divided into immediate actions that require immediate service and delayable actions that can be accumulated and processed when convenient for the host processor (thereby turning them into synchronous events). With this separation of data and control events we only need resources sufficient to bypass the non-host processor events rather than all events.

Conventional communication protocols (especially for LANs) are not designed to facilitate fast separation of incoming messages into these various event types, increasing the processing required at the destination. Instead, we can exploit the sender's knowledge, shifting more of the burden to the sender, in order to simplify the message processing at the destination. In effect, the sender uses its knowledge to pre-demultiplex the messages. By doing so, we posit three benefits: greater accuracy in separating events at the destination, simpler and cheaper communication co-processors, and reduced host processor load.

Table 1 shows a classification of messaging based on the division of the burden between sender and destination. *dest* represents destination state, *sender* represents sender state transmitted with the message, and $intr_{msg}$ is the interrupt status on message arrival. In *sender-based addressing* demultiplexing is trivial: a message contains an address, supplied by the source, into which the message is directly deposited. In *destination-based addressing*, the source has no direct input on the final address of a message: a message identifies a buffer into which the message is stored at some implicit location, e.g. by sequencing a pointer.

Destination-based addressing is the conventional approach in distributed systems and for send-receive models in parallel machines. Sender-based addressing is common in parallel machines, though usually at fine granularities e.g. words. Recently, sender-based addressing has become popular for optimizing messaging passing, as in SHRIMP [Be94] which uses virtual memory mapped communication. Three recent works promote sender-based addressing in a LAN environment. [Wil92] describes a high level design of an interface called Hamlyn; [SP90] describes an interface design called Axon; and [TLL93] describes a in-kernel implementation of a remote read/write model (a follow-on of Spector's work [Spe82]). The Meiko CS-2 multicomputer also supports a remote read/write model, though with custom co-processors and proprietary network.

Sender-based addressing allows random access transfers and thus is attractive for parallel computing. However, sender-based addressing suffers from two problems in our context of a LAN distributed system. The first problem is lack of isolation. Pure sender-based addressing releases considerable information to the sender and requires the destination to trust the sender. The second problem is the inefficiency of certain operations. For example, three messages are required to add to the end of a shared queue. Whereas this is acceptable in a fine grained global address parallel machine, it is not acceptable in a LAN environment with its higher communication costs.

Hybrid-based addressing solves these problems while retaining the advantage of sender-based addressing. The storage address is partially a function of an address the sender may not know, either for isolation, or so we can add to the end of a shared queue in one message. Active Messages provide hybrid addressing. An "active message" contains the name of an interrupt handler, arguments, and data. The destination executes the interrupt handler to apply the message action [von92]. This provides great flexibility in combining sender and destination information. However, to get low latency these interrupt handlers execute on the host processor and execute in the context of the interrupted task. Thus Active Messages must be combined with a protection mechanism to be useful in a multiuser environment. [DLM94] provides a limited form of hybrid addressing in which a message controls whether it is deposited at an address contained in the message or an address stored at the destination.

Hybrid-based interrupts are also useful. Pure destination-based interrupts, like in Active Messages, cause an interrupt on every message arrival. On the other hand, pure sender-based interrupts, as used in Hamlyn, limit the interrupt semantics: it becomes very awkward, for instance, to priority schedule interrupts at the destination. Thus, as with addressing, interrupts should be a function of both sender and destination information. This allows the interrupt status to be partially a function of message priorities contributed by processes that the sender may not know exist. [TLL93] and [DLM94] describe very limited approximations to hybrid interrupts in which the sender indicates an interrupt in the message and the destination can simply enable or disable all interrupts for a given buffer.

Our work combines sender and destination information for both addressing and interrupts. This gives flexibility to exploit various points in the sender-destination spectrum, but with full protection unlike with Active Messages. This full hybrid functionality, for both data and control, distinguishes our work from previous work.

3 HYBRID DEPOSIT MESSAGING

[Se93] introduced the "deposit model" for a form of sender-based addressing used by a compiler in a message passing parallel computing paradigm. We generalize this appropriately suggestive term to mean demultiplexing messages and depositing them directly where they are needed e.g. depositing data directly in application memory and delivering (significant) control events to the host processor (similar to the ideas in [Se94]). To this notion we add hybrid addressing and interrupt generation.

3.1 Overview

The application view of hybrid deposit messaging is two "endpoint" buffers linked by a "connection". As shown in Figure 1, messages originating in the source endpoint can bypass the conventional operating system and host processor route to the network and either be delivered directly to any location in the destination endpoint or conditionally delivered to the destination operating system. Messages contain both control and data information.

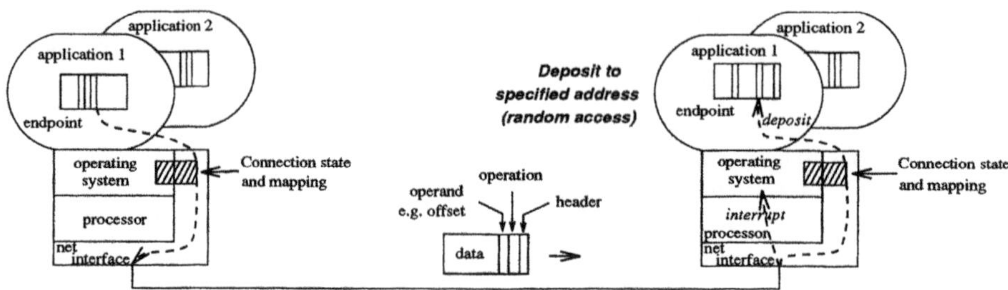

Figure 1: Hybrid deposit operational model

The destination decodes the control information and chooses an action and/or destination location based on **both** the source state contained in the message and destination state. Although these actions may be quite general, we focus on a set of primitive actions representing common operations that can be implemented simply without host processor or operating system intervention. We leave more complex actions to the host processor. This is the RISC philosophy of processor architecture applied to communication.

3.2 Endpoints and Connections

Endpoints and connections are allocated and deallocated with kernel calls. An endpoint (buffer) is any page-aligned, contiguous region of virtual memory. Consequently, the host virtual memory page protection can be used within endpoints. A connection is a channel authorizing communication between a pair of endpoints (connections can also be duplex and multicast). Some state and protection information is associated with the source and destination end of each connection.

An endpoint may have multiple originating connections and/or multiple terminating connections. All connections associated with a given endpoint use the same virtual memory mapping and protection information. Endpoint buffers can be overlapped or nested to effect different degrees of sharing and isolation between connections (like fbufs [DP93]). Different protection schemes can also be realized by mapping the physical pages behind an endpoint buffer to virtual address ranges with different page protections.

3.3 Messages

Messages contain a connection ID (which implicitly identifies the endpoint), control information consisting of an operation and some operands, and data. Each operand is an address, immediate data, or the name of some destination state. Addresses are encoded as an offset from the destination endpoint. The offset is essentially a network logical address that insulates the sender and destination from the addressing details (e.g. address space size, virtual to physical mappings, and page size) at the other. This separation promotes modularity and accommodates node heterogeneity. Furthermore, an offset typically does not need the full dynamic range of a virtual or physical address and thus can be encoded in fewer bits within

Address generation	effaddr= $operand$	(direct addressing)
	effaddr= $< addreg_i >$	(indirect addressing)
	effaddr= $< addreg_i > + operand$	(indexed addressing)
Register operations	$addreg_i \leftarrow operand$	
	$addreg_i \leftarrow$ **unary-op** $\{ < addreg_j >$ or $operand \}$	
	$addreg_i \leftarrow < addreg_i >$ **binary-op** $\{ < addreg_j >$ or $operand \}$	
Conditional operations	if ($< addreg_i >$ **compare-op** $operand$) then generate interrupt at end	
	if ($< addreg_i >$ **compare-op** $< addreg_j >$) then generate interrupt at end	

Figure 2: Example set of primitive operations

a message.

3.4 Operations

The simplest operations are pure sender-based direct read and write data transfers. For a direct write, the sender specifies the source data by its offset from the source endpoint base and the destination location by the offset from the destination endpoint base. For reads, the source sends a message with a direct write request to the destination, along with the offset in the destination and the deposit offset in the source (and a reply connection if the connection is not duplex).

To enable operations which are a function of both sender and destination state, the destination end of each connection has some state which message operands can name. To simplify matters (and to foreshadow our implementation in Section 6), we assume this state is contained in specially addressable locations which we call "address registers". (This state could also be held in general memory locations.) Thus in the full model, message actions are a function of an operation specified by the sender, operands representing sender state, and the contents of the address registers.

The operations must be simple enough to complete in one cell time without host processor or operating system intervention. Figure 2 shows an example set of such primitive operations. The address generation subset calculates the effective address (effaddr) at which to read or write for data transfers. Indirection is useful for queue manipulation and more generally for isolating sender and destination. (<X> denotes the contents of location X.) The conditional operation subset provides the primitives for conditional interrupts to implement delayable or immediate actions. Finally, the register operation subset allows manipulation of registers for address generation (e.g. postincrementing) and conditional interrupts (e.g. interrupt masks). The message operation controls whether a read or write occurs (or otherwise), the primitives selected, and their order. (The conditional test may occur at any time but the interrupt occurs at the end of the compound operation.)

The few primitive operations in this example set allow a rich set of powerful and flexible compound operations. For example, a store indirect with postincrement can be synthesized with an indirection followed by a register operation:

$$< addreg_i > \leftarrow MSG$$
$$addreg_i \leftarrow < addreg_i > + operand \quad \text{(Or } addreg_i \leftarrow < addreg_i > + < addreg_j > \text{)}$$

Done on a per-cell basis, this amounts to DMA with stride equal to the increment value. However, note that varying $operand/< addreg_j >$ yields variable strides. As another detailed example, we can synthesize priority queueing and interrupts as follows:

$< addreg_p > \leftarrow MSG$
$addreg_p \leftarrow < addreg_p > + < addreg_s >$
`if` ($operand$ `greater than` $< addreg_i >$) generate interrupt at end
$addreg_i \leftarrow < addreg_i >$ `bitwise-or` $operand$

operand indicates the priority of the message, $addreg_p$ points to the end of the queue to which this priority message should be added, $addreg_s$ contains the size of MSG, and $addreg_i$ holds the priority level at the destination[2] ($p \neq i \neq s$). The message specifies *operand* and register indices p, i, and s. Complex compound operations like this priority queueing may require multiple compound operations. For example, two compound operation messages would be required in this case if the destination executes one register operation per message.

In a variation of this priority queueing example, we could append messages to one of several different queues without generating interrupts and maintain a bit vector of non-empty queues. This mechanism can be implemented in the same way as priority-based interrupts, but with the most significant bit of $addreg_i$ set to block interrupts.

As other examples, various atomic operations such as fetch-and-increment, read-modify-write, and compare-and-swap can be implemented by devoting one or more of the address registers for the target location and using the register operations for incrementing and comparing. We can also implement barrier synchronization this way. When a process reaches a barrier point, it toggles a bit in a specified address register of all processes in the "barrier set" and then waits for a conditional interrupt when all the bits are set (or cleared).

3.5 Protection and Isolation

There are two levels of protection: access to and from the network is authorized via connection IDs and the source and sink of endpoint data is authorized by virtual memory mapping and page protections. Together with address register indirection, these mechanisms provide protection and isolation between sender and destination manipulation of endpoints. Towards the destination-based end of the addressing spectrum, there is also the need for protection and isolation between sender and destination manipulation of the address registers. For example, in the priority queueing example in Section 3.4, many senders might send messages to the same endpoint within the operating system. Then it would be necessary for the integrity of the priority queueing (and possibly for the operating system) that a sender not be able to overwrite a queue pointer or the message size. We may also not want a sender to be able to read any of this information. Finally, the destination needs to be able to prevent unwanted interrupts, such as from malicious or errant senders.

To allow a destination to control the information a sender can read and modify, we add address register protection. We assume that each address register can be marked as readable and/or writable by the sender, or accessible only by the destination. This adds protection but does not increase isolation since the sender must still name all the operands.

[2] This assumes fewer queues than bits in $addreg_i$.

A sender could still give inconsistent operands (e.g. an operand priority and priority queue that do not match) or specify the wrong registers. Also, a malicious sender could still force interrupts at the destination. To solve these problems, we need to isolate operand names to the destination.

To implement this classic destination-based addressing, the destination could decode the operation to find the destination operands or simply interrupt the host processor. The latter option usually incurs significant overhead but can be useful as a flexible escape mechanism. For the former option, we take the following approach to minimize complexity. We extend the hybrid deposit model to allow the operation in a message to be replaced by a pointer to an "instruction" comprising an operation and operands in the destination. This allows the destination operation to directly reference destination operands (in the address registers) without the sender naming those operands, thereby isolating the destination state from the sender. The destination can refuse all non-instruction pointer style messages (on a per connection basis) to give isolation. However, messages can still provide immediate operands and name destination operands, though the destination may choose not to use these operands (as necessary to maintain isolation). This extension is merely another way to name the operations and operands at the destination; there is no change in the set of primitive operations at the destination.

Now, to implement the priority queueing example given earlier, the sender only needs to specify the instruction pointer for priority enqueueing, the priority, and the data.

Although the instruction pointer approach may appear to subsume the direct operation approach, it really just presents a different cost/benefit point. Some of the increased costs of the instruction pointer approach are a setup phase — the storage of any destination operands and conveyance of an appropriate instruction pointer back to sender — and increased complexity on the part of the destination.

3.6 Summary

Combining sender and destination state in determining message action makes the hybrid deposit model powerful and flexible. We have the full superset of capabilities of the conventional, destination-based approach and sender-based addressing. We have the flexibility to vary the mix of sender and destination information — on a per message basis — to accommodate different requirements on what the sender knows, or alternatively, different requirements on the isolation of knowledge between sender and destination. We can form compound operations from primitive operations. Finally, we can form even more complex operations by combining the result of multiple compound message operations. The hybrid deposit model is more than end-to-end DMA since consecutive messages can be stored in non-sequential locations and non-data operations like conditional interrupts and register operations are supported.

Different application areas are likely to use the hybrid deposit model in different ways. In parallel computing, the endpoint buffers are likely to be large and applications will probably mostly use sender-based addressing, reflecting the cooperation and trust placed in application "partners" on each node. In distributed computing, the endpoint buffers are likely to be small and applications will probably mostly use indirection for isolation and protection.

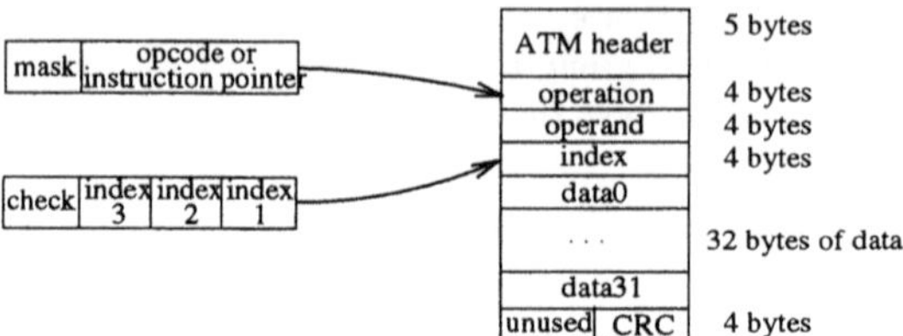

Figure 3: Example format of 53 byte ATM cell for hybrid deposit model

Taking this to the extreme, the hybrid deposit model can model the conventional approach to networking by causing an interrupt on every message (or by having an endpoint within the kernel and using conditional interrupts). In real-time computing, the emphasis will likely be on using conditional interrupts to control asynchronous event delivery. The hybrid deposit model is flexible enough to span all these application areas.

4 SPECIALIZING TO ATM

In the rest of this paper we specialize the hybrid deposit model to ATM networks — it can certainly be specialized to other networks as well.

Figure 3 shows an example cell format (one of many possible variants) that we will use throughout the rest of the paper. The connection number is encoded in the VCI/VPI field (not shown) in the header. The payload is divided into 32 bytes of data — a size selected to match memory and cache (sub)block sizes — and 16 bytes of control. The opcode/instruction pointer field either directly specifies the operation to be performed at the destination or provides a pointer to the operation and operands at the destination; operand is a 32 bit immediate source operand (offset or data). Destination operands are specified via three separate register indices encoded in the index field. Reads and writes occur in 32 byte blocks. The mask field can be used to deselect the reading or writing of 4 byte words within such a block. (Bit i in the mask controls whether data word i is read or written.) This feature is useful to update a location without changing the values (e.g. variables) in neighboring locations in a block. The check field contains a checksum over the prior control fields so that decoding of these fields can begin before the entire cell arrives.[3] The two byte CRC covers the data field. To take advantage of CRC hardware for AAL5 format cells [Onv94], the CRC field could be extended to 4 bytes and cover the entire payload.

The example in Figure 3 is for a write. For a read request to a remote node, the data section contains control fields for the read reply message.

There is also a multiple cell message format for block transfer. In this format, the first cell is a "control" cell in the write format in Figure 3 and the following cells are standard AAL5 cells. To avoid complexities involving cell boundaries and the trailer (length and

[3]An 8 bit checksum for 11 control bytes gives more protection than the 10 bit CRC in AAL3/4 cells.

CRC) in the last AAL5 cell, all block transfers are multiples of 16 bytes.

5 SOFTWARE IMPLEMENTATION

To experiment with the hybrid deposit model, especially with different protocol variations and applications, we built a software implementation of the hybrid deposit model described in Section 3.[4] We also wanted to understand the implementation alternatives and performance limits on a stock machine with a primitive network interface to help build a case for hardware assistance in the network interface. In this section we give a quick overview of this implementation and present some preliminary experimental results.

The implementation hardware platform is a DECStation 5000/240 with a 140Mbps Fore Systems TCA-100 ATM interface card. The operating system is Mach 3.0. We modified the Mach kernel to send a cell via an illegal instruction trap, partly optimized the ATM interface interrupt path, and added hybrid deposit emulation in the kernel. This is similar to the software implementation described in [TLL93] except they modified Ultrix and only support simple remote read and write (using only direct addressing). We added more functionality (address indirection and register operations) and modified the cell format (to use 32 bytes instead of 40 bytes of data) in order to be a better match for host computer systems.

We measured the round trip latency for a 32 byte remote write with two DECStations. This is the time to send 32 bytes from user level on the source to user level on the destination, discover via polling that the message arrived, and send the same 32 bytes back to the sender. On the first such cycle, the round trip latency was greater than 1msec, due to TLB and cache misses. For back-to-back connected workstations (i.e. no switch), the average round trip time on successive cycles was 49μsec, yielding a best case one way send to receive time of 24.5μsec. The best case remote read time was 45μsec. Connecting the workstations to our Fore Systems ASX-100 ATM switch added about 5.5μsec per switch transit to these numbers.

Since the TCA-100 ATM interface does not have DMA, all accesses to the interface to send and receive cells must use programmed I/O. Unfortunately, programmed I/O is quite slow on the DECStation TURBOchannel I/O bus resulting in nearly one third of the latency.

For 155Mbps, we estimate that point to point data transfer latencies in the 10 to 20μsec range are achievable using next generation stock hardware, conventional interface cards with DMA, and software implementation. Using Fore's ATM switch as a guide, 15μsec to 26μsec is a reasonable range in a uncongested single switch LAN.

6 HARDWARE IMPLEMENTATION

While the software implementation in Section 5 shows that stock hardware can get quite low latencies, this comes at the cost of host processor load and bandwidth. Furthermore, the latency is highly dependent on hard-to-control factors such as cache and TLB misses and page faults. The worst case can be 10 times the best case due to memory system (mis)behavior. Predictable latency is important for real-time systems. Consequently, hardware implemen-

[4] Conditional interrupts have not been implemented yet.

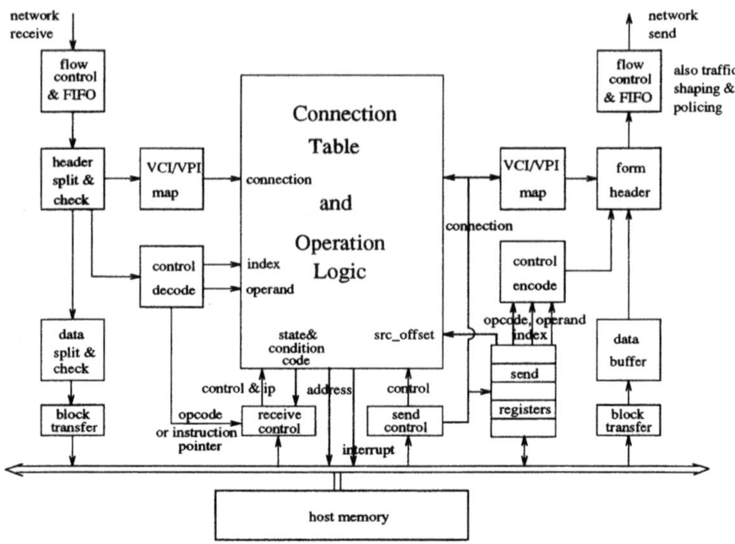

Figure 4: DART block diagram

tations — which can achieve constant low latency and high bandwidth with minimal loading on the host — can be useful even at 155Mbps. Such hardware is essential at 622Mbps for both low latency and high bandwidth. While software implementations might be adequate to achieve 10μsec latency in first generation 155Mbps ATM LANs, they will not be adequate to achieve latencies under 3μsec in second generation 622Mbps ATM LANs.

We present an architecture called DART which implements the hybrid deposit model described in Section 3. Figure 4 shows a general block diagram of the common DART components. This paper focuses on the Connection Table, Operation Logic, and Send Registers. The Receive and Send Control blocks are finite state control machines. The remaining function blocks are either rather standard for ATM interfaces or uninteresting (for this paper).

6.1 DART Overview

The Connection Table contains state for each active connection (incoming and outgoing) as shown in Figure 5. Each entry contains an endpoint number, address register information (explained later), some connection state, and a reply connection number. The endpoint number indexes into the Endpoint Table which contains the buffer base and bounds information for each endpoint buffer. The endpoint information is in a separate table so that multiple connections to the same endpoint buffer can share the information. Finally, the reply connection indicates the connection number to use to reply in responding to an incoming message. The Connection Table is mapped into uncached processor address space and managed by the operating system.

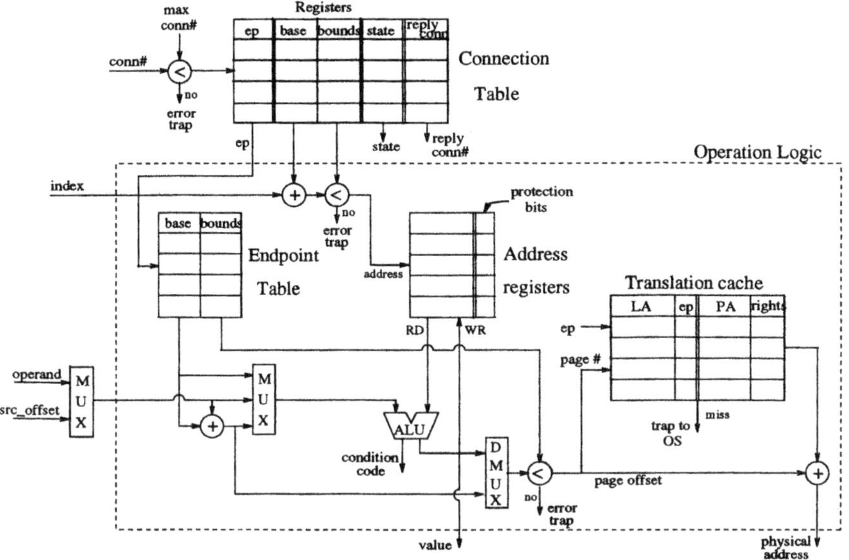

Figure 5: Connection Table and Operation Logic

The Send Registers are a set of six registers for each active outgoing connection. These registers are used for forming send messages. Each connection's send registers can be mapped (and marked uncached), via an operating system call, into the first six locations in a page in the application address space. Thereafter the application can access the send registers without operating system interaction.

The Operation Logic in Figure 5 performs protection, mapping, and the primitive operations listed in Figure 2. For simplicity, we make the following initial assumptions: Address register operations are restricted to at most one address register read and write operation per cell; endpoint pages must be pinned in physical memory while an endpoint buffer is active; remote reads must be handled by the host processor (via an interrupt); and the instruction pointer message variant is not supported. We also omit latches and control wiring for clarity.

6.2 DART Receive Side

Incoming messages index into the Connection Table using a connection number derived from the cell VCI/VPI. Each connection has a number of local memory locations for address registers. This introduces the awkwardness of a separate name space, but avoids main memory accesses in the critical path (for indirection). Each connection has a number of contiguous locations to form a register "window". For convenience and flexibility, we allow each connection to dynamically allocate the window size at connection set up time. The address register base and bounds fields in the connection table entry point to the beginning and end of this window respectively. This scheme allows the overlapping and nesting of register windows to effect different sharing and protection (as described for endpoints in Section 3.2). The protection bits on each address register control read, write, and access privileges as described in Section 3.5.

Because the endpoint pages are pinned in this architecture, the Endpoint Table contains the physical address of the buffer base. However, since physical pages are not necessarily allocated contiguously, we use a TLB-like address translation cache to contain mappings from the logical address of the endpoint base address + offset to the appropriate physical address. Note that the TLB must match on bits identifying the endpoint as well as on the physical address (PA) since multiple endpoints may have the same PA but different protection. A TLB miss causes an interrupt to the host processor. This allows full software flexibility in managing the storage of endpoint mappings. One way to reduce the number of TLB misses is to use a hybrid scheme which stores the first N endpoint buffer page mappings directly in the Endpoint Table and manage any overflow mapping entries with the TLB. A small number, like $N = 2$ is probably sufficient for most endpoints.

To implement a multiple cell format for block data transfers, data cells following an initial "control" cell (as described in Section 4) are deposited in successive memory locations after the data in the control cell. The state field in the Connection Table entry indicates whether an arriving cell for that connection should be interpreted as a control cell or a data cell.

The Address Registers, and TLB are all mapped, via additional data and control paths, into the processor address space so the operating system can manage their contents.

6.3 DART Send Side

Both control information and data must be provided to send a cell. The control information comes from the set of Send Registers associated with an outgoing connection and mapped into a known location in the application address space. The first three registers in this set are the control information — the opcode, operand, and index — for the cell. The data comes from the endpoint associated with the connection. The fourth register, named "go" controls the initiation of a send.

The send side reuses the Operation Logic to map from the application address space to physical endpoint addresses, i.e. we multiplex the Operation Logic between the receiver side and the sender side. To send a block of data at *src_offset* from an endpoint base, we simply write *src_offset* into the "go" Send Register for the appropriate connection attached to that endpoint. This write causes a data block at the endpoint offset indicated by *src_offset* (32 byte block aligned) to be read and composed into a cell with the control information in the opcode, operand, and index registers and sent via the associated connection.

The set of Send Registers for each connection also contains a size register and a mode register. The size register controls the number of 16 byte data blocks (minimum of 2 blocks) sent starting from the offset written into the "go" Send Register. This register is set to 0 when the last cell has been sent to the flow control and traffic shaping unit. The mode register enables two variants of the send procedure just described. The first is a shortcut: the operand is taken from the value actually written to the "go" register. The second causes an exception if an attempt is made to send when the status field is non-zero.

6.4 Exception processing

Exceptions arising due to error traps, TLB misses, and unimplemented operations cause an interrupt to the host processor. (Error traps can also be configured to discard the offending cell without causing an exception.) Any state operated on by the cell, e.g. write to address registers, does not get updated until after the point of the last possible exception point for that cell. We retain the cell in the input FIFO and block processing of further cells from that connection until re-enabled by the host processor: cells are only removed from the input FIFO when a cell "commits" after the last exception point. (We assume per connection buffering and flow control at the destination.) This in turn might cause the flow control mechanism for that connection to be invoked. Cells belonging to other connections can be processed once the exception condition is saved (even if these connections share the same endpoint as an exceptioned connection). We cannot process cells belonging to the exceptioned connection, even if they are not affected by the exception, since some applications may depend on the per-channel ordering guarantee of ATM cells.

Cell processing can also be blocked globally across all connections. The operating system uses this feature to get atomic access to DART state.

6.5 Relaxing the assumptions

To allow the execution of up to four primitive operations per message (one address generation, two register operations, and one conditional)[5] we clock the Operation Logic through multiple primitive operations per message, feeding back immediate values as necessary via the "value" path shown in Figure 6. A main opcode controls the selection and ordering of the primitive operations. Example opcodes are read, read multiple, write, write multiple, and software exception which causes an interrupt to the host processor. The instruction format allows up to three different register operands to be named in addition to an immediate operand. To retain the simple roll-back exception model in the face of multiple register operations, address register writes are cached and only written back after the cell processing commits.

Removing the restriction on pinning endpoint pages is straightforward: the Endpoint Table now contains virtual addresses and the TLB maps from virtual addresses to physical addresses. This enhancement also introduces a new category of exceptions: page faults from references to paged out endpoint pages. These are treated as another class of exceptions and are serviced by the host processor (which maintains the main virtual mapping tables). The operating system is now responsible for keeping the mapping information in DART consistent with the main memory state.

Handling remote reads without the interrupting the host processor just requires a more complicated receive controller.

To support the instruction pointer message variant, we modify the Connection Table and add an Instruction Memory on the receive side as shown in Figure 6. The Connection Table now contains a base and bounds entry, similar to that for address registers, for access to the

[5]So we can do priority queueing and interrupts.

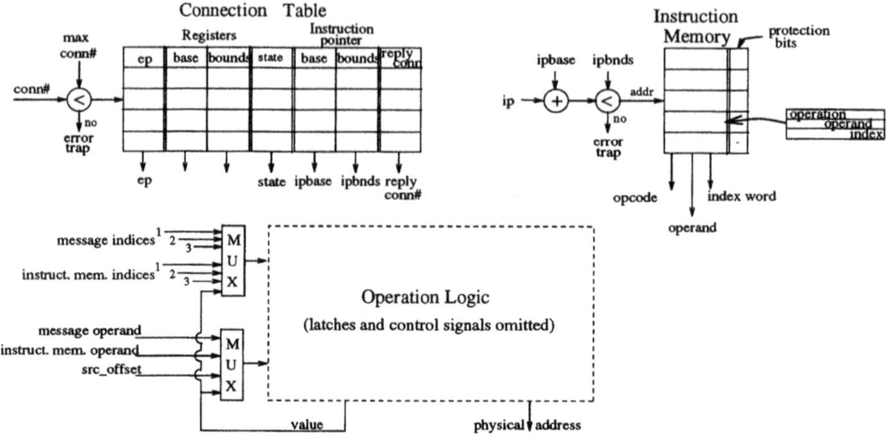

Figure 6: Modifications for multiple primitive operations and instruction pointer format

Instruction memory. The instruction pointer (ip) from the receive control logic and the ipbase and ipbnds information from the Connection Table combine to index into the Instruction Memory. To keep the scheme simple, each instruction is composed of an operation and operands in exactly the same format as in an ATM cell (as in Figure 3). The operation controls from which location — the instruction memory in the destination or the message — operands and indices are taken. Protection bits, like those for the address registers, allow the destination to control which instructions serve as entry points to the sender. One very useful addition is a sequencer to step the destination through several instructions per cell. It is tempting to add further enhancements, such as conditional sequencing operations, but we avoid them to keep the interface simple: more complex functionality can be obtained by trapping to the host processor.

6.6 Status

At this point we have only completed a high level design of DART. We omitted some of the detail (such as data masking) from the architectural sketch just presented; many more details and implementation options require further evaluation. Some of the major open issues are programmable control for destination operand decoding, the division between hardwired and microcontrolled (e.g. using a small microprocessor) operations, and the duplication of the Operation Logic for sender and receiver. We are planning to interface DART to the PCI bus.

The only difficult timing constraint is completing whatever bus operations are required for data transfers before the control information in the next cell arrives. The address generation, register, and conditional operations can be overlapped with the data and CRC arrival which takes 1.8μsec at 155Mbps and 450nsec at 622Mbps. Since register operations take one cycle to access each register plus one cycle to writeback, the worst case of an address operation, two binary register operations, and a conditional operation on registers takes nine operation logic cycles. This yields a generous clock period of 200nsec at 155Mbps and 50nsec at 622Mbps.

7 CONCLUSIONS AND FURTHER WORK

We presented a new model for low overhead communication. This hybrid deposit model increases the role of the sender in order to simplify the role of the destination. Using both sender-supplied information and destination information, messages are routed so as to minimize the impact on the host processor: data messages are deposited directly in memory, delayable action messages are queued for later processing when convenient, and immediate action messages are sent directly to the host processor. Moreover, the sender and destination information can be combined in a flexible way to span the spectrum from purely sender-based communication, as in some parallel computing systems, to conventional, purely destination-based communication. Thus, our hybrid deposit model is applicable across a wide range of applications (as well as a wide range of networks) in parallel, distributed, and real-time computing.

We specialized this hybrid deposit model to ATM LANs and exploited the synergy between the relatively small, fixed size data units in ATM and common memory block sizes. Our preliminary results show that a purely software implementation on stock workstation hardware can do fairly well, to a point. However, hardware is required to get consistent low latency, high bandwidth, and reduced processor load. We sketched an architecture called DART which hardware demultiplexes incoming cells directly to where they are required, on a per-cell basis. DART provides a small number of demultiplexing primitives in hardware, along with some simple ways to perform compound multiplexing operations. More complex operations are handled via software exception to the host processor.

The per cell processing architecture of DART is similar in principle to the message-driven processor (MDP) [De87]. However, we use DART in a filtering role for depositing messages, rather than for direct computation and we provide full protected multiuser communication. The closest related work we know of for LANs are Hamlyn and Axon. Hamlyn [Wil92] adopts a similar emphasis on increasing the participation of the sender to minimize the required functionality at the destination. However Hamlyn differs in some major ways: it supports only direct addressing, one delayed action queue per node (there can be any number in DART), and it pins endpoint pages. Like DART, Axon [SP90] has self describing packets for direct deposit at the destination. However, Axon lacks hardware support for flexible sender and destination-based addressing, hybrid interrupt control, and register operations.

To date we have developed the model and designed plausible implementations. Now we intend to turn our attention mostly to the other half of the hybrid deposit "hypothesis": examining how higher level protocols such as TCP/IP, services such as RPC, and applications can benefit and what services they require in an network interface. We plan to use our existing software implementation as the basis for this work.

REFERENCES

[AI87] Arvind and R. Ianucci. Two Fundamental Issues in Multiprocessing. In *Proc. of DFVLR - Conf. on Parallel Processing in Science and Eng.*, June 1987.

[Be94] M. Blumrich and et al. Virtual Memory Mapped Network Interface for the SHRIMP Multicomputer. In *Intl Symposium on Computer Architecture*, April 1994.

[BP93] D. Banks and M. Prudence. A High-Performance Network Architecture for a PA-RISC Workstation. *Journal of Selected Areas in Communications*, pages 191–202, February 1993.

[Dav93] B. Davie. The Architecture and Implementation of a High-Speed Host Interface. *Journal of Selected Areas in Communications*, pages 228–239, February 1993.

[De87] W. Dally and et al. Architecture of a Message-Driven Processor. In *Intl Symposium on Computer Architecture*, 1987.

[DLM94] C. Dubnicki, K. Li, and M. Mesarina. Network Interface Support for User-Level Buffer Management. In *Parallel Computer Routing and Comm. Workshop, Univ. of Washington*, May 1994.

[DP93] P. Druschel and L. Peterson. Fbufs: A High-Bandwidth Cross-Domain Transfer Facility. In *Proc. of the Sympos. on Operating System Principles*, December 1993.

[Ke94] P. Keleher and et al. TreadMarks: Distributed Shared Memory on Standard Workstations and Operating Systems. In *Proc. of Winter Usenix Conf.*, January 1994.

[NPA92] R. Nikhil, G. Papadopoulos, and Arvind. *T: A Multithreaded Massively Parallel Architecture. In *Intl Symposium on Computer Architecture*, pages 156–169, May 1992.

[Onv94] R. Onvural. *Asynchronous Transfer Mode Networks: Performance Issues.* Artech House, 1994.

[Se93] J. Subhlok and et al. Programming Task and Data Parallelism on a Multicomputer. In *Proc. of ACM Sympos. on Principles and Practice of Parallel Programming*, pages 13–22, May 1993.

[Se94] T. Stricker and et al. Decoupling Communication Services for Compiled Parallel Programs. Technical Report CMU-CS-94-139, CMU, 1994.

[SP90] J. Sterbenz and G. Parulka. Axon: A High Speed Communication Architecture for Distributed Applications. In *Proceedings of IEEE INFOCOM*, 1990.

[Spe82] A. Spector. Performing Remote Operations Efficiently on a Local Computer Network. *Communications of the ACM*, pages 246–260, April 1982.

[SR91] J. Stankovic and K. Ramamrithm. The Spring Kernel: A New Paradigm for Real-Time Systems. *IEEE Software*, 8(3), May 1991.

[TLL93] C. Thekkath, H. Levy, and E. Lazowska. Efficient Support for Multicomputing on ATM Networks. Technical Report TR93-04-03, Dept. of Computer Science, Univ. of Washington, April 1993.

[TS93] C. Traw and J. Smith. Hardware/Software Organization of a High-Performance ATM Host Interface. *Journal of Selected Areas in Communications*, pages 240–253, February 1993.

[von92] von Eicken et al. Active Messages: A Mechanism for Integrated Communication and Computation. In *Intl Symposium on Computer Architecture*, pages 256–266, May 1992.

[Wil92] J. Wilkes. Hamlyn: An Interface for Sender-based Communication. Technical Report HPL-OSR-92-13, HP Labs, November 1992.

ACKNOWLEDGMENTS

MERL colleagues Chia Shen, Richard Waters, John Howard, Hugh Lauer, and Qin Zheng gave helpful suggestions on the interface ideas and paper presentation. John Kubiatowicz of the MIT Lab for Computer Science gave valuable feedback on the design ideas and presentation in an earlier draft. The development of the hybrid deposit model benefitted from discussions with Peter Steenkiste and Thomas Gross at CMU. Takushi Kawada and Vu Le Phan of Mitsubishi Electric helped install Mach 3.0.

PART SEVEN

Posters

18

A Multimedia Document Distribution System Over DQDB MANs*

Luis Orozco-Barbosa[a] and Michel Soto[b]

[a] Department of Electrical Engineering, University of Ottawa,
161 Louis Pasteur, Ottawa, Ont., K1N 6N5 Canada

[b] Institut Blaise Pascal, Laboratoire MASI, Université Pierre et Marie Curie
4 Place Jussieu, 75252 PARIS Cedex 05, France

In this paper, we present a system architecture for supporting Real-time Multimedia Document Distribution Applications. This architecture is based on an extended version of the ODA (Office Document Architecture) standard using as communications support the IEEE 802.6 MAN Standard : an integrated services local area network (ILAN). Multimedia document delivery is performed by dynamically allocating the network bandwidth. This allocation strategy is based on the bandwidth required by the various media composing the documents. A performance analysis of the proposed bandwidth allocation scheme is given.

Keyword Codes: C.2.2, C.2.4, C.2.5
Keywords: Network Protocols; Distributed Systems, Local Networks

1. INTRODUCTION

Multimedia communications may be simply defined as the exchange of highly-structured documents including text, graphics, voice, audio and video sequences. Three main requirements could be identified for multimedia communications systems [1] :

- Multimedia document standards : The spread of multimedia applications makes possible the interchange of multimedia documents. This can only be achieved by means of well accepted international standards.
- Multimedia workstations : The use and integration of new peripherals capable of supporting a wide variety of multimedia applications, such as video-conferencing (digital telephony and video).
- High speed networks : High speed communications networks will have to provide the means to support a wide variety of communications services : bandwidth on demand, synchronous and asynchronous services, guaranteed quality of service (low response times and low blocking probabilities).

In the near future, large storage capacities and transmission technologies able to provide sufficient data bandwidth transfer will be both necessary to handle multimedia documents. This will be needed particularly for handling multimedia documents comprising digital video sequences. Several research axes on technologies, such as optical storage, file handling

* This work was supported in part by the Telecommunications Research Institute of Ontario (TRIO) and the Natural Sciences and Engineering Research Council of Canada under grant number OGPIN 334.

techniques and non-volatile cache memory as applied to multimedia applications are being carried out by research laboratories and universities [1].

Multimedia information implementations are built up by combining a wide variety devices, technologies and procedures. For example, a multimedia information system in a medical environment will require the use of very large capacity storage devices, e.g., optical disks, high speed computer network, special acquisition devices and highly reliable protocols [2]. This kind of systems, multimedia medical information systems, will support digital imaging applications requiring high resolution images and lossless compression techniques. Higher compression rates may be used in other environments, such as multimedia office information system. In this last case, the amount of information to be stored may be reduced substantially.

Multimedia information systems often rely on a distributed architecture where several servers are used for storing multimedia documents. It is now a reality that working environments are already heterogeneous because they are composed of different types of devices going from fully equipped workstations to more or less powerful personal computers with low storage capacities.

Another important point is the need of a structuring and interchanging multimedia documents standard. These standards should define the links and synchronization of static and time-based information as well as the interchange of multimedia information. One of the most promising standard is ODA, the Office Document Architecture [3]. Some other relevant standards such as HyTime [4] and MHEG [5] may also be considered. However, these two later standards are in a committee draft and in a working document stage.

In this paper, we present a system architecture for supporting Real-time Multimedia Distribution Application. By Real-time Multimedia Distribution Application, we mean a system capable of providing the support for storage and on-line retrieval of multimedia documents from a multimedia server. As multimedia documents will be much bigger in size than simple ASCII text files, the retrieval of these documents are much likely to be carried on-line. In this paper, we propose a system architecture for supporting the real-time delivery of multimedia documents. This architecture is based on the CCITT Office Document Architecture (ODA) Standard and it uses as transmission support an integrated services local area network (ILAN).

The paper is organized as follows. Section 2 reviews the general architecture of the ODA Standard. In Section 3, we provide a functional description of bandwidth allocation strategy. Section 4 contains a performance analysis of the proposed strategy aiming to assess the bandwidth required for supporting multiple multimedia connections. Finally, in Section 5 our conclusions and future research directions are given.

2. MULTIMEDIA DOCUMENT ARCHITECTURE

2.1 Standards

The definition of the multimedia document architecture is a fundamental element in a multimedia communications system. Current effort are being carried to define a standard able to specify the different types of data needed for use in multimedia systems [6][7]. It has been recognized that one of the best way to approach this problem will be to define a standard that could be easily extended to accomodate additional formats [6]. Another approach has been to extend current standards [8]. In our approach, we have preferred to look into current standards. We believe that current standards, such as ODA provide the modelling basis required for the definition of a multimedia documents architecture. Furthermore, it has been recognized that whatever standards will be adopted, multimedia communications systems having different characteristics will have to be able to interwork. We therefore look into the particular case of a multimedia document architecture based on an extended version of ODA [3].

ODA has been studied in the context of office applications. The types of documents that may be represented in the ODA document model are those frequently used in the general office work such as reports, letters, forms, invoices, memoranda. In practice such office documents may contain text, graphics and pictures within a single document: a so called "Multimedia Document". The purpose of the document model is to provide a method of describing the electronic representation of these documents, including the type of information found in their content. The structured description of the documents is termed the "architecture" of the document. A document architecture can be subdivided in three categories, namely, logical structure, layout structure and content .

The logical structure provides a method for organizing the content of the document and it is intended to closely correspond to those aspects of the document structure related to the functional semantic of the document. The layout structure provides a method for organizing the content of the document into pages and areas within pages. The details on the document architecture are given in [3].

Multimedia document may contain information such as text, graphics and images. This information may be organized as follows:

- a header which contains information related to the originator and recipient of the document such as name, document title and date .
- a control part containing the scenario, i.e., the description of the existing links between the various document's components.
- the body of the message. Each body part will contain one or various formats.

The goal of the document architecture is to propose means by which the document generation and reading scenarios can be represented and described with respect to the logical structure and the synchronization requirements between body parts.

2.2 An ODA-based Multimedia Document Architecture

The multimedia document architecture defines the method for generating and reading documents. The reading of the document by the recipient is performed in an animated fashion. The term animation is used to express the simultaneous treatment (e.g. display, playback) of the different body parts.

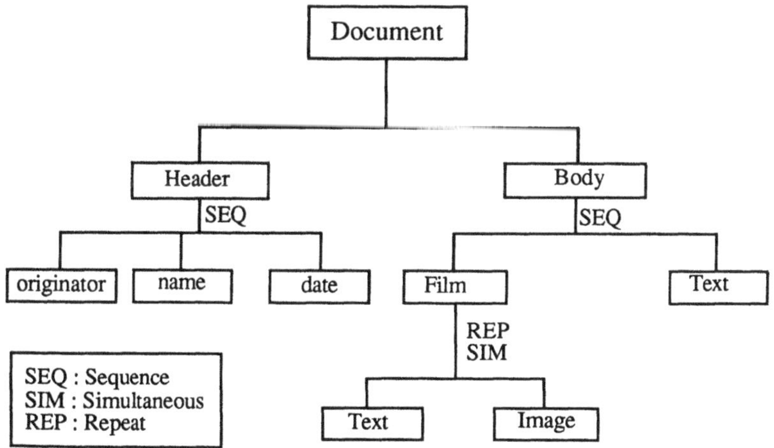

Fig. 1 Generic logical structure of a multimedia document

The generic logical structure of the document specifies the set of rules by which the document is to be constructed. The generic structure is very similar to the concept of a class in object oriented programming languages and indeed the generic structure of a document is referred to as its class. Fig. 1 shows the generic structure for a typical multimedia document. This structure, however, is given as an example of possible document organizations.

In Fig. 1 the generic structure defines a document as a sequence containing a Header and a body. The header in turn is composed of a originator identifier, name, date of the document. This kind can be further extended to contain other relevant information. The body of the document is a set of parts. The figure does not show the control part. As already stated, this control part will contain the information regarding the links among the different body parts. In particular, it will include all the operators defining the relation among the various body parts.

The film section consists of a text and an associated image. The text may be related to the image as a whole. In this case the text is used for this purpose. If a more detailed association is necessary, a window section can be used, first to indicate the particular portion of the image to be referred (the position may be located, in the image, by means of a reference to the particular portion for example).

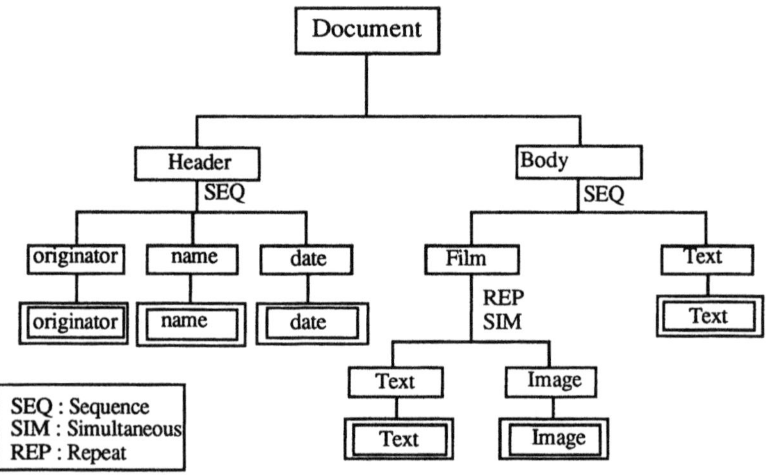

Fig. 2 Generic logical structure of a multimedia document

The specific structure of the document is derived from the generic structure. Fig. 2 represents an instance of a specific logical structure generated, with respect to the generic logical structure. Content portions are located at the lowest level components of the tree. Content architecture defined in the current version of ODA may be used for text, geometric and raster graphics. However for bulky data, such as video sequences further extensions to the current ODA standard are needed. Some experimental implementations have overcomed this limitation by placing bulky data out of the document architecture by replacing them by references [9].

3. REAL-TIME MULTIMEDIA DOCUMENT DISTRIBUTION

The various elements of a real-time multimedia system architecture have to interact in order to properly carry out the distribution of documents. In the following, we give a description of the functions to be carried by network bandwidth allocation scheme. In this description, we also explain the way the allocation scheme makes use of the information describing the organization of the multimedia document (scenario). This is a central element enabling the proper allocation of the bandwidth following the requirements of the application (transfer of multimedia documents). We also describe the way these elements interact with all the other service elements and the communication services provided by DQDB.

3.1. Scenario

As already mentioned, the scenario describes the organization of a multimedia document. This description provides a temporal as well as a spatial relation of the different elements of a multimedia document. Figure 3 depicts an example of a scenario of a multimedia document composed by three different media, namely text, voice and video. The scenario contains all the required information to *playback* the multimedia document. For the particular example depicted in Fig. 3, the scenario corresponds to a multimedia document which starts by a video segment of 3 minutes, followed by a segment of voice and text of 3 minutes, a pause of 2 seconds and ending with a voice and video segment of 5 minutes. From the example, we can distinguish the two main temporal relation depicted in the example of Figures 1 and 2: simultaneous and sequential. The scenario also provides us with the bandwidth requirements of the various multimedia segments. This information is particularly useful for playing back the multimedia document as the retrieval of the document from a multimedia database server takes place. The main challenge is to allocate the bandwidth to make possible the timely transfer of the different elements of the document. This is the purpose of defining a bandwidth allocation scheme using the information provided by the scenario.

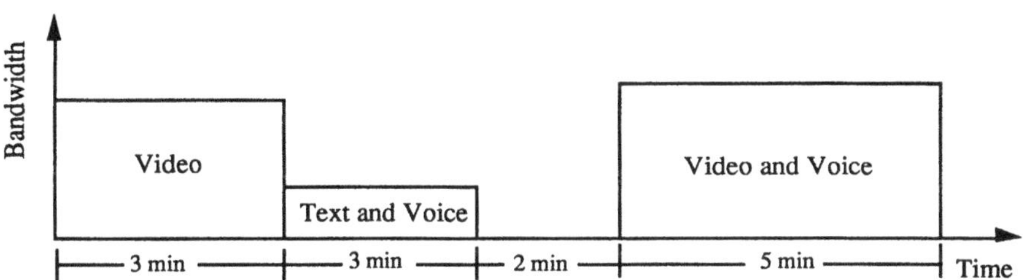

Fig. 3 A Typical Scenario

3.2. Network Bandwidth Allocation Scheme

In our architecture, the allocation of the network bandwidth is dynamically performed. One of the major challenges is therefore to define a Network Bandwidth Allocation Scheme (NBAS) suitable to handle in real-time the requests for allocating as well as freeing the network bandwidth. The operation of NBAS should be closely related to the services provided by the network. In this study, we have focused on the use of the IEEE 802.6 standard [10]. DQDB (Distributed Queue Dual Bus) has been selected by the IEEE 802.6 working group as the standard for Metropolitan Area Networks (MANs). The timing structure of DQDB is based on frames of 125 μs comprising a given number of fixed slots. In turn, each DQDB slot comprises 48 (voice) channels with a nominal bit rate of 64 kbps per channel. Slots are accessible by two methods :

- In the pre-arbitrated (PA) access, slots are marked by the node that is head of the bus. These slots carry isochronous service octets at defined offsets, every 125 μs.
- In the queues arbitrated (QA) access, slots are assigned by means of a queueing mechanism for outstanding slots. A queue is maintained in a distributed manner through all the active nodes within the MAN. Three levels of priority are supported by this method.

In order to adapt the two aforementioned access methods to the higher levels the following Convergence Functions (CF) have been included in the DQDB standard:

- Isochronous Convergence Function (ICF)
  This function has specially been designed for connection-oriented services requiring timing relation between the source and the destination, such as constant bit rate.
- MAC Convergence Function (MCF)
  This function provides the basis for connectionless communications. The destination and source addresses are included in the header of every MAC protocol data unit.
- Connection Oriented Convergence Function (COCF)
  This function has not been completely defined in the standard. However provisions are contained for its future definition.
- Other Convergence Functions (OCF)
  The inclusion of these functions will allow to define new services.

In order to provide the service required to dynamically allocate the bandwidth for retrieving, we define a new convergence function. The main reason behind our choice of defining a new convergence is twofold, 1) under this scheme, a multimedia document is progressively transmitted by means of multimedia segments; 2) the definition of a dynamic bandwidth scheme looks suitable for optimizing the bandwidth utilization. In the next section, we will demonstrate the advantages of using this scheme.

Fig. 4 shows the protocol architecture relating Network Bandwidth Allocation Entity (NBAE) to the new OCF. The main objective of the NBAE is to coordinate the operation of the application (not shown in the figure) and the convergence function. The functions to be performed by the NBAE are : to interpret the multimedia document scenario and upon this interpretation to drive the convergence function in allocating the bandwidth needed for conveying the document accordingly.

In the following, the operation of the proposed bandwidth allocation scheme will be explained by using the scenario depicted in Fig. 3. In order to convey all the information needed to indicate to the Head Of Bus (HOB) how to mark the slots, the NBAE will use, through the OCF, the PA slot with Virtual Channel Identifier (VCI) all ones defined in the DQDB to convey all the required signaling information [10]. We will refer to this slot as the S-slot and in order for the scheme to operate properly, the S-slot should precede the QA slots of the current cycle. The 48-bit payload of the S-slot is divided into three fields, namely F-0, F-1 and F-2. These fields are initially set up to all zeros by the HOB. As the S-frame passes by the nodes, the NBAE will request the OCF to reserve a given number of slots according to the rules depicted in Table 1. The number of slots is reserved by incrementing the fields F-0, F-1 and F-2 which corresponds to the one of three levels of priority with priority 2 being the highest one. According to this scheme, NBAE requests with the lowest priority the allocation of corresponding number of slots indicated by F-0. The priority one is used at the beginning of a new segment. For instance, at the beginning of the multimedia segment consisting of text and voice, the NBAE will direct OCF to allocate the required number of slots with a priority one. Finally, the highest priority, number 2, is used when a session is engaged in transferring a multimedia segment. Upon receiving this information and for a given cycle, the HOB will start by allocating the slots with priority 2, followed by 1 and finally 0.

The rationale to define three priorities can be simple explained as follows. As the number of slots per cycle is finite, the number of active sessions may exceed the number of slots available. Therefore, by assigning the highest priority to the session currently engage in the transfer of segment, we attempt to provide to the user with a synchronous transfer of the required information. The second level of priority will give an advantage to the session already in progress but initiating a new segment which may be slightly delayed without affecting the quality of service provided to the user. Finally, the third level of priority will make sure that new sessions will be initiated only if enough bandwidth to provide the quality of service required is available.

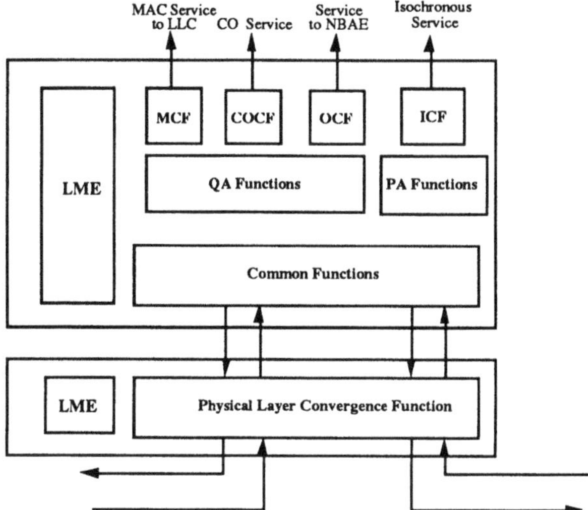

Fig. 4 DQDB Node Functional Architecture

Table 1 Bandwidth Allocation Priorities

Priority	Condition
0	New Session
1	Beginning of New Segment
2	Segment in Progres

4. PERFORMANCE STUDY

The architecture suggested in this paper aims to optimize the use of the bandwidth of the underlying communication network by taking into consideration the quality of service required by the real-time multimedia application. It is particularly important to evaluate the amount of bandwidth and the blocking probability required for providing service to a large number of users. Blocking may result during the delivery of a document mainly by lack of bandwidth. From the user point of view, it will be very annoying to be blocked in the middle of a document delivery. It is therefore essential for the service provider to guarantee a low blocking probability.

4.1. Analytical Model

The bandwidth request process can be modeled by using a closed, multi-class queueing network (Fig. 5). The service rates, μ_{NODE_i} of the infinite servers $NODE_i$, i = 1, 2, ... N, represent the interarrival rates of the connection requests. Servers SIGNAL and HOB describe the signalling channel and the Head Of the Bus, respectively. By signalling channel, we mean the PA slot used for conveying the requests for slots. The communication active period is modelled by infinite servers whose service rates, $\mu_{TRAFFIC_i}$, are set according to the length (holding time) of the various HBS's. By using this model, it is possible to carry out an analysis of the bandwidth allocation scheme under various conditions. The model can be solved by applying the numerical method Convolution[11].

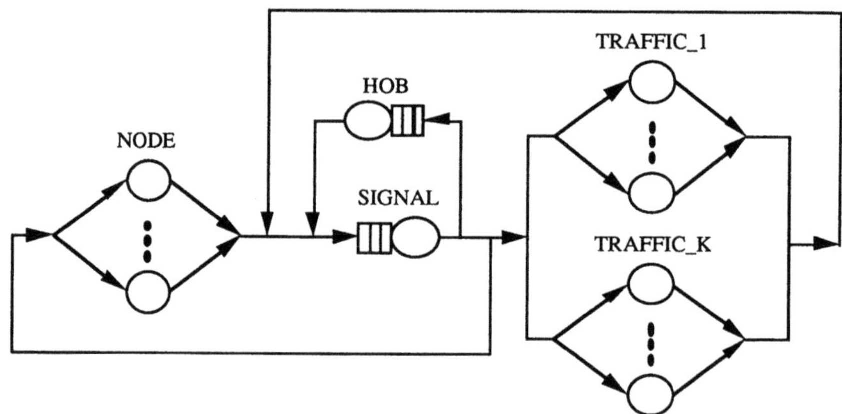

Fig. 5 Queueing Model of the Bandwidth Allocation Scheme

The number of channels used for the transmission of a multimedia document will vary from a segment to another. The blocking probability, p_B, can be expressed by :

$$p_B = p(y_i > K) \qquad (1)$$

where y_i is the number of segment in service and is given by :

$$y_i = (n_{i1}, n_{i2}, \ldots n_{iR}) \tag{2}$$

where R and n_{ir} are the total number of different classes of segment and the number of segment of class r in service center i = TRAFFIC, respectively. The parameter K is given by :

$$K = \left\lceil Q / \sum_{j=1}^{R} s_{ij} \right\rceil \tag{3}$$

where s_{ij} and Q are the number of channels used per type of HBS at server i = TRAFFIC, and the total number of channels in the system, respectively. $\lceil T \rceil$ is defined as the smallest integer not less than the real number T.

The blocking probability due to lack of bandwidth can be calculated by first determining the equilibrium state probabilities at server TRAFFIC. These probabilities, according to the queueing theory [11], are given by :

$$P(S = (y_1, y_2, \ldots y_N)) = G\, g_1(y_1)\, g_2(y_2) \ldots g_N(y_N) \tag{4}$$

where S is an aggregate state of the system and G is a normalizing constant chosen to make the equilibrium state probability sum to 1, respectively. For the case of a infinite server (IS) service center :

$$g_i(y_i) = \prod_{r=1}^{R} (1 / n_{ir}!)\, [e_{ir} / \mu_{ir}]^{n_{ir}} \tag{5}$$

where e_{ir} and μ_{ir} are the relative arrival rate of class r customers to service center i and the mean service rate of class r customers at service center i, respectively.

The blocking probability is therefore given by :

$$P_B = \sum_{S(N,i)} P(S = (y_1, y_2, \ldots y_N)) \tag{6}$$

where the state space, S(N,i), for N customers at the ith queue is given by :

$$S(N,i) = \{y_i = (n_{i1}, n_{i2}, \ldots n_{iR}) \mid \sum_{j=1}^{R} s_{ij}\, n_{ij} > Q\} \tag{7}$$

4.2. Numerical Results

In this section, we illustrate the performance of the proposed bandwidth allocation scheme. Our main aim is to show the required bandwidth for supporting multiple multimedia document deliveries, as well to evaluate the blocking probability which may result during the document delivery.

In order to illustrate the performance, we have assumed that the documents consist of two media, namely, audio and video. Audio sources require a sustained bit rate of 772 kbps, that is to say 12-bytes at every DQDB frame [11], while video sources require a bit rate of 1.544 Mbps. This latter value is in compliance with the current CCITT's Recommendation H.261 (also called pX64) for visual telephony (videophone and vide-conferencing) [12]. It is assumed that the multimedia documents are transmitted by using two different classes of segment, namely SEG_{mm} and SEG_{audio}. The former corresponds to an HBS comprising both media, while the latter only consists of audio. We assume a nominal bit rate of 44.736 Mbps for DQDB, corresponding to 14 DQDB slots (672 voice channels). This value has been suggested as a prime candidate for a network speed compatible with DS3 circuits [10]. Assuming the aforementioned values and a static bandwidth allocation scheme is easy to verify that the maximum number of concurrent multimedia document deliveries is limited to 18.

By now, it is difficult to predict the ratio of the various media composing a multimedia document. It is however expected that the development of new technology will make common practice the use of visual (video and images) services [1][13][14]. In order to better evaluate the performance of the proposed bandwidth allocation scheme, we have divided our study into two parts. First, the use of video has been limited to half of the multimedia document contents, while in the second part the extensive use of video has been assumed. Furthermore, it is worth to notice that typically the service times for the different segments are several orders of magnitude higher than the signalling and processing to be carried out by the head-end node.

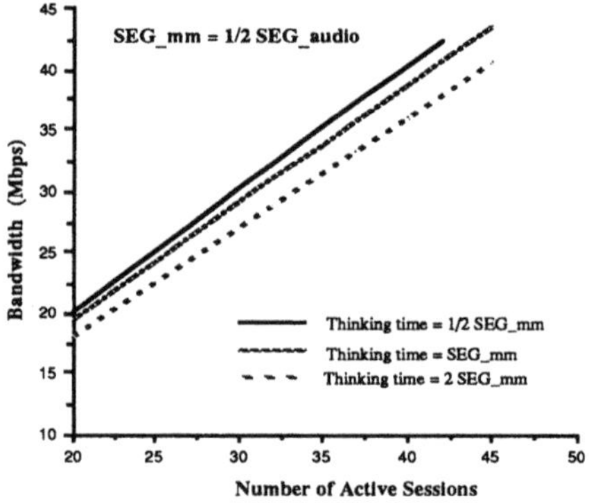

Fig. 6 Bandwidth vs. Number of Active Sessions

Fig. 6 shows the bandwidth required for supporting the concurrent retrieval of multimedia documents. In this case, the extensive use of audio service is assumed, while the use of video only accounts for a third of the total document length. This bandwidth has been evaluated for three different thinking times (time between two successive requests). These values range from half to twice of the service (holding) time of the SEG_mm. For all three cases, the figure shows an important improvement in the number of concurrent active retrieval sessions as opposed to the static bandwidth allocation scheme of 18. For all three cases, the improvement is in the order of 150%.

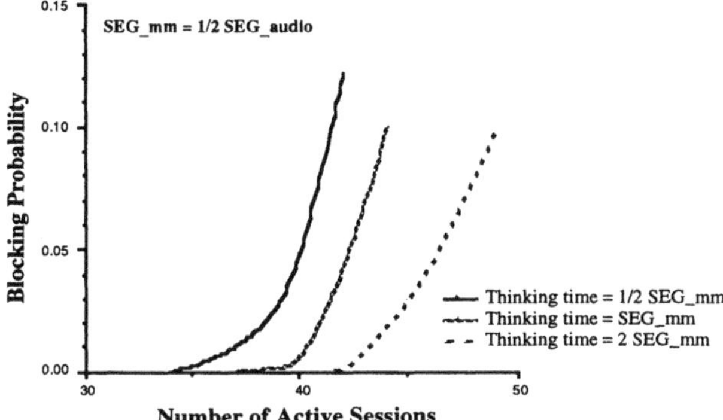

Fig. 7 Blocking Probability vs. Number of Active Sessions

Fig. 7 shows the blocking probability as a function of the number of active sessions. As expected, the figure shows that the blocking probability increases less steeply as the ratio between thinking time/service (holding) time of the multimedia segment (SEG_mm) increases. It is very important to provide flexibility in the services to be provided by a multimedia distribution system. One of the main performance measure of interest to the users is grade of service that the system may provide, i.e. loss probability. Real-time applications based on audio-visual services may tolerate some losses. The grade of service required will be determined by the application, as well as the coding techniques used [13][14].

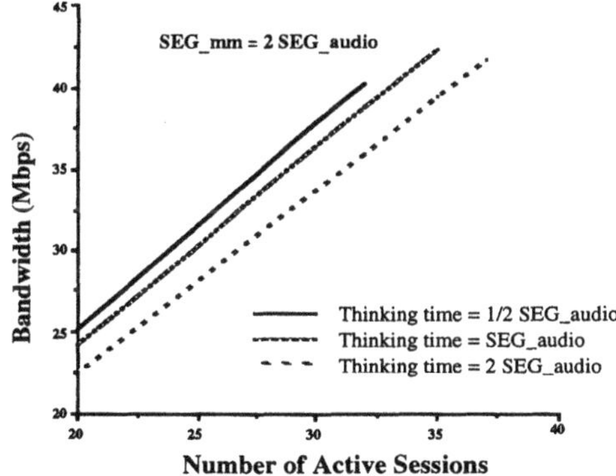

Fig. 8 Bandwidth vs. Number of Active Sessions

In the second part of our analysis, we have been interested in evaluating the impact of introducing a higher degree of visual services. In this second case, the composition of the document is assumed to consist of multimedia segment with a holding time twice the holding time of an audio segment. Fig. 8 shows the bandwidth required as a function of concurrent active sessions. As expected, a higher degree of visual composition of the multimedia document severely affects the performance of the system. This results from the fact that, in general, visual services have a more stringent communications needs than audio communications. As for the previous case, Fig. 8 shows the bandwidth required for three different thinking times. The values of the three thinking times considered range from half to twice of the shortest segment namely the segment of type audio. In this case, an improvement of 100% in the number of concurrent sessions is obtained by using the proposed bandwidth allocation scheme.

Finally, Fig. 9 shows the blocking probability as a function of the active sessions. As in the previous case, this blocking probability gives a measure of the expected probability of being blocked once a user is engaged in the retrieval of a multimedia document. In the case of a real-time application, the fact of offering a degraded service (blocking probability) may be considered as an option for providing service to a larger number of concurrent users.

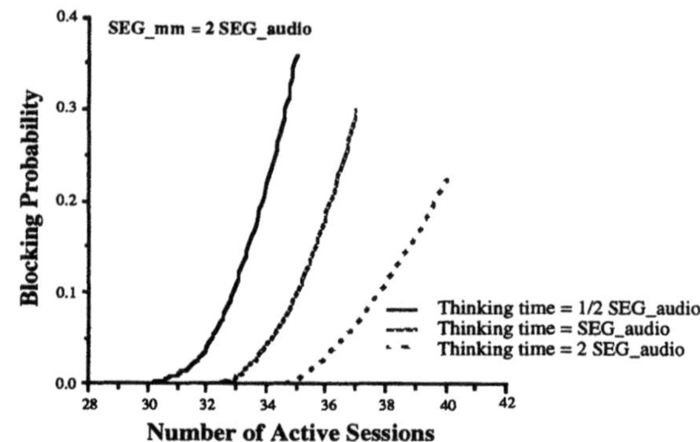

Fig. 9 Blocking Probability vs. Number of Active Sessions

5. CONCLUSIONS AND FURTHER WORK

In this paper, we have presented a novel protocol architecture suitable for multimedia document distribution systems. The design of this protocol architecture has been based on the CCITT Standard for Office Document Architecture It has also been assumed the use of the IEEE 802.6 DQDB MAN as the transmission support. The introduction of new protocol elements has been necessary so as to cope with the stringent requirements of a real-time multimedia delivery system, namely : short delays and intermedia synchronization. A performance analysis of the bandwidth allocation scheme has been carried out. Our results show that the blocking probability can be limited for a larger number of sessions as compared to the number of sessions supported by using an static allocation scheme.

The proposed bandwidth allocation scheme can also be considered to be used with other local or metropolitan computer networks providing integrated services, such as FDDI network environments [15][16]. By making use of the multimedia document description (scenario), the scheme can drive the medium access control procedures to allocate the bandwidth according to the requirements of the different media. However, in order to extend this scheme to high speed long haul networks, such as ATM [17], this mechanism can be integrated as part of the procedures used for the establishment of the connection. This last issue requires further study.

REFERENCES

[1] L. Orozco-Barbosa and N. D. Georganas, "Multimedia Services and Applications", *European Trans on Telecommun. and Related Technologies*, vol. II, pp. 5-19, No. JAN/FEB 1/1991.

[2] R. Kositpaiboon, P. Tsingotjidis, L. Orozco-Barbosa and N.D. Georganas, "Packetized Radiographic Image Transfers over Local Area Networks for Diagnosis and Conferencing", *IEEE Journal on Select. Areas on Commun*, vol. 7, no. 5, pp. 842-856, June 1989.

[3] ISO, Information Processing - Text and Office Systems. Office Document Architecture (ODA) and interchange Format (ODIF), Part one to eight, International Standard 8613, March 1988.

[4] ISO/IEC - Committee Draft International Standard 10744 - Information Technology - "Hypermedia/Time-based Structuring Language (HyTime)", April 1991

[5] ISO MHEG Working document "S" - ISO/IEC JTC1/SC29/WG12, Jan. 1992.

[6] N. Borenstein and N. Freed, MIME (Multipurpose Internet Mail Extensions) Mechanisms for Specifying and Describing the Format of Internet Message Bodies, RFC 1341.

[7] ISO-10021(1) and CCITT-X.400. "Message Handling Systems - Information Processing Systems - Text Communication - MOTIS", March 1988.

[8] N. Borenstein and C. Thyberg, "Power, Ease of Use and Cooperative Work in a Practical Multimedia Message System", International Journal of Man-Machine Studies, April 1991.

[9] A. Karmouch, L. Orozco-Barbosa, N.D. Georganas and M. Goldberg, "A Multimedia Medical Communications System,", *IEEE Journal on Select. Areas in Commun*, vol. 8, no. 3, April 1990, pp. 325-339

[10] IEEE 802.6-1990, IEEE Standards for Local and Metropolitan Area Networks : "Distributed Queue Dual Bus (DQDB) - Subnetwork of a Metropolitan Area Network (MAN)", 1991.

[11] S.S. Lavenberg (Ed.), *Computer Performance Modeling Handbook*, Academic Press, Inc, San Diego, Cal, 1983.

[12] R. K. Jurgen, "Digital Video", *IEEE Spectrum*, vol. 29, no. 3, March 1992, pp. 24-30.

[13] J.M. Cioffi, J.W.M. Bergmans, H.K. Thapar, and J.K. Wolf (Eds), *IEEE Journal on Select. Areas on Commun*, Special Issue on Signal Processing and Coding for Recording Channels, vol. 10, no. 1, January 1992.

[14] N. A. Davies and J.R. Nicol, "Technological Perspective on Multimedia Computing", *Computer Comms*, vol. 14, No. 5, June 1991, pp. 260-272.

[15] ISO 9314-1, Information Processing systems - Fibre Distributed Data Interface (FDDI), Part 1 : Token Ring Layer Protocol (PHY), April 1989.

[16] ISO 9314-2, Information Processing systems - Fibre Distributed Data Interface (FDDI), Part 2 : Token Ring Media Access Control (MAC), May 1989.

[17] T1S1.5/90-001 R2, Broadband Aspects of ISDN, Baseline Document, June 1990.

19

From SDL Specifications to Optimized Parallel Protocol Implementations

Stefan Leue[a] and Philippe Oechslin[b]

[a] Institute for Informatics, University of Berne, Länggassstrasse 51, CH-3012 Berne, Switzerland

[b] Computer Network Lab LTI, Swiss Federal Institute of Technology, DI-LTI EPFL, CH-1015 Lausanne, Switzerland

Keyword Codes: C.1.2; C.2.2; D.2.1; D.2.2
Keywords: Computer-Communication Networks, Network Protocols; Software Engineering, Requirements/Specification; Software Engineering, Tools and Techniques

Abstract

We propose a formalized method that allows to automatically derive an optimized implementation from the formal specification of a protocol. Our method starts with the SDL specification of a protocol stack. We first derive a data and control flow dependence graph from each SDL process. Then, in order to perform cross-layer optimizations we combine the dependence graphs of different SDL processes. Next, we determine the common path through the multi-layer dependence graph. We then parallelize this graph wherever possible which yields a relaxed dependence graph. Based on this relaxed dependence graph we interpret different optimization concepts that have been suggested in the literature, in particular lazy messages and combination of data manipulation operations. Together with these interpretations the relaxed dependence graph can be be used as a foundation for a compile-time schedule on a sequential or parallel machine architecture. The formalization we provide allows our method to be embedded in a more comprehensive protocol engineering methodology.

1. Introduction

Optimized protocol implementation has become an important field of research as network speed has increased much faster than computer processing power over the last decade. We present a method for the mainly automated derivation of efficient implementations of protocol stacks, starting from formal specifications. The rigor in the formalization is useful when implementing our method as a tool, which we are currently doing. In the paper we formalize and generalize optimization approaches that can be found in the literature, in particular in the literature on optimal protocol implementation.

Overview. In Figures 1 and 2 we present a (partial) view of the SDL specification of a two layer protocol stack consisting of two processes named `N` and `N+1`. It will serve as a running example in our paper and we will refer to it as TLS. It is the purpose of our optimization and implementation method to transform specifications similar to TLS into parallelized and optimized implementations. In Section 2 we discuss the sort of layered SDL specifications we consider in the paper. Here, we also argue why a direct and faithful implementation of SDL specifications would lead to inefficient implementations. This is mainly due to the structuring of SDL specifications into per-layer processes and the resulting inter-layer asynchronous queue based communication mechanism.

First, we construct a dependence graph representing control-flow and data dependences among statements in an SDL specification. This leads us to so-called *Transition Dependence Graphs.* Their construction is explained in Section 3.1. For the dependence graphs for example TLS see Figures 1 and 2. The dependence graph construction is an application of methods known from the domain of compiler optimization and parallel compilation as they are for example described in [10] and [3]. Control flow dependences relate directly successive statements (e. g. S2 and S3 in Figure 1) whereas data dependences relate statements where the depending statement uses a variable that is defined in the other statement (e. g. S and D1 in Figure 1).

In the second step of our method we perform an optimization and parallelization of the operations which are caused by the processing of a packet. We consider the way the packet takes from the point where it enters the protocol stack to where it exits. Therefore we have to combine transition dependence graphs belonging to different SDL processes. We do so by eliminating the inter-layer communication statements, e. g. the statements S4 and S9 in the example TLS. The result is a Multi-Layer Dependence graph. We describe the construction in Section 3.1, for an example see Figure 4.

Third, we identify the path a packet takes through the protocol stack in the so-called common case, from the root node representing the point where a packet is accepted from the environment to the exit node, where the packet is conveyed to the environment. For example, we assume that in the example TLS decision D1 has one common and one uncommon branch, whereas decision D2 has two common branches. This resulting graph is called *common path graph*, for an example see Figure 6. We will apply our later optimizations only to the common case part of the specification.

Fourth, we relax dependences on the common path graph in the following steps.

- *Anticipation of the common case:* In this step we ignore that certain statements depend on a decision, namely for those decisions where we assumed a common outcome. Henceforth we treat these decision nodes as if no other node depends on their execution. An example is decision D1 (see Figure 6).

- *Parallelization:* We construct a relaxed dependence graph by taking the data flow dependence relation of the CPG and by adding additional dependences which ensure that a node is never executed before the last decision node on which it depends in the control flow dependence relation has been executed (see Section 6.2). For example node S10 is not data flow dependent on decision node D2, but still both nodes may not be executed in any order, because the execution of S10 depends on the evaluation of D2. However, S10 and S11 are not data dependent and may thus be executed in parallel (meaning in any order).

Finally, in Section 7 we show how suggestions that have been made in the literature to optimize the implementation of communication protocols can be interpreted based on the relaxed dependence graph. We refer to the concepts of *Lazy Messages* (see [20]), and, in particular, *Grouping of Data Manipulation Operations* (see [7], [8] and [1]).

The optimized and parallelized graph now serves as a foundation for an implementation on either a sequential or a parallel machine architecture. The discussion of implementation aspects such as scheduling is outside the scope of this paper. We refer the reader to [17] and [15] for further discussion.

Related work. Aspects of hardware and software architecture that increase an implementation's efficiency are discussed in [7], [8], [20], [9] and [23]. Hardware implementations for high speed protocols have been proposed in [13]. Special attention has been paid to the parallelization of protocol implementations, so for example in [5] and [24]. However, the parallelization proposed in these papers depends entirely on the intuition of the designer and thus its efficiency may be non-optimal. Therefore automated support for the parallelization is desirable. An approach based on the scheduling of parallel tasks generated by an Estelle compiler is presented in [11]. In [19] the determination of data-flow dependence graphs for parallel implementations of stream processing programs on transputers are described. Others ([22], [14]) analyze the data- and message flow dependences between communicating processes, whereas we restrict ourselves to the analysis of dependences inside processes.

Precursors. Precursors of our work appeared in [17] where we describe the application of our method to a IP/TCP/FTP protocol stack. Further results will appear in [16]. More technical detail can be found in [15].

The role of SDL. The formal specification technique we consider is the CCITT standardized *Specification and Description Language* SDL [6]. We chose this language not because we particularly advocate its suitability as an implementation language, but rather because it enjoys wide acceptance in the protocol engineering community. For an overview of SDL see for example [4]. The choice of a formal description technique as starting point connects our method to existing techniques and methods in the domain of protocol engineering (see for example [18]). We may for example assume that as result of a previous verification step the specifications on which we base our optimization are dead- and livelock free. Also, conformance tests developed based on the formal specification can be directly applied to the implementation. Part of our method (dependence analysis and construction of multi-layer dependence graphs) are specific to features of SDL. However, we claim that for many other procedural specification methods an easy adaptation is possible. The later steps (starting with the CPG construction and down to the optimization steps we describe) are independent of the specification method on which the dependence graph is based.

2. A Discussion of SDL Specifications

2.1. SDL Specifications of Protocol Stacks

SDL is a Formal Description Technique frequently used in the specification of telecommunications systems, in particular for the layered specification of communications protocols. An SDL specification of a protocol stack can usually be structured into different concurrent processes, each one representing the functionality of one protocol layer. A process is structured into transitions which describe its dynamic behaviour. Processes communicate via asynchronous signal queues. In SDL the mapping of the output signals of the sending process to corresponding input signals of the receiving process is done using a relatively complicated mapping of signal names to signal routes, where the signal routes carry the sender and receiver identification information. For reasons of conciseness of the presentation we abstract away from this mechanism and identify sender and receiver of messages simply by identity of the message type. Thus, in the Example TLS the message `Y` sent out by process `N` is consumed by process `N+1`.

2.2. Inadequacy of 'Faithful' Implementations

By the term *faithful* we refer to an implementation which follows in its structure and in the sequence of operations exactly the original SDL specification from which it is derived. This may for example mean that the SDL specification is directly compiled so that every statement in the SDL specification is mapped to a statement in the implementation, that every SDL process corresponds to a process in the implementation, and that the processes in the implementation communicate using the SDL asynchronous communication mechanism via infinite queues. However, as we argue in the following, such a faithful implementation is potentially inefficient.

- *No explicit parallelism:* Although SDL processes run concurrently the processing inside an SDL process is strictly sequential. This means that the structuring of the specification into processes, which in many cases is influenced by general design decisions, determines the degree of parallelism of a specification. It also means that without optimizations the sequential processing of operations inside a process may be inefficient compared to a parallel execution.

- *Structuring of the specification into processes:* The structure of the specification often means that there is one process per protocol layer peer entity of the protocol (see for example the specifications presented in [4]). The design of communication protocols is often governed by the principle that *'a good specification is a highly modular and layered specification'*. Though from a structured-design point of view a layered design may be desirable, we stipulate that in order to derive efficient parallel protocol implementations such a layered design is obstructive. This is mainly due to the fact that the parallel scheduling and combined execution of operations belonging to different protocol layers, which can lead to a considerable gain in efficiency, are inhibited by the layer-wise structuring of the specification. Similar arguments can be found in [9].

- *Asynchronous inter-layer communication via infinite queues:* An efficient implementation of a protocol stack for one peer entity will usually be a non-distributed system. Apparently it is very inefficient to implement the exchange of data in a non-distributed system via asynchronous queues. Instead, the protocol data will be stored in a local memory and the communication between the processes will be by shared variables.

The objectives of our method are therefore to remove the boundaries between processes, to remove the asynchronous communication between processes, and to analyze dependences between statements in order to enable parallel and combined execution of statements belonging to different processes.

3. Dependence Analysis for SDL Processes

In this section we explain how a Dependence Graph can be obtained by syntactical analysis from an SDL specification. For a definition of the mathematical notation we use here and in later Sections see the Appendix. First we will explain how transitions as basic building blocks of SDL process specifications can be formalized and then how entire protocol stacks can be represented as graphs, based on the graphs representing the transitions.

3.1. Transitions in SDL Specifications

Syntactic structure. A *transition* in an SDL specification is a construct which describes the transition of an SDL process from one *symbolic state* into a successor symbolic state. The body of a transition consists of a collection of statements which we group in the set of statements S. We only consider a limited subset of SDL-statements, namely `INPUT`, `TASK`, `DECISION` and `OUTPUT` statements, and we identify one of these four statement types with every element of S. For a complete description of the syntax of SDL transitions see [4]. The syntactic subset we have chosen is a concise subset of the full SDL syntax. It allows for the analysis of standard protocol specifications as presented in [4]. For the sake of conciseness we have limited our considerations to the language subset described above but we conjecture that an adequate treatment of most of these constructs is possible when extending our method.

3.2. Control Flow and Data Flow Dependences

The syntactical analysis of the SDL specifications that we describe in this Section yields a graph structure over the set of statements S. This so-called dependence graph represents the two types of dependences between the statements of a specification, namely control flow and data flow dependences.

Statements, which according to the syntactical and semantical rules of SDL are direct successors, are part of the *control flow dependence* relation *cfd* over the set S. A statement of type `DECISION` has two or more directly succeeding statements, all pairs of a `DECISION` statement and it successor statements are part of the *cfd* relation. The execution of a statement directly succeeding a `DECISION` statement depends on the run-time evaluation of the decision predicate. This is represented by a branching of the *cfd* graph.

A statement usually describes operations on process variables in which these are usually referenced in two different ways.

- We say that a statement S_n *uses* a variable x iff it references the variables current value without modifying it. Note that in one statement more than one variable may be used. A typical use of a variable would be to reference its value in the expression on the right hand side of an assignment statement.

- We say that a statement S_n *defines* a variable x iff it assigns an initial or new value to the variable without referencing its previous value. A typical example is the definition of a variable on the left hand side of an assignment statement.

A pair of statements (s_1, s_2) is in the *data flow dependence* relation *dfd* if (s_1, s_2) is in the transitive closure of the *cfd*-relation[1] and s_2 *uses* a variable which is *defined* in s_1. For simplicity we assume that no re-definition of variable names inside transitions occurs[2]. Also, we assume that every variable name used in a transition is defined inside of the transition, therefore no data dependences from statements in other transitions exist. Function calls are assumed to have no side-effects and to return a single value. Assignments to structured variables are decomposed into component-wise assignments. An `INPUT(X)` statement is a *define* statement with respect to a variable named X, an `OUTPUT(Y)` statement is a *use* statement with respect to variable named Y[3].

3.3. Transition Dependence Graphs (TDG)

Definition Transition Dependence Graph. Let S, STT and X denote pairwise disjoint sets, the elements of which we call *statements*, *statement types* and *variables*. Formally, we define a Transition Dependence Graph (TDG) as a tuple $T = (S, STT, X, sttype, cfd, dfd)$ where $cfd \subseteq S \times S$, $dfd \subseteq cfd^+$, $STT = \{input, decision, task, output\}$, $sttype \subseteq S \times STT$ is a functional relation (relating a statement to a statement type), $use \subseteq S \times \mathcal{P}(X)$ is a functional relation (relating a statement to the set of variable names which are being *used* in it), and $define \subseteq S \times X$ is a partial functional relation (relating a statement to the variable name which is being *defined* in it), satisfying the following conditions:

1. (S, cfd) is a tree.

2. $\forall s \in S$ the following conditions hold: $(sttype(s) = \{input\}) \leftrightarrow (\mid \{s\} \triangleleft cfd \mid = 1 \wedge root(S, cfd) = \{s\})$ (an `INPUT` statement has exactly one successor, and it is the root of the tree), $sttype(s) = \{decision\} \rightarrow \mid \{s\} \triangleleft cfd \mid \geq 2$ (every `DECISION` node has at least two successors), $sttype(s) = \{task\} \rightarrow \mid \{s\} \triangleleft cfd \mid \leq 1$ (every `TASK` node has at most one successor), and $sttype(s) = \{output\} \rightarrow s \in leaves(S, cfd)$ (an `OUTPUT` statement is a leaf of the tree).

3. $(\forall (v, w) \in dfd)(define(v) \subseteq use(w))$.

3.4. Example SDL processes and TDGs

In the following examples we give the SDL specification of a transition in graphical representation (SDL-GR) on the left hand side, and equivalent specification in textual form (phrase representation, SDL-PR) in the middle of the chart, and a resulting dependence graph on the right hand side. We add labels `Sn` and `Dn` to help us to identify regular and decision statements, respectively. However, these labels are not part of the specification.

The example in Figure 1 shows a partial view of the specification of a process `N` of which we show only two transitions. The dependences are as follows. The *control flow* dependence follows the linear sequence of the statements `S1`, `S2`, `S3` and `D1` and then branches to either `S4` or `S5`. The `DECISON` statement `D1` has possible successor statements

[1]Thus our definition of the data dependence implies that an 'earlier' statement in the control flow cannot be data dependent on a 'later' one.

[2]This avoids additional *output* dependences, see [21].

[3]The data dependences we consider are purely local to the processes, we do not consider data dependences between processes caused by message flows.

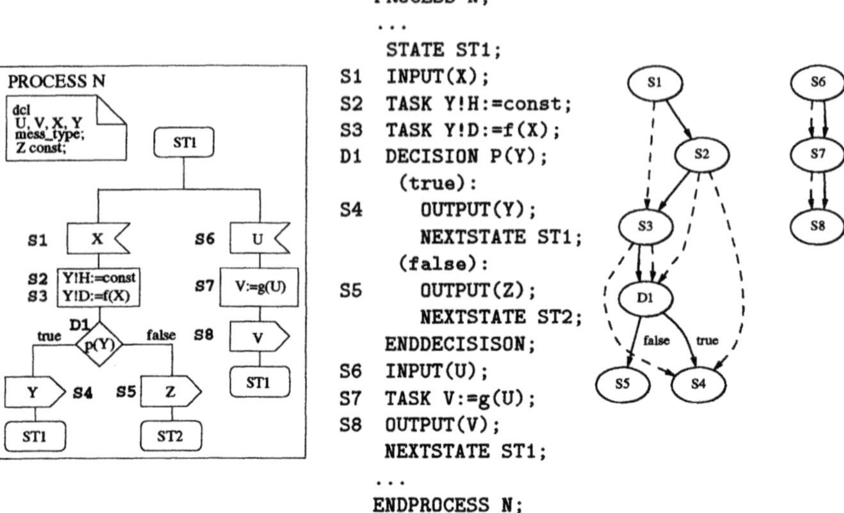

Figure 1. SDL-PR (left) and SDL-GR (middle) specifications and TDGs (right) for process N

S4 and S5, the respective control flow dependence edges are labeled for illustrative purposes by *true* and *false.* The data flow dependences are so that S3 depends on S1 because of variable X, whereas D1 and S4 both depend on S2 and S3 because of the use of variable Y[4]. Figure 1 presents a graphical representation of this TDG which we call T_1, namely on the right hand side by the graph starting in node $S1$. *Solid* line arrows represent control flow dependencies, thus elements of *cfd*, and *dashed* line arrows represent elements of *dfd.* The dependences for the transition starting in statement S6 are obvious.

For our later argumentation, which aims at combining multiple processes to one process, we need a further example of an SDL process, namely the process named N+1 presented in Figure 2. The syntactical analysis leads to the transition dependence graph T_3 shown on the right hand side of Figure 2. The structure of the dependence graph is quite similar to the structure of the dependence graph for process N. It should also be noted that the decision D2 does not have a boolean evaluations, instead it evaluates to strings A1 or A2.

4. Dependence Graphs for Protocol Stacks

As we saw in Section 2 protocol stacks are usually specified by a set of independent concurrent processes. Each of these processes consists of a number of transitions. In Section 3 we described how to syntactically analyze each of these processes in order to

[4]One may envisage Y!H to stand for the header and Y!D for the data part of a protocol data unit or a packet.

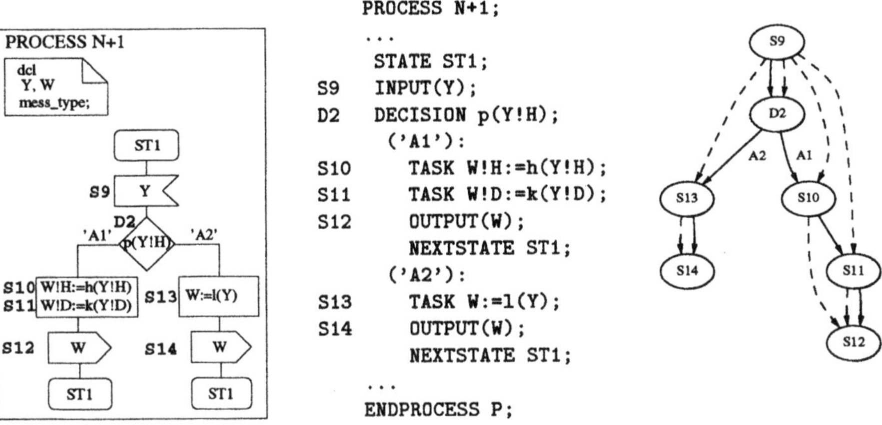

Figure 2. SDL-PR (left) and SDL-GR (middle) specifications and TDGs (right) for process N+1

derive a set of transition dependence graphs for each process. As we argued in Section 2.2 it is advantageous to remove the boundaries between layers of SDL processes and to eliminate the inter-layer communication via infinite queues. In this section we describe the necessary steps to combine the transition dependence graphs of different SDL processes and to remove the communication between them. Technically, we perform this in two steps. First, we label all TDGs of all processes by so-called *input/output labels.* These labels are the names of the signals exchanged by the `INPUT` and `OUTPUT` statements at the beginning and at the end of each transitions. Second, we combine all TDGs with matching input/output labels, eliminate the `OUTPUT(X)/INPUT(X)` statement pairs, and perform a cross-layer data dependence analysis. We may do this because we assume that every `OUTPUT` statement can be mapped to a unique `INPUT` statement of another process. The result is a graph which we call *Multi-Layer Dependence Graph.*

4.1. Input/Output labeled Transition Dependence Graphs (IOTDGs)

We assume that all transitions we consider for the combination process start with an `INPUT` statement accepting a data packet from an adjacent layer process, and end with an `OUTPUT` statement which delivers the processed packet to the next adjacent layer process. Hence, we assume that all the processing for a packet in a layer process is carried out in the course of *one* transition, and that no looping inside a transition occurs. Thus, our dependence graphs are always trees. Different transitions starting in different states in one process may exist, but they only represent the process to be in different states (e. g. state *waiting* and state *transmission*). Furthermore, we assume that the packet passing is unidirectional, either from the medium towards the user or vice versa.

Formal Definition of Input/Output labeled TDGs. Based on the above stated assumptions on the structure of the SDL Transitions we formalize the concept of labeling of root and leave nodes of TDGs by the appropriate signal names as follows. Let $T = (S, STT, X, sttype, cfd, dfd)$ denote a TDG and let SIG denote a set disjoint from any other set in sight, the elements of which we call *signal names*. Furthermore, let $insig \subseteq ((S \cap root(T)) \times SIG)$ and $outsig \subseteq ((S \cap leaves(T)) \times SIG)$ denote functional relations. We define an Input-Output labeled Transition Dependence Graph (IOTDG) as a tuple $I = (S, STT, X, SIG, sttype, cfd, dfd, insig, outsig)$ for which the following conditions hold: $sttype(root(I)) = input$, and $(\forall x \in leaves(I))(sttype(x) = output)$.

Example IOTDG In Figure 3 we show the three IOTDGs representing the TDGs for Example TLS.

4.2. Multi-layer Dependence Graph (MLDG)

What we have obtained so far is a set $\mathcal{T} = \{T_1, \ldots, T_n\}$ of IOTDGs. $\mathcal{T}$ represents the dependences of all transition of the specification that we analyze. In this section we describe an algorithm that transforms $\mathcal{T}$ into a set $\mathcal{M}$ of Multi-Layer Dependence Graphs (MLDG). Each MLDG represents the dependences of the processing of one packet or protocol data unit in adjacent layers of the protocol stack. We are interested in following the processing of one packet from the code location where it enters into the protocol stack to the location where it exits. In our example this means that we will derive a connected control flow dependence graph from statement `S1`, where the packet `X` enters the processing in process `N`, to the statements `S12` and `S14`, where it exits the stream of processing in process `N+1` as a message of type `W`. Thus we have to compose the individual IOTDGs in $\mathcal{T}$. The criterion for composing two IOTDGs will be that they exchange a message with identical names, e. g. one IOTDG ends with an `OUTPUT(Y)` statement and another IOTDG begins with an `INPUT(Y)` statement. We assume that the names of the types of the messages exchanged are unique at the interfaces between two processes, and that the direction of the message flow is uniquely determined by the message type names. Also, we assume that every `OUTPUT` statement can be mapped to a unique `INPUT` statement. Note that SDL transitions are deterministic on `INPUT` signals, i. e. in one state the future behavior is uniquely determined by the type of the message that is consumed next.

MLDG Construction Algorithm. The algorithm is as follows. First, a set $\mathcal{T}'$ of initial IOTDGs is selected (step I.). This set contains all those IOTDGs that do not input a message that is output-ed by another IOTDG. The algorithm then loops over all these IOTDGs (III.). The set $\mathcal{Z}$ (V.) contains all those IOTDGs that can be appended to a leaf node of an IOTDG from $\mathcal{T}$. The next loop (VI.) performs the merging of two IOTGs (VII. to XVI.) for all elements of $\mathcal{Z}$. The merging of two IOTDGs comprises the elimination of the two nodes x and $root(Z)$ by which the two graphs are merged (IX.), this corresponds to the elimination of the `OUTPUT/INPUT` statements. XIII. describes the construction of the new *cfd* relation. Every node that depended on $root(Z)$ is made dependent on every node from which x depended. The construction of the new *dfd* relation (XIV.) is very similar, but we additionally check whether a node on which x depended defines a variable which is used in a node that depended on $root(Z)$. XVII. constructs the result, a set $\mathcal{M}$ of MLDGs.

Algorithm 1

I. SELECT $\mathcal{T}' = \{T'_1, \ldots, T'_m\} \subseteq \mathcal{T}$ *SO THAT*
$(\forall T'_i)(\forall T_j)(\text{insig}(\text{root}(T'_i)) \cap \bigcup_{j \neq i} \text{outsig}(\text{leaves}(T_j)) = \emptyset)$;

II. $\mathcal{M} := \emptyset$;

III. FOR ALL $T'_i \in \mathcal{T}'$

IV. $M := T'_i$;
V. $\mathcal{Z} := \{T \in \mathcal{T} \mid \text{outsig}(\text{leaves}(M)) \cap (\text{insig}(\text{root}(T))) \neq \emptyset\}$;
VI. WHILE $\mathcal{Z} \neq \emptyset$
VII. FOR ALL $Z \in \mathcal{Z}$
VIII. SELECT $x \in \text{leaves}(M)$ *SO THAT*
$(\text{outsig}(x) \in \text{insig}(\text{leaves}(\text{root}(Z))))$;
IX. $S'_M := S_M \cup S_Z - \{x\} - \text{root}(Z)$;
X. $X'_M := X_M \cup X_Z$;
XII. $\text{sttype}'_M := S'_M \triangleleft (\text{sttype}_M \cup \text{sttype}_Z)$;
XIII. $\text{cfd}'_M := \text{cfd}_M \cup \text{cfd}_Z - (\text{cfd}_M \triangleright \{x\}) - (\text{root}(Z) \triangleleft \text{cfd}_Z)$
$\cup \{\text{domain}(\text{cfd}_M \triangleright \{x\}) \times \text{range}(\text{root}(Z) \triangleleft \text{cfd}_Z)\}$;
XIV. $\text{dfd}'_M := \text{dfd}_M \cup \text{dfd}_Z - (\text{dfd}_M \triangleright \{x\}) - (\text{root}(Z) \triangleleft \text{dfd}_Z)$
$\cup \{(v, w) \in \{\text{domain}(\text{dfd}_M \triangleright \{x\}) \times \text{range}(\text{root}(Z) \triangleleft \text{dfd}_Z)\} \mid \text{define}(v) \subseteq \text{use}(w)\}$;
XV. $M := (S'_M, STT_M, X'_M, \text{sttype}'_M, \text{cfd}'_M, \text{dfd}'_M)$
XVI. $\mathcal{Z} := \{T \in \mathcal{T} \mid \text{outsig}(\text{leaves}(M)) \cap (\text{insig}(\text{root}(T))) \neq \emptyset\}$;
XVII. $\mathcal{M} := \mathcal{M} \cup M$

As a result we obtain a set of MLDGs $\mathcal{M}$. Each $M \in \mathcal{M}$ is a multi-edged labeled tree $(S, STT, X, SIG,$ *sttype*, *cfd*, *dfd*$)$. Note, however, that not all of the conditions we required for IOTDGs still hold. For example it is not true any more that a node of type *input* has no predecessor in the *cfd* relation. For a MLDG M we say that a node in *root*(M) is an *entry* node, that a node in *branchnodes*(M) is a *branching* node, and that a node in *leaves*(M) is an *exit* node. An entry node represents a statement where a message (in most cases a packet or protocol data unit) is accepted from the environment, and an exit node refers to a statement in the code where a message is delivered to the environment.

Example MLDG Figure 4 shows the set $\mathcal{M}$ which we obtain by applying our algorithm to the IOTDGs of our example TLS. It contains two MLDGs, one with root `S1` and one with root `S6`. In order to illustrate the input/output labeling we also retained the respective labels at the nodes. Note that the *cfd*-relation forms the skeleton of the MLDGs. The nodes `S4` and `S9` have been eliminated, reflecting the elimination of the `OUTPUT(Y)` / `INPUT(Y)` statement pair. The additional *cfd* pair $(D1, D2)$ has been added. Furthermore, data dependences between statements of the two merged graphs have been added, so for example $(S2, D2)$.

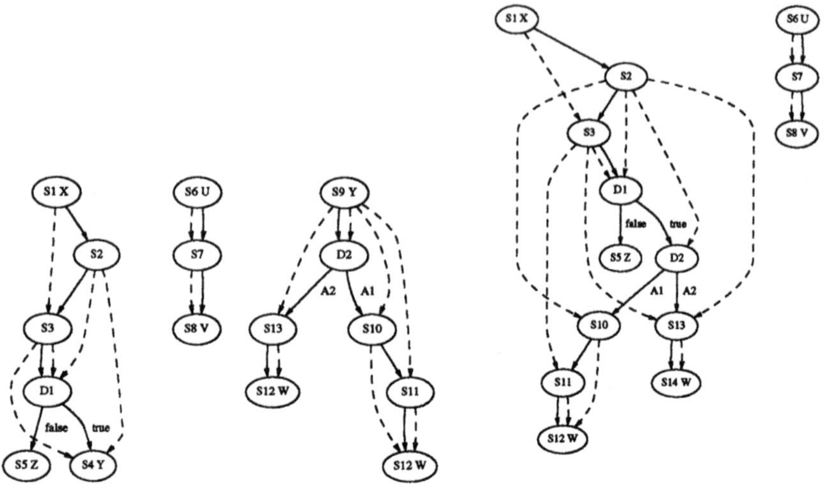

Figure 3. IOTDGs for Example TLS.

Figure 4. MLDGs for Example TLS.

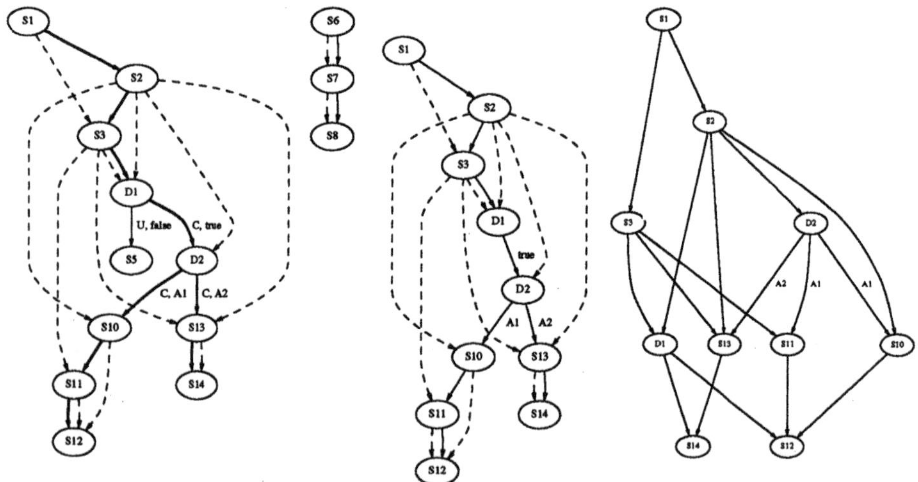

Figure 5. Labeled MLDG for Example TLS.

Figure 6. Common Path Graph for Example TLS.

Figure 7. The Relaxed Dependence Graph for Example TLS

Justification for MLDG construction. When building the MLDG we modified the original SDL specification in two ways. Firstly, we ignored the asynchronous queue communication mechanism, and secondly, we eliminated the corresponding `OUTPUT / INPUT` statement pair. The justified question arises whether these modifications preserve the correctness of the original specification. We argue that ignoring the queue can be justified because this is a refinement step which preserves two essential queue properties, namely 1. the *safety* property that it is always true that if something is received it must have been sent before, and 2. the *liveness* property that it is always true that if something is sent it will eventually be received. The safety property is trivially satisfied because the order of the `OUTPUT(X)` and `INPUT(X)` statements is preserved. The liveness property is satisfied if we assume our implementation to be live, namely that every transition which is continuously enabled will eventually be taken. The elimination of the `OUTPUT / INPUT` statement pair can be justified by the fact that we preserved all control flow and data flow dependences. Thus, the above argument concerning the safety properties now holds for all those statements which are direct predecessors or successors of the `OUTPUT / INPUT` statements that we eliminated. Concluding we can say that out of the many interleavings of events which are possible according to the original specification we only implement one possible representative, namely the interleaving where a packet is accepted at one end of the protocol stack, entirely processed, and finally handed over at the other end before the next packet is accepted for processing. Also, as opposed to work reported in [12] in the context of ESTELLE we do not eliminate the asynchronous queue communication mechanism between adjacent layer processes by replacing these processes by one product automaton because this would induce an extreme blow-up in the complexity of the implementation.

5. Determination of the Common Path Graph

The later steps of our optimization method rely on the assumption that we optimize the processing of a packet only for the 'common case' (we will come to a clearer understanding of this expression in this Chapter). In Chapter 6 we introduce optimization steps that anticipate certain common decision results according to a common case assumption. We consider our common path determination a generalization of the *Common Path* optimization as advocated in [7].

Protocols usually have the task of hiding imperfect behavior of lower layer services from upper layer users. This means that a major part of their functionality aims at detection and treatment of many kinds of exceptions and errors. Exceptions and errors, however, are usually uncommon, in particular in typical high speed communication environments. On the other hand, optimizing the common case implies that we need to take care of uncommon cases using alternate non-optimized error-case implementations. But, as we argued above, because of the low probability of these error handling cases we can tolerate the non-optimized processing of these error cases without risking a considerable degradation of the performance of the protocol. However, not all branching in the control flow can be classified so that *one* branch is common and *all others* are uncommon. It may as well be the case that more that one alternative is a common choice, namely when the branching does not aim at handling exception cases.

Now, what does the term *common case* mean technically? We distinguish the decision edges (outgoing *cfd*-edges of a node with outdegree > 1) of the *cfd* relation of an MLDG M disjointly into those which are taken with a probability above a certain value (the *common* ones, labeled with 'C') and those for which the probability is below a certain value (the *uncommon* ones, labeled with 'U'). The labeling of the decision edges is described in Section 5.1. It defines a *common path graph* which is a subgraph of the *cfd* graph. Hence, our further optimization will only address the common way a packet takes through the protocol stack, along a common path, and not the uncommon cases. In order to obtain what we call the *Common Path Graph* (CPG) we drop those subgraphs of M which start with an edge labeled as uncommon from every decision node (see Section 5.2).

5.1. Labeling of MLDGs

Common/Uncommon Labeling of MLDGs. Let M denote an MLDG and let $C = \{C, U\}$ a set disjoint from any other set in sight. Furthermore let $cul \subseteq (branchedges(S, cfd) \times C)$ a functional relation. We say that *cul* is a *common/uncommon labeling* of the MLDG M.

Example Common/Uncommon Labeled MLDG. Figure 5 presents and example of a common / uncommon labeled MLDG. Note that the labeling of the branching edges yields a tree which represents the common path of the processing of a packet. This tree, which is indicated by bold solid line *cfd* edges in Figure 5, is obtained by traversing an MLDG so that no decision edge with label `U` is traversed.

Discussion. Whether a decision edge is common or uncommon depends in part on the environment in which a protocol is running. The common/uncommon attributes can thus not be automatically derived from the protocol specification. The attribution has to be provided by the implementor as an input for our method. One way of finding out which decisions are uncommon is to analyze a working implementation using for example a tool like it has been proposed in [2][5]. In case such analyses are not available it may be necessary to use simulation techniques or estimations in order to determine whether a particular decision edge belongs to the common or the uncommon case.

5.2. Common Path Graph (CPG)

The very easy algorithm for the construction of the CPG can be found in [15]. It simply prunes all those subtrees of the labeled MLDG which start with an edge labeled 'U' in a decision node.

Example CPG. In Figure 6 we present the CPG derived from the common / uncommon labeled MLDG in Figure 5. The subgraph that has been removed is the graph starting with the edge $(D1, S5)$. The subgraphs starting in node $D2$ have both been retained as they both represent common decisions.

6. Construction of the Relaxed Dependence Graph

In the previous chapters we have shown how a common path graph (CPG) can be derived from an SDL specification based on a control flow and data flow dependence analysis. In this Section we will construct a relaxed dependence graph (RDG) based on which

[5] The authors describe a tool called *Chitra* which analyzes program execution sequences yielding a semi-Markov chain model representing the time behavior of a program.

the original specification can be implemented. We conjecture that the implementation of the RDG will be functionally equivalent to the faithful implementation, but will execute faster. We propose the following steps to generate a RDG.

- *Anticipation of the common case:* In this step we ignore that certain statements depend on a decision, namely for those decisions where we assumed a common outcome. Henceforth we treat these decision nodes as if no other node depends on their execution.
- *Parallelization:* We construct a relaxed dependence graph by taking the data flow dependence relation of the anticipated CPG and by adding additional dependences which ensure that a node is never executed before the last decision node on which it depends in the control flow dependence relation has been executed. Two statements can be executed in parallel iff in the RDG they do not dependent on each other.

6.1. Anticipation of the Common Case

The CPG may contain decisions with only one outcome in the CPG. As we will see in the next transformation, decisions enforce an execution order, namely that a node must be executed after all decisions it depends on have been taken. Decisions thus limit potential parallelism. To enhance potential parallelism we anticipate the outcome of decisions that have only one outcome in the CPG. We treat such decisions as if they represented tasks instead of decisions. In [15] we discuss the handling of the uncommon cases in an implementation and argue that there is always a way to handle them consistently. Anticipation of the common case is applied to the CPG changing the type of those decision nodes which have only one successor in the *cfd* relation of the CPG to *task*. The algorithm can be found in [15].

In our example, anticipating the common case results in changing the statement type of D1 from*decision* to *task*.

6.2. Relaxation of Dependences

In this transformation we remove dependences from the CPG graph to allow its parallel execution. More precisely we remove all dependences and replace them by a smaller set of *relaxed* dependences. There are three precedence relations that the relaxed dependence graph must enforce. *Data flow dependences*: a node using a variable may not be executed before a node which defines that variable. *Control flow dependences*: a node which is (directly or transitively) control flow dependent of a decision node may not be executed before this decision has been taken. *Final execution of exit nodes:* Exit nodes must be the last nodes to be executed because they are the point where a protocol interacts with its environment and makes the result of the processing visible. Thus all statements which are no exit nodes must be executed before executing an exit node. The result of the transformation is a relaxed common path graph (RDG) in which the *cfd* and *dfd* relations have been replaced by a relaxed dependence relation *rxd*. We create the *rxd* relation in three steps. First we include all elements of the original CPG's *dfd* relation in *rxd*. This will ensure that data dependences are respected. Then we examine each node of the RDG to see if it already depends (directly or transitively) from its nearest preceding decision node in the *cfd* relation. If not, we add a dependence between the examined node and the nearest decision node. This ensures that a node is not executed before the last decision it depends on. Finally we check that all exit nodes reachable from a given node in the CPG are also dependent of that node in the RDG. If this is not the case, we add relaxed dependences between the given node and the concerned exit nodes.

Algorithm 2 is an algorithm for the construction of the RDG. Starting with a given, possibly anticipated CPG C it uses the cfd_C and dfd_C relations to create the rxd relation over $S_C \times S_C$ of the resulting RDG. The algorithm first selects a set D of all decision nodes of the graph plus the root of the graph (I.). It includes all elements of dfd_C into rxd (II.). Then, for every node x of the graph, it finds the nearest node in D from which x is transitively dependent in the cfd_C relation (III.). If in the rxd relation x is not yet transitively dependent of that node, a new dependence is added. Next, all nodes except the exit nodes of the graph are examined. A dependence is added (V.) between an examined (VI.) node y and each exit node which is transitively dependent of y in the cfd_C relation of the CPG but not in the rxd relation of the RDG (VII. and VIII.).

Algorithm 2

I. SELECT $\mathcal{D} = \{D_1, \ldots, D_m\} \subseteq S_C$ *SO THAT*
$(\forall D_i)(sttype(D_i) = decision \vee D_i = root(C))$

II. $\mathrm{rxd} := \mathrm{dfd}_C$

III. FOR ALL $n \in S_C - \mathrm{root}(C)$

IV. SELECT D_n *SO THAT*
$\{s \in \mathcal{D} \mid (s, n) \in \mathrm{cfd}_C^+ \wedge (D_n, s) \in \mathrm{cfd}_C^+\} = \emptyset$

V. IF $(D_n, n) \notin \mathrm{rxd}^+$ *THEN* $\mathrm{rxd} := \mathrm{rxd} \cup \{(D_n, n)\}$

VI. FOR ALL $m \in S - \mathrm{leaves}(C)$

VII. FOR ALL $x \in \mathrm{leaves}(C)$

VIII. IF $(m, x) \in \mathrm{cfd}_C^+ \wedge (m, x) \notin \mathrm{rxd}^+$ *THEN* $\mathrm{rxd} := \mathrm{rxd} \cup \{(m, x)\}$

We call the resulting directed graph $R = (S_C, rxd)$ the relaxed dependence graph for CPG C. It should be noted that R is not a tree. Figure 7 shows the RDG for the CPG in Figure 6. We see that S2 and S3 depend on S1 but not on each other. This means that once S1 has been executed S2 and S3 can be executed in parallel.

7. Optimizations based on the Relaxed Dependence Graph

An implementation will be based on the RDG. In particular, the scheduling of the operations on a given hardware architecture is a central task of any implementation, be it at compile- or at run-time. When scheduling the operations the scheduler may take advantage of the relaxation of dependencies in the RDG. The execution of an operation may be scheduled at a different point of time compared to its execution according to the sequential SDL specification. In particular, the scheduler may schedule the processing of certain operations whenever it seems optimal, for example when the required resources and data are available. A further gain in efficiency can be achieved by combining the execution of so-called *Data Manipulation Operations* (DMOs).

Grouping of Data Manipulation Operations. We call data manipulation operations (DMOs) operations that manipulate entire data parts of protocol data units. Combining two such operations into one that does two manipulations at the same time saves an extra storing and fetching of all the data and thus executes much faster. This has already been demonstrated in [7]. It is also central to the work reported in [8] and [1]. Particularly, it has been shown in [20] that in presence of decisions along the path of execution of a protocol, it is better to wait with the execution of DMOs until all decisions have been taken. At that point the set of DMOs to be executed is known and the DMOs can be combined. The technique is referred to as *lazy messages.* Our algorithm is a generalization of this technique. In order to enable the joint execution of DMOs the RDG has to be modified. It has to be taken into account that when grouping the execution of two DMOs so that one operation depends on a decision higher up in the RDG than the other operation, the higher operation must be executed along every possible path through the RDG.

An algorithm for grouping of DMOs. We propose a recursive algorithm that starts at the root of the RDG. Let B be the name of the node the algorithm is applied to. The algorithm distributes the DMOs that depend of B over each decision that depends of B, called B', iff other DMOs exist which can only be executed after B'. The algorithm is then recursively applied to all decisions B' that depend on B. Algorithm 3 does the operation described above. It is applied to the root node of a RDG C. It also takes as input the *cfd* relation of the CPG from which C was derived. The node it is applied to is called B. For each DMO D which depends of B (I.) a second DMO D_2 depending on an other decision node is searched in the subset of nodes that may be executed if D is executed (II.). If such a DMO exists, the decision node B' depending of B and leading to D_2 is found (III.). Then D is removed from the graph (IV.-VI.) and several copies of D, called D'_i, are created, one for each possible evaluation of B' (VII.-X.). Dependences are added from D'_i to all exit nodes which can be reached from B' with the corresponding evaluation of the predicate (XI.). Once all DMOs depending of B have been treated, the algorithm is applied to all decision nodes depending of B (XII. & XIII.). The algorithm stops when B has no more successors which are decision nodes.

Algorithm 3
RecursiveCombine(B)

I. FOR ALL $\{D \in \mathcal{DMO} \mid (B, D) \in \mathrm{rxd}\}$

II. IF $\exists D_2 \in \mathcal{DMO} \mid (D, D_2) \in \mathrm{cfd}^+$ *AND* $\{s \in S_C \mid (D, s) \in \mathrm{rxd}^+ \wedge (s, D_2) \in \mathrm{rxd}^+\} = \emptyset$

III. $B' = s \in \mathrm{branchnodes}(C) \mid (B, s) in \mathrm{rxd} \wedge (s, D_2) \in \mathrm{rxd}^+$

IV. $S_C := S_C - \{D\}$

V. $\mathcal{DMO} := \mathcal{DMO} - \{D\}$

VI. $\mathrm{rxd} := \mathrm{rxd} - \{\{D\} \lhd \mathrm{rxd} \cup \mathrm{rxd} \rhd \{D\}\}$

VII. FOR ALL $N_i \in \{B'\} \lhd \mathrm{cfd}$

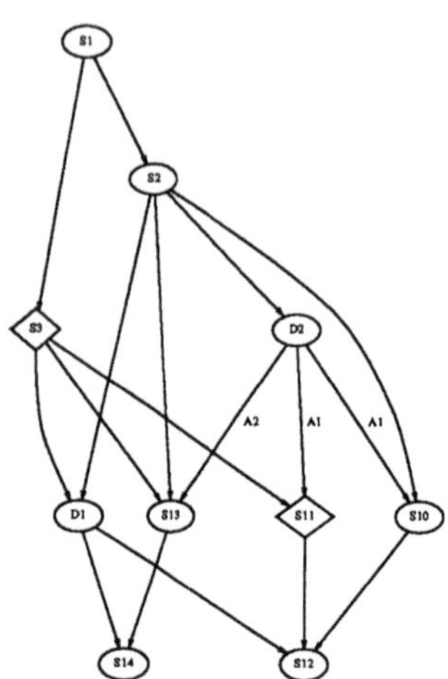

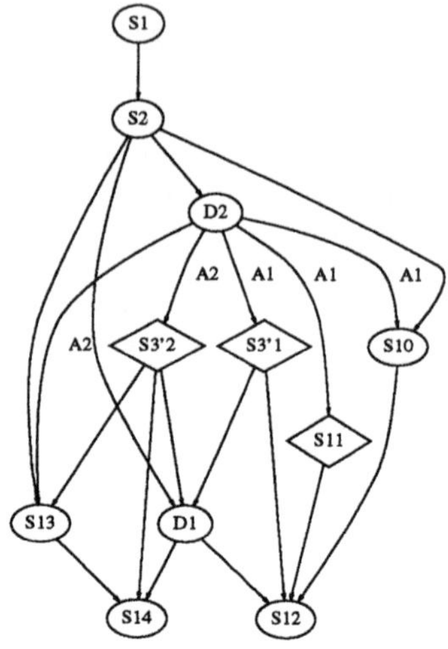

Figure 8. CPG with marked DMOs (diamonds)

Figure 9. Grouped DMOs

VIII. $S_C := S_C \cup \{D_i'\}$
IX. $\mathcal{DMO} := \mathcal{DMO} \cup \{D_i'\}$
X. $\mathrm{rxd} := \mathrm{rxd} \cup (B', D_i')$
XI. $\mathrm{rxd} := \mathrm{rxd} \cup \{(D_i', x), x \in leaves(C) \mid (N_i, x) \in \mathrm{cfd}^+\}$

XII. FOR all nodes $newB \in B \lhd \mathrm{rxd}$ ***and*** $\mathrm{sttype}(newB) = decision$

XIII. call recursiveCombine($newB$)

Example. The application of the algorithm to our example is shown in Figure 8 and Figure 9. The two DMOs identified are S3 and S11. S3 is replicated for each evaluation of D2, yielding S3'1 and S3'2. If D2 evaluates to 'A1' then a combined DMO S3'1/S11 can be executed. If D2 evaluates to 'A2', then S3'2 is executed alone. In the final implementation the schedule of the operations has to be such that depending on the evaluation of the decision predicate D2 either S3'1 or S3'2 is executed before D1, but not both.

8. Conclusions

In this paper we presented formalizations and algorithms for the derivation of optimized protocol implementations from SDL specifications. We started with a syntactical dependence analysis for SDL processes. We then showed how multiple dependence graphs can be combined to multi-layer dependence graphs. Next we determined the common path graph, a subgraph of a multi-layer dependence graph which represents the common case of processing of a packet in the protocol stack. This graph was the basis for an optimization by anticipating the evaluation of some decision statements in the CPG, and then by relaxing the dependences. This essentially meant to omit control flow dependences and to only consider data flow dependences and dependences that express the dependence of a statement from the evaluation of a decision predicate. We called the result a relaxed dependence graph. When scheduling the operations on a given hardware architecture the scheduler may take advantage of the relaxation of dependencies in the RDG in particular by scheduling certain operations at a different point of time compared to the sequential execution in the SDL specification. In particular we showed how the optimization concepts of lazy messages and grouping of Data Manipulation Operations can be interpreted based on the Relaxed Dependence Graph.

We are currently developing a toolset for the support of our method. The toolset will consist in an SDL parser which generates dependence graphs, and a set of graph optimizing routines. The graph optimizing algorithms have already been implemented, the SDL parser is currently under development. Furthermore, we have implemented a prototype tool to support the scheduling aspect of the implementation. The fact that we have provided a rigorous formal description of our method clearly supports the implementation of such a toolset. It also connects our method well to other formally supported steps of an overall protocol engineering methodology, like testing and validation.

Acknowledgments. The work of both authors was supported by the Swiss National Science Foundation. We would like to thank Peter Ladkin for very helpful commentary on an earlier draft of this paper.

REFERENCES

1. M. Abbott and L. Peterson. Increasing network throughput by integrating protocol layers. *IEEE/ACM Transactions on Networking*, 1(5), October 1993.
2. M. Abrams, N. Doraswamy, and A. Mathur. Chitra: Visual ananlysis of parallel and distributed programs in the time, event, and frequency domains. *IEEE Transactions on Parallel and Distributed Systems*, 3(6):672–685, November 1992.
3. U. Banerjee, R. Eigenmann, A. Nicolau, and D. Padua. Automatic program parallelization. *Proceedings of the IEEE*, 81(2):211–243, Feb 1993.
4. F. Belina, D. Hogrefe, and A. Sarma. *SDL with Applications from Protocol Specification*. Prentice Hall International, 1991.
5. T. Braun and M. Zitterbart. Parallel transport system design. In A. Danthine and O. Spaniol, editors, *Proceedings of the 4th IFIP conference on high performance networking*, 1992.
6. CCITT. Recommendation Z.100: CCITT Specification and Description Language (SDL). CCITT, Geneva, 1992.
7. D. D. Clark, V. Jacobson, J. Romkey, and H. Salwen. An analysis of TCP processing overhead. *IEEE Communications Magazine*, 27(6):23–29, June 1989.
8. D. D. Clark and D. L. Tennenhouse. Architectural considerations for a new generation of protocols. In *Proceedings of the ACM SIGCOMM '90 conference*, Computer Communication Review, pages 200–208, 1990.
9. J. Crowcroft, I. Wakeman, Z. Wang, and D. Sirovica. Is layering harmful? *IEEE*

Network Magazine, pages 20–24, january 1992.
10. J. Ferrante, K. J. Ottenstein, and J. D. Warren. The program dependence graph and its use in optimization. *ACM Transactions on Programming Languages and Systems*, pages 319–349, July 1987.
11. S. Fischer and B. Hofmann. An Estelle compiler for multiprocessor platforms. In R. L. Tenney, P. D. Amer, and M. Ü. Uyar, editors, *Formal Description Techniques, VI*, IFIP Transactions C, Proceedings of the Sixth International Conference on Formal Description Techniques. North-Holland, 1994. To appear.
12. B. Hofmann and W. Effelsberg. Efficient implementation of Estelle specifications. Technical report Reihe Informatik, Nr. 3/93, University of Mannheim, Mannheim, Germany, 1993.
13. A. S. Krishnakumar and K. Sabnani. VLSI implementation of communication protocols - a survey. *IEEE Journal on Selected Areas in Communications*, 7(7):1082–1090, September 1989.
14. P.B. Ladkin and B.B. Simons. Compile-time analysis of communicating processes. In *Proceedings of the Sixth ACM International Conference on Supercomputing*, pages 248–259. ACM Press, 1992.
15. S. Leue and Ph. Oechslin. A formal approach to optimized parallel protocol implementation. Technical report, University of Berne, Institute for Informatics, Berne, Switzerland, 1994.
16. S. Leue and Ph. Oechslin. Formalizations and algorithms for optimized parallel protocol implementation. In D. Lee et al., editor, *Proceedings of the 1994 International Conference on Network Protocols ICNP-94*. IEEE Computer Society Press, 1994. To appear.
17. S. Leue and Ph. Oechslin. Optimization techniques for parallel protocol implementation. In *Proceedings of the Fourth IEEE Workshop on Future Trends in Distributed Computing Systems*, Lisbon, Sep. 1993. To apear.
18. M. T. Liu. Protocol engineering. In M. C. Yovitis, editor, *Advances in Computers*, volume 29, pages 79–195. Academic Press, Inc., 1989.
19. A. Mitschele-Thiel. On the integration of model-based performance optimization and program implementation. In *4th Workshop on Future Trends of Distributed Computing Systems*, 93.
20. S. W. O'Malley and L. L. Peterson. A highly layered architecture for high-speed networks. In M. J. Johnson, editor, *Protocols for High Speed Networks II*, pages 141–156. Elsevier Science Publishers (North-Holland), 1991.
21. D. A. Padua and M. J. Wolfe. Advanced compiler optimizations for supercomputers. *Communications of the ACM*, 29(12):1184–1201, Dec 1986.
22. W. Peng and S. Purushothaman. Data flow analysis of communicating finite state machines. *ACM TOPLAS*, 21(3):399–442, 1991.
23. Y.H. Thia and C.M. Woodside. High-speed OSI protocol bypass algorithm with window flow control. In B. Pehrson, P.Gunningberg, and S. Pink, editors, *Protocols For High-Speed Networks III C*, volume C-9, pages 53–68. IFIP, NORTH-HOLLAND, 1993.
24. C. M. Woodside and R. G. Franks. Alternative software architectures for parallel protocol execution with synchronous IPC. *IEEE/ACM Transactions On Networking*,

1(2):178–186, April 1993.

Appendix

Notation and Definitions

Relations. Let $f \subseteq R \times R$ denote a binary relation over a set R, let $x, y \in R$ and S a set. We define the following *restrictions* and *operators* on a relation f.

$$f \rhd S \triangleq \{(a,b) | (a,b) \in f \ \wedge \ b \in S\}$$

$$S \lhd f \triangleq \{(a,b) | (a,b) \in f \ \wedge \ a \in S\}$$

$$\mathit{domain}(f) \triangleq \{a \mid (\exists b \in R)((a,b) \in f)\}$$

$$\mathit{range}(f) \triangleq \{b \mid (\exists a \in R)((a,b) \in f)\}$$

$$\mathit{field}(f) \triangleq \mathit{domain}(f) \cup \mathit{range}(f)$$

A relation f is *functional* if and only if each element in its domain is related to a unique element in its range. For a functional relation f and an $x \in R$ we sometimes write $f(x)$ to denote $\mathit{range}(\{x\} \lhd f)$. We use f^+ to denote the transitive closure of a relation f, and f^* to denote the transitive reflexive closure of f.

Digraphs and Trees. Let V denote a set and let $E \subseteq V \times V$, then we call $T = (V, E)$ a *digraph*. We call T a *tree* if and only if the following additional conditions hold:

- $(\exists v \in V)((E \rhd \{v\} = \emptyset)) \wedge (\forall w \in V, w \neq v)(E \rhd \{w\} \neq \emptyset))$ (we call v the *root*),
- $(\forall v, w \in V)((E \rhd \{v\} = \emptyset) \rightarrow (v, w) \in E^+)$ (all nodes are reachable from the root),
- $E^+ \cap E^* = \emptyset$ (there are no cycles), and
- $(\forall v \in V)(| \{v\} \lhd E | \leq 1)$ (every node except for the root has exactly one predecessor).

Furthermore, for a tree $T = (V, E)$ we define: $\mathit{root}(V, E) \triangleq \{v \in V \mid E \rhd \{v\} = \emptyset\}$, $\mathit{leaves}(V, E) \triangleq \{v \in V \mid \{v\} \lhd E = \emptyset\}$, $\mathit{branchnodes}(V, E) \triangleq \{v \in V \mid (| \{v\} \lhd E |) > 1\}$, and $\mathit{branchedges}(V, E) \triangleq \mathit{branchnodes}(V, E) \lhd E$.

Multi-edged and Labeled Trees.

- Let $E_1 \dots E_n \subseteq V \times V$ for $n \geq 1$. Then we call $T = (V, E_1 \dots E_n)$ a *multi-edged tree* iff (V, E_1) is a tree.
- Let $T = (V, E_1 \dots E_n)$ a multi-edged tree. Let $D_1 \dots D_n$ denote sets which are pairwise disjoint from any other set in sight. Let $L_1 \dots L_n$ denote functional relations with $L_i \subseteq (E_i \times D_i)$. Then we call $T = (V, E_1 \dots E_n, D_1 \dots D_n, L_1 \dots L_n)$ a *multi-edged labeled tree.* We shall slightly abuse notation in that we extend the notations $\mathit{root}(T)$ and $\mathit{leaves}(T)$ to multi-edged labeled trees, in the obvious way.

20

Partial-Frame Retransmission Scheme for Data Communication Error Recovery in B-ISDN

Ichiro INOUE and Naotaka MORITA

Communication Switching Laboratories,
Nippon Telegraph and Telephone Corporation,
3-9-11 Midori-cho, Musashino-shi, Tokyo 180, Japan

Abstract

An increasing number of computers and workstations will require data communications like that of LAN-LAN interconnections. The requirement for high end-to-end throughput will be satisfied by B-ISDN, in which the ATM technique is the key technology. In B-ISDN, user data is segmented into fixed-length short packets (cells) and transferred at very high speed. The greater the amount of user data to be transmitted at one time, the more efficient the transmission has to be. Efficient error recovery is needed for the higher transmission efficiency as well as for a shorter response time, which is also a key data communication requirement.

This paper describes the functions needed for data communication in B-ISDN and classifies the error recovery methods that can be used by the end-to-end protocol for data communication. A partial-frame retransmission scheme is proposed and its performance is compared to that for selective-frame retransmission from the viewpoints of implementation complexity, efficiency, total throughput, and cell-loss probability.

A functional block model for partial-frame retransmission is proposed and it is shown that partial-frame retransmission can be implemented with little addition to the model for selective-frame retransmission.

The evaluation results show that partial-frame retransmission can provide higher efficiency for a wide variety of parameter values compared with selective-frame retransmission. They also show that partial frame retransmission reduces the cell-loss rate to less than 10% of that for selective-frame retransmission. In particular, when the incoming traffic load is relatively high, the proposed partial-frame retransmission scheme can avoid the congestion that selective-frame retransmission allows.

Keyword Codes: C.2.2; C.2.0; C.2.1
Keywords: Network Protocols; Computer-Communication Networks, General; Network Architecture and Design

1. Introduction

With the continuing speed of processing capability in personal computers, workstations, and servers, conversational data communication between computers and distributed processing among processors increases the requirements for inter-computer communication. The expansion of the area in which computers communicate with each other strengthens the requirements for inter-local-area data communication, like LAN-LAN interconnections, as well for wide-area data communication, like wide-area networks (WANs). Applications for such inter-computer communication include the traditional LAN applications (file transfer, remote procedure call, transaction processing, and LAN management message exchange), the WAN signaling message exchange, and the large file transfer for WAN management and operation, e.g., Telecommunications Management Network (TMN[1]).

Data communication traffic is generally very bursty, with a peak rate as high as several Mbps. Typical traffic in a LAN or LAN-LAN-interconnect environment is the exchange of control data units for managing and maintaining the network configuration. Signaling traffic is the exchange of the relatively short messages that occurs for call control (set up, release, etc.). The TMN traffic includes both large file exchange and short operational message exchange. The required average rate for this kind of traffic may be relatively low since a large file transfer with a high peak rate does not often occur. On the other hand, when a typical application in such an environment, e.g., file transfer, occurs, the peak rate can be as high as the physical medium's speed.

Data communication is currently handled with leased lines, X.25 services, or dedicated networks (in the case of No. 7 signaling). None of these alternatives, however, meets the requirement for high speed (bit rate) with efficient transmission (cost).

B-ISDN will provide multimedia communication services for voice, images, motion pictures, data, etc. Asynchronous transfer mode (ATM[2][3]), which is based on the asynchronous transfer of fixed-size packets (cells) provides a wide range of communication rates, as well as variable-rate communication, while providing statistical multiplexing capability. Therefore, ATM is expected to be the common transfer technique in B-ISDN. ATM-based B-ISDN will provide both high-bit-rate, up to hundreds of Mbps, and efficient transmission.

B-ISDN will initially provide connection-oriented services suitable, for example, for LAN-LAN interconnections. In LAN-LAN interconnections there can be many types of data communication applications, such as file transfer, transaction processing, image referencing, electronic mailing, and computer-aided design. However, many problems remain to be solved before such applications can economically use B-ISDN's high performance.

One problem to be solved is the protocol for data communication. Voice, motion picture, image, and data communication services require different transfer characteristics, but ATM provides only uniform transfer characteristics. Thus, an ATM adaptation layer (AAL[4][5]) protocol is needed to adapt the ATM layer capability to provide service-specific characteristics. The AAL protocol for data communication should provide flow control for high-speed data transmission, error control for error correction, and congestion control to avoid network congestion and to prevent user buffer overflow.

This paper clarifies the requirements for the AAL protocol for high-speed data communication in B-ISDN in Section 2. Section 3 shows examples of error recovery methods and proposes a partial-frame retransmission method for error recovery; a block diagram shows that the proposed method can be implemented easily. Section 4 gives the theoretical and simulated evaluation results for the current selective-frame retransmission method and the proposed partial-frame retransmission method. The results show that the proposed partial-frame retransmission method meets the requirements for data communication. Section 5 summarizes the paper and mentions remaining problems.

2. **Data communication in B-ISDN**

A typical point-to-point B-ISDN data communication model (Fig. 1) has two end terminals that communicate by segmenting and reassembling the data units (frames) into and from the ATM cells, which have a fixed 53-byte length.

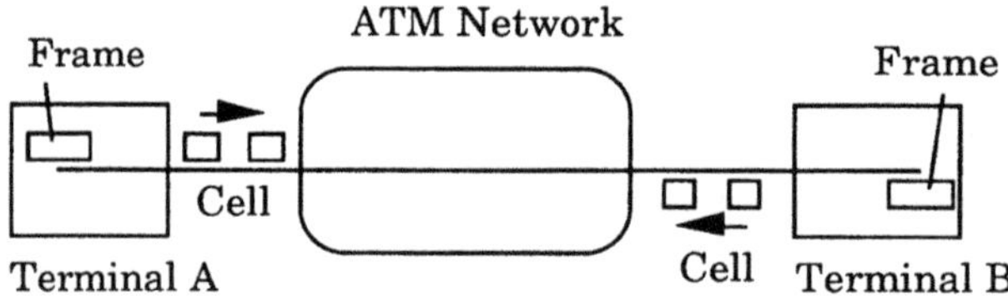

Figure 1. B-ISDN data communication model.

The B-ISDN environment can be characterized by its high-speed linking over virtual channels (VCs), by its low to moderate cell-loss rate depending on the required multiplexing effect, and by its small to large round-trip delay depending on the distance between the source and destination entities. The data communication requirements in B-ISDN are high speed, a low end-to-end error rate, high efficiency, and a quick response time. To meet these requirements, a simple functional configuration, proper flow control, and proper error control are needed. Efficient error recovery is especially needed to achieve low overhead and reduce traffic volume.

Two possible protocol stacks for B-ISDN data communication are shown in Fig. 2. Combination (1) has a service-specific convergence sublayer (SSCS), which may include a service-specific coordination function (SSCF) and a service-specific connection-oriented protocol (SSCOP[6]) over an AAL-common part (CP) (types 3/4 or 5) over the ATM and physical (PHY) layers as shown in Fig.3. Combination (2) uses the existing protocol over RFC 1483[7] over the AAL-CP5, ATM, and PHY layers. In both cases, the upper layer relies on the underlying transmission provided by the AAL, ATM, and PHY layers.

Figure 3 shows the data communication protocol structure recently standardized by the ITU-T. The figure focuses on the internal structure of AAL Types 3/4 and

5. In addition to the functions shown for each sublayer, the SSCS provides end-to-end frame error recovery if necessary.

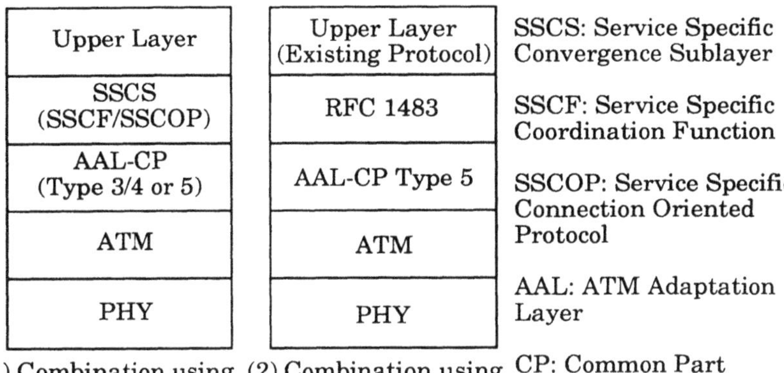

(1) Combination using SSCS (2) Combination using RFC1483

Figure 2. Possible protocol stacks for B-ISDN data communication.

Layer				Example functions
Upper Layers				
AAL	*CS*	*SSCS*	*SSCF*	Primitive mapping, retrieval invocation for signalling
AAL	*CS*	*SSCS*	*SSCOP*	Error control, flow control, link reset, keep alive, retrieval
AAL	*CS*	*CPCS*		CS-PDU validation, buffer allocation (3/4), CRC (5)
AAL	*SAR*			Cell<->AAL PDU mapping, CRC (3/4), multiplexing (3/4)
ATM Layer				Cell transfer, header error check, loss priority control
Physical Layer				Bit transfer, synchronous multiplexing (SDH)

SSCF: Service-Specific Coordination Function
SSCOP: Service-Specific Connection-Oriented Protocol
SSCS: Service-Specific CS
CPCS: Common Part CS
CS: Convergence Sublayer
SAR: Segmentation and Reassembly Sublayer
AAL: ATM Adaptation Layer
ATM: Asynchronous Transfer Mode
SDH: Synchronous Digital Hierarchy
CRC: Cyclic redundancy check code and associated actions
(3/4): Function specific to Type 3/4
(5): Function specific to Type 5

Note: The SSCF and SSCOP sublayers are only defined for the AAL type 5 at present.

Figure 3. Framework of AAL Types 3/4 and 5 for data communication.

The ATM layer provides cell-level transfer and cell-header error detection. The AAL-CP is the AAL-CPCS and AAL-SAR. It provides segmentation and reassembly of the AAL-CP service data unit (SDU) into and from the cell. It also provides cell-level error detection; the detection method varies for each AAL type. AAL Type 3/4 uses the cyclic redundancy check code (CRC), the cell sequence number, the message identifier, the length indicator, and the frame length. AAL Type 5 uses both the CRC and the length indicator at the frame level. The AAL service-specific part provides frame-level error detection and recovery, as well as frame-level flow control. Existing protocols, including TCP[8]/IP[9], XTP[10], VMTP[11], and high-speed data link control procedure (HDLC[12]), provide frame-level flow control and error recovery.

If RFC 1483 is used rather than SSCOP, frame-level error recovery can only be provided by the existing protocol above the RFC. If additional error recovery functions are needed, the existing protocol or RFC 1483 has to be amended to provide them.

Errors in B-ISDN may be classified into those at the bit level, those at the cell level, and those at the frame level. Bit-level errors, which may be caused by link bit errors, are detected by ATM or AAL-CP error detection. Cell-level errors are defined as cell loss due to buffer overflow or cell misinsertion. These errors can be detected by AAL-CP as well as by the AAL service-specific part (AAL-SS). They can also be detected as frame corruption or frame loss by the AAL-CP and the AAL-SS. If the bit error and cell-loss rate are not too high, even if an error occurs, the corruption may be restricted to within a part of the corrupted frame.

Although ATM and AAL-CP provide error detection, end-to-end error correction is needed for data communication because ATM cell loss cannot be recovered with only error detection. While B-ISDN generally provides a very low cell-loss rate, about 10^{-9}, if the multiplexing gain is large, the utilized cell-loss rate may temporarily be higher[13]. In such a case, an efficient error recovery scheme is needed that minimizes the total amount of data to be transmitted and re-transmitted and the probability of congestion.

3. Error recovery schemes

3.1. Preventive recovery (Forward Error Control: FEC)

3.1.1. CRC

(a) Overview

The cyclic redundancy check code (CRC) detects a variety of bit errors but corrects only same of them: it is targeted at bit-level error control. In the sense that no retransmission is not required for error recovery, the CRC can be defined as a kind of preventive error recovery method. The CRC is generated by processing the data according to a specific formula. When the data is received, this same formula is applied to the data and the result is compared with the CRC value contained in the received data unit.

(b) Hardware implementation

Since the CRC only uses simple addition with modulus 2 and a fixed-length delay, it is easily implemented. The processing speed can be high enough so that processing time can generally be neglected compared with the end-to-end round-trip delay.

(c) Possible Problem

Even if the CRC is used, the data communication user may still require further error recovery by retransmission. One reason for this is that the CRC's error correction capability is limited. It can only correct a limited variety of errors (a limited number of the errored bits). Loss of data, which typically occurs due to cell loss, cannot be recovered by the CRC. Another reason is that the efficiency degrades: the CRC has to be lengthened in order to correct a greater variety of errors. Each data item needs a CRC, which leads to low efficiency, especially when the error rate is not very high.

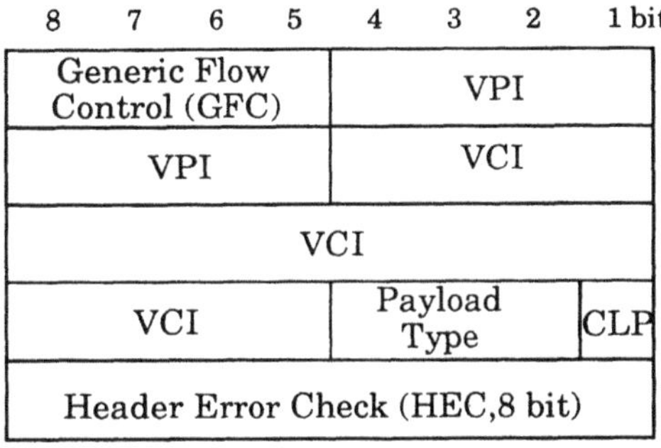

VPI: Virtual Path Identifier
VCI: Virtual Channel Identifier
CLP: Cell Loss Priority

Note: GFC is replaced with a part of a VPI at network node interface (NNI).

(a) Header structure of cell at UNI

2	4	10		6	10 bits
Segment Type	Sequence Number	Message Identifier	SAR-Payload	Length Indicator	CRC

(b) SAR-PDU format for AAL Type 3/4

	0-47	1	1	2	4 Bytes
CPCS-PDU Payload	PAD	CPCS-UU	Common Part ID	Length	CRC

CPCS: Common Part Convergence Sublayer; :PAD: Padding;
UU: User-User data; ID: identifier

(c) CPCS-PDU format for AAL Type 5

Figure 4. CRC used in ATM and AAL layers.

(d) Application

Although CRC error detection can be easily implemented without any delay due to retransmission, further retransmission is needed for error recovery. These characteristics make the CRC more suitable for the lower layers, for example, ATM header error check (HEC) and the SAR CRC, as well as AAL5-CP CRC32, each of which assumes higher layer retransmission if necessary. Figure 4 shows how these CRCs are used by these protocols. The CRC may therefore not be suitable as the end-to-end error recovery method in a B-ISDN environment.

3.1.2. PCR

(a) Overview

Preventive cyclic retransmission (PCR[14]) autonomously retransmits the previously sent data when no new data is to be sent or the amount of already sent data reaches the preestablished default value. Therefore, if the received data is errored, the receiver can expect to receive the retransmitted data soon after receiving the original data. Using PCR to recover data errors avoids the round-trip delay that comes with the ordinary retransmission mechanism in which the receiver must request retransmission of the errored data. It is thus effective in environments with a large round-trip delay or a relatively high error rate, such as satellite links.

(b) Hardware implementation

Since PCR simply retransmits the previously sent data, implementation is not complex. It only requires that the sender and receiver have a relatively large buffer for the retransmission and the reordering, respectively.

(c) Problem

Though PCR recovers errors, including any data loss, without much delay, data is retransmitted even when it is unnecessary, which increases costs because it uses more bandwidth.

(d) Application

Since PCR provides quick recovery, but at low efficiency, it is good for leased-line users or for internal network information transfer, which are not sensitive to efficiency. It is also good when there is a very large round-trip delay, as with a satellite line. It is not suitable as the error recovery method for B-ISDN.

3.2. Retransmission Recovery (Automatic Repeat Request: ARQ)

Since preventive error recovery methods do not recover lost data or handle some kinds of errors, reactive recovery, i.e., retransmitting necessary information, is needed. Recovery by retransmission can be classified according to the unit of retransmission.

3.2.1. Go-back-N frames retransmission

This retransmission scheme retransmits all the data after an error has occurred. Therefore, it is effective in networks with moderate to low error rates, relatively small round-trip delay, and moderate to low speed. The traditional data link protocol, HDLC, uses this Go-back-N (GBN) scheme. HDLC is the basis for existing data link protocols, including the packet network datalink layer (X.25 LAPB[15]), the ISDN D-channel (LAPD[16]), and the frame relay control part (Q.922[17] control, LAPF). All of these protocols are based on GBN retransmission. The GBN error recovery scheme is also used by both TCP and XTP. In these two protocols, data is identified by both its length (bytes) and its segment sequence number.

(a) **Procedure**

In GBN retransmission, a sequence number is maintained in both the sending and receiving entities. This sequence number can be based on either the number of the data unit (frame) or the amount of data (bytes). The sender holds a copy of the sent data in its buffer until the acknowledgment for the data is received. When an error is detected, the sender retransmits all data units with sequence numbers larger than that of the errored unit. The retransmission may be invoked by a notification sent by the receiving side, which checks the continuity of the sequence numbers. This notification may also be sent by a control data unit (S-frame in HDLC, STAT PDU in SSCOP).

Figure 5 shows a sequence chart for GBN error recovery as used in HDLC. The information frame with a sequence number of 0 is lost and the next frame is received. The receiver detects the sequence number gap and requests retransmission by sending a reject frame (REJ). The received information frames are discarded before the error is recovered. In some protocols, like TCP, the sending entity also retransmits the data if the timer for the sent data expires before an acknowledgment is received.

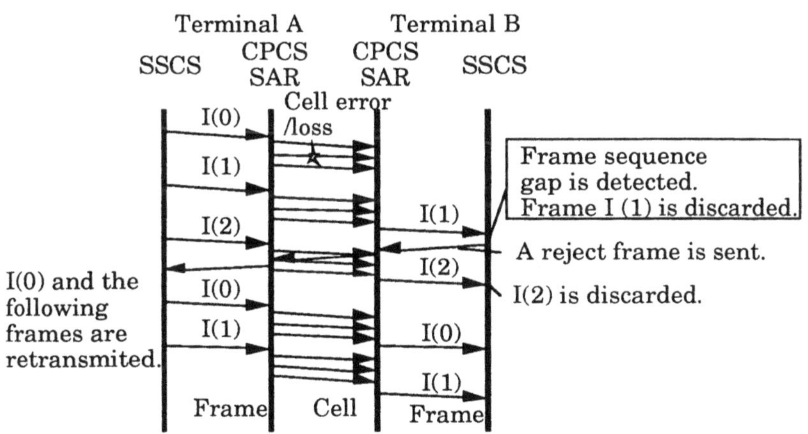

Figure 5. Sequence chart for GBN error recovery.

(b) **Problem**

With the GBN method, the sender simply retransmits all the data after the error and the receiver simply discards all the data after the error until the error is recovered. Implementation is easy because the sender needs to buffer only a copy of the unacknowledged data and the receiver needs to buffer no more than one frame. Although this method is simple and easy, it leads to unnecessarily retransmitting data that has been correctly received; this may lead to large transmission overhead, especially in environments with a large bit-rate round-trip delay product. B-ISDN has a high bit rate as well as varying round-trip delay. In addition, it uses a small transmission unit, a 53-byte cell, and error or data loss occurs only in this cell. Since GBN error recovery in B-ISDN would retransmit a large number of cells unnecessarily, using it would greatly degrade efficiency.

3.2.2. **Selective-Frame Retransmission**

(a) **Procedure**

In the SSCOP and in some options of the TCP, XTP, VMTP, and HDLC, errored data is selectively retransmitted to recover the error. This scheme is called selective-frame retransmission (SFR); it aims for efficient error recovery in environments with moderate to small error rates, moderate to high transmission speed, and varying round-trip delay. As in GBN retransmission, the data units are identified by a sequence number or flag by both the sending and receiving entities. The sequence number may be a pointer based on the amount of data (bytes). As in GBN retransmission, the sender holds a copy of the unacknowledged data in its buffer and the receiver checks if the sequence number of the received data is continuous or not. When an error occurs, the sender retransmits only the errored or lost data.

The retransmission is invoked either by the receiver's notification using a control frame (an S-Frame in HDLC or a USTAT PDU in SSCOP) or by a timeout at the sender. When the receiver receives the retransmitted data, it combines it with the data in the receiving buffer and reorders it. With this scheme, the sender has to have a buffer for the unacknowledged data and the receiver has to have a buffer for the correctly received data. The buffers are larger than those needed for the GBN scheme.

Figure 6 shows a sequence chart for SFR. A cell in frame I(1) is lost and the frame is discarded at the receiving side. When the next frame is received, the receiver detects the sequence gap and requests retransmission of the lost frame by sending a selective reject (SREJ) frame containing the sequence number of the lost frame. All the frames received after the error are contained at the receiver for reordering.

(b) **Problem**

The SFR scheme requires a reordering procedure and a larger buffer space at the receiver, which means more overhead than with GBN. However, SFR goes increase transmission efficiency in terms of the total amount of data that needs to be retransmitted. SFR, therefore, reduces the average overall time for sending.

With short frames, the transmission efficiency may not be improved much because of the relatively large overhead for each frame. Large files can be sent more efficiently by segmenting them into longer frames. With longer frames, however, the frame loss rate increases for the same cell error/loss rate. The number of correctly received cells will also grow, but they will more likely be retransmitted unnecessarily because they are contained in errored or missing frames.

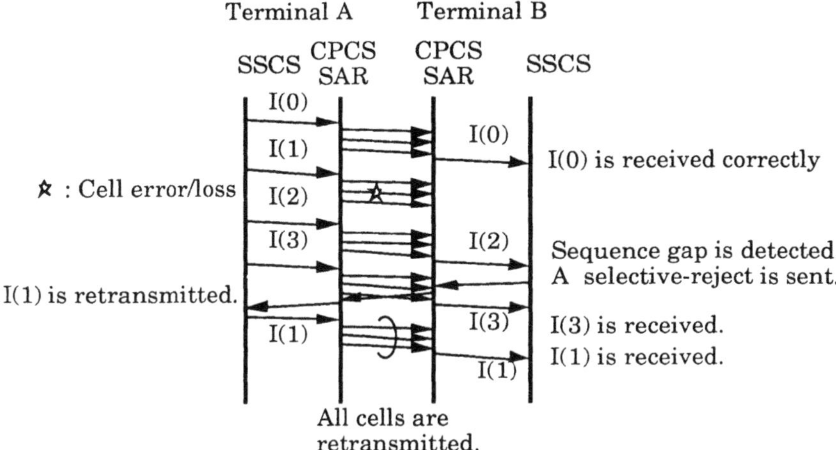

Figure 6. Sequence chart for SFR error recovery.

Although the SFR scheme improves efficiency and total mean response time, as well as total throughput, compared with GBN, its error recovery retransmission performance has to be further improved. This is because in some cases[13], the B-ISDN cell-loss rate of may be not as small as 10^{-9}.

3.2.3. **Partial-Frame Retransmission[13][18][19][20]**

The partial-frame retransmission (PFR) method for error recovery, which depends on B-ISDN cell-based transfer, is proposed. It aims to increase transmission efficiency compared to SFR. The basic concept is to retransmit only errored or lost cells; the receiver then combines the cells which were initially received correctly with the retransmitted cells into a complete frame.

(a) **Procedure**

With PFR, the frames are given frame-level sequence numbers, as in GBN and SFR. The receiver constructs a frame even if all the constituent cells have not been received correctly. In doing so, the receiver creates a list indicating which cells were received correctly and which were not. The length of the list is minimized by using a bit sequence to indicate each cell's reception result. We call this list a bitmap. To identify the lost or errored cells, an identifier is needed to show each cell's location in the frame. AAL-CP Type 3/4 gives a cell-level sequence number (SN) to each cell payload; this SN can be used as an identifier. Although AAL-CP Type 5 does not provide such an identification mechanism, if no VC multiplexing is performed within a user's VP, the VCI value can be used for identification. In other words, the cell's VCI would be incremented each time a new frame is sent.

This bitmap and frame-level sequence number identify the cells to be retransmitted. Although identification by using only the cell-level sequence number would be possible, since this sequence number needs to be long, it would be a greater burden than the proposed identification mechanism.

When a frame is determined to be errored, the receiver notifies the sender of the sequence number of the errored and/or lost frame with its bitmap. The correctly received cells are kept in the receiver buffer until the retransmitted cells are received; the complete frame is then reconstructed and sent to the upper part. When the sender receives an error notification with its cell-level reception result (bitmap), the sender constructs a new frame containing only the requested cells as well as a copy of the bitmap.

When the receiver detects frame losses by gaps in the sequence number, it sends a notification of the losses with the bitmap showing that all the cells were not received. When the receiver receives the retransmitted frame, it reconstructs the complete frame by combining the held cells with the retransmitted cells according to the bitmap contained in the retransmitted frame. These error notifications can be done with control frames (SREJ PDUs in HDLC and USTAT/STAT PDUs in SSCOP). Figure 7 shows a sequence chart for PFR. The second cell of the frame I(0) is lost; it is requested to be retransmitted by a SREJ with a bit map. The sender sends back a short frame containing the same bitmap and the second cell.

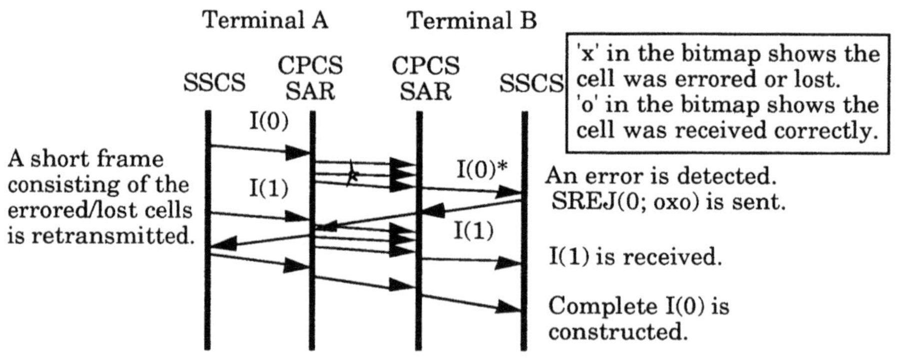

Figure 7. Sequence chart for partial frame retransmission.

Though XTP and VMTP provide selective retransmission, their retransmission units are basically user data units (packets etc.) and are much longer than the payload length of an ATM cell. In particular, VMTP provides no transmitter procedure for polling the receiver's status and retrying the retransmission if necessary.

(b) **Hardware Implementation**

Figures 8 and 9 show the functional blocks needed to provide SFR and PFR. AAL Type 3/4 is assumed. SSCS is assumed to be the SSCOP and the SSCF. The SSCF and higher parts, as well as the ATM and lower parts, are not shown because they are common to the two retransmission schemes. In general, these functional blocks add AAL-SS (frame) overhead and AAL-CP overhead to the data, and segment the data into SAR units with AAL-SAR overhead and an ATM header is added. The data is kept in the transmitter's buffer (Tx buffer) for retransmission. Received cells are processed in reverse order. When the received data is user data, it is sent to its local user. When it is a PDU requesting retransmission, the requested PDUs in the Tx buffer are retransmitted. When the data is polling the receiver status information, the information is sent back in the STAT PDU.

Blocks additional to the ones for SFR are the described below; they are dashed in Fig. 9 and add little overhead.

(1) SD generation based on the received bitmap

This block generates a short SD PDU according to the received bitmap by using the SD PDU in the transmission (Tx) buffer. This function can be performed by the Tx buffer manager; it can also be done with a processor and the proper software.

(2) Creation of bitmap

This function creates a bitmap for every received SD PDU; it can be performed with a register that has the same length as the maximum bitmap length. For example, the approximate number of needed gates would be five [gates/bit] times the bitmap length plus the fixed number of gates.

(3) Check for the initially received SD PDU

This function determines if the received CPCS PDU is an initially received SD PDU or not. If not, the created bitmap is discarded. This function can be performed by a small number of gates.

(4) Receiver partial-frame buffer

This buffer stores the received SD PDUs, even if they are not complete. When it receives a retransmitted partial frame, it reconstructs a complete frame from the stored and the received (partial) frames. This block would be the same size as the SD generation function block.

These additional blocks can be implemented with several thousand gates, using a several-hundred bits bitmap, which corresponds to a maximum frame length of several tens of kilo-bytes. Considering that an AAL Type 5 can be made with several tens of kilogates, this addition is not significant.

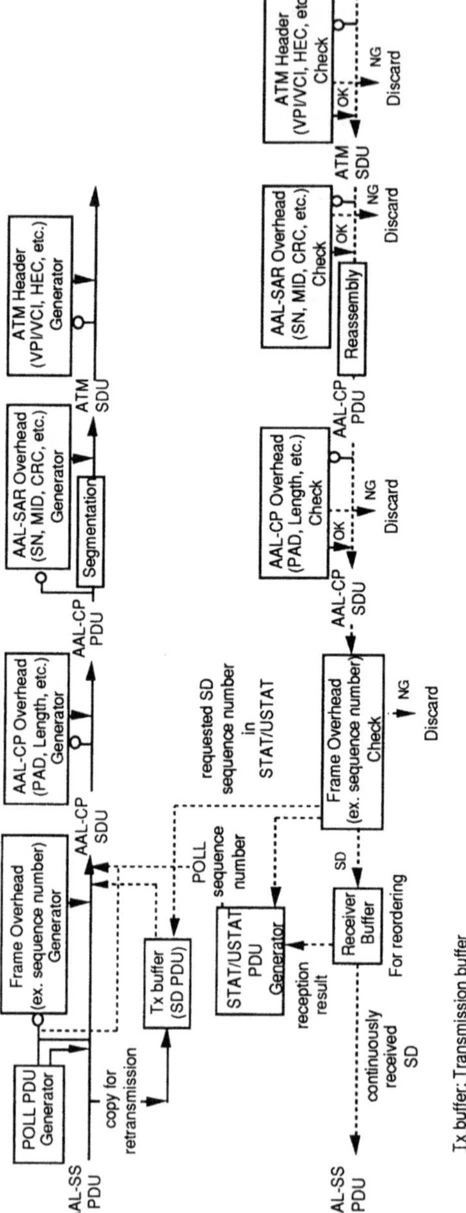

Figure 8. Functional blocks for SFR.

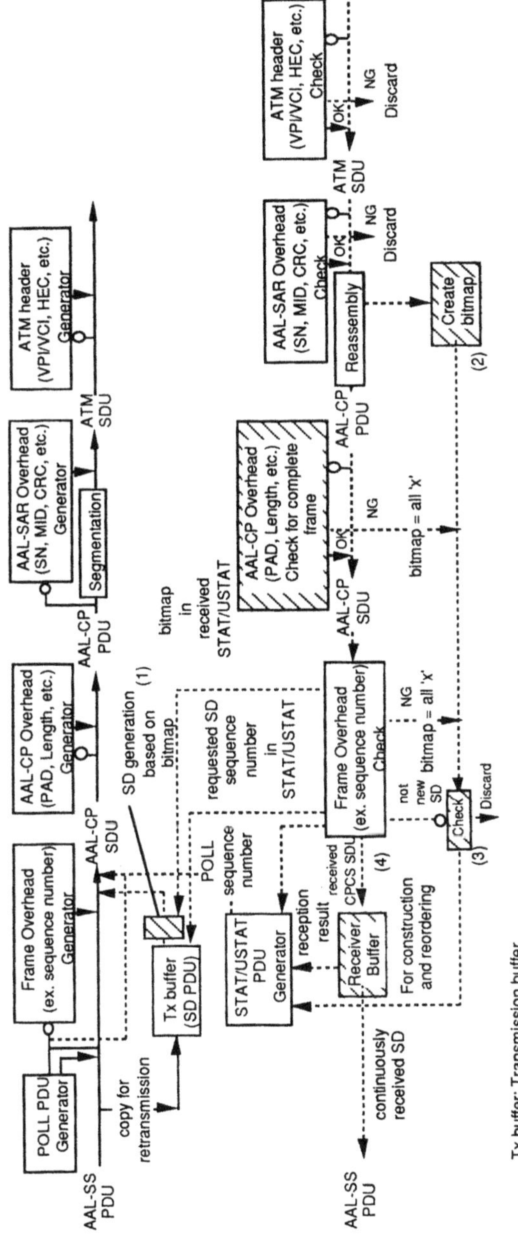

Figure 9. Functional blocks for PFR.

PFR can thus be performed with only a little more burden than for GBN or SFR; it would therefore not reduce the processing speed. Additionally, the required receiver buffer space is almost the same size as for SFR if the other conditions are the same. This is because the required buffer space is determined approximately by the product of the link throughput and the delay.

(c) Expected Benefits

The proposed PFR method will reduce the amount of retransmitted traffic. This will reduce the response time for each frame's acknowledgment and reduce the total traffic volume (initially sent data plus retransmitted data). It will also lower the total cell-loss rate as well as the congestion probability, both of which are the result of the improved efficiency.

4. Evaluation[18][19][20]

In this section we evaluate the PFR method in terms of data communication efficiency and total error-rate reduction. These two criteria are coupled with the data communication application requirements for B-ISDN because efficient and low-error-rate communication reduces response time, congestion probability, and charges (when usage-sensitive charging is used).

4.1. Transmission Efficiency

Transmission efficiency is defined as:

the amount of correctly transmitted user data divided by the total amount of data for the transmission, including retransmitted data and the header overhead.

To focus on the retransmission scheme's effect on efficiency, the following assumptions are made. The cell-loss rate is assumed to be fixed. The receiver buffer is assumed to be infinite. The cell-loss rate for each cell in the network is independent. The network is modeled as two communicating terminals with a point-to-point link that has a fixed cell-loss rate and a fixed throughput capacity. The sender continuously sends data at a fixed rate.

Following these assumptions, the efficiency can be obtained from the formula below.

$$\text{Efficiency for SFR} = \frac{\text{AAL-CS SDU length}}{\text{Total data amount per AAL-CS SDU} \times \dfrac{1}{1 - \text{frame-error rate (FER)}}}$$

Total data amount per AAL-CS SDU:
AAL-CS SDU length +
AAL-CS overhead +
ATM overhead x (AAL-CS PDU length / AAL-SAR payload length)

This is because after a frame-error occurs with a probability of FER, another frame is retransmitted ; the total number of frames is thus multiplied by (1+ FER + FER x FER + ...).

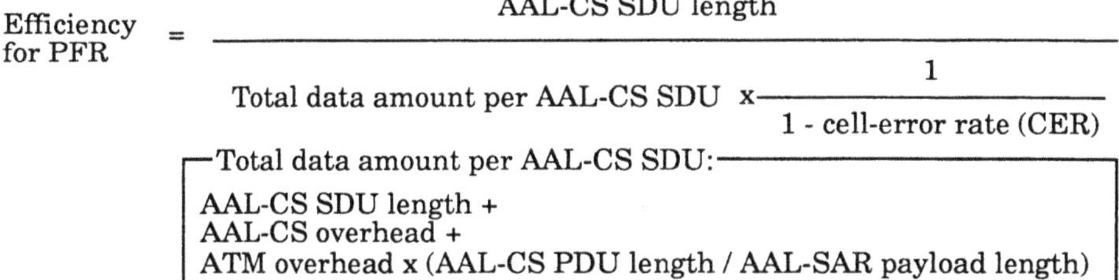

$$\text{Efficiency for PFR} = \frac{\text{AAL-CS SDU length}}{\text{Total data amount per AAL-CS SDU} \times \dfrac{1}{1 - \text{cell-error rate (CER)}}}$$

Total data amount per AAL-CS SDU:
AAL-CS SDU length +
AAL-CS overhead +
ATM overhead x (AAL-CS PDU length / AAL-SAR payload length)

This is because if a cell error occurs, only that cell is retransmitted. If the same cell is again errored, one more cell is retransmitted with a probability of CER. Therefore, the total number of retransmitted cells is multiplied by (1+ CER + CER x CER +...).

In both cases, maximum efficiency is obtained when the error rate is zero.

$$\text{Maximum Efficiency} = \frac{\text{AAL-CS SDU length}}{\text{Total data amount per AAL-CS SDU}}$$

Total data amount per AAL-CS SDU:
AAL-CS SDU length +
AAL-CS overhead +
ATM overhead x (AAL-CS PDU length / AAL-SAR payload length)

Here,

$$\text{FER} = 1 - (1 - \text{CER})^{\left[\frac{\text{AAL - CS PDU length}}{\text{AAL - SAR SDU payload length}}\right]}$$

[#] is the maximum integer no larger than #.
AAL-CS overhead : 8 bytes
ATM overhead : 5 bytes
SAR payload : 44 bytes

The evaluation results (Fig. 10) show that the efficiency provided by PFR can be higher than that of SFR for a wide variety of cell-loss rates, for a variety of frame lengths.

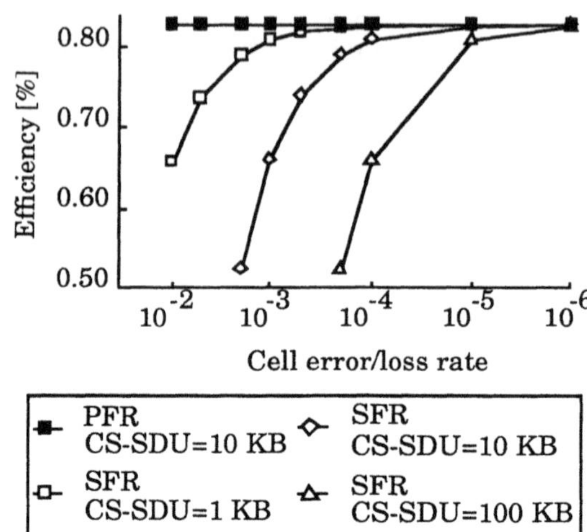

Figure 10. Error recovery efficiency depending on the cell error/loss rate.

4.2. **Total Traffic (Error Rate) Reduction**

To evaluate the transmission efficiency, the cell-loss rate was assumed to be fixed. Actual cell loss, however, depends on the overflow of the finite-size buffers in the nodes, e.g., switches, in the network; it varies depending on the buffer space as well as on the amount traffic of coming into the network. The proposed PFR can reduce the of amount retransmitted traffic compared with GBN and SFR. Therefore, when the terminals try to send a fixed amount of data under the same conditions (link rate, buffer space frame length, etc.), the cell-loss rate may vary depending on the retransmission scheme, i.e., SFR or PFR. In particular, PFR will lower the cell-loss rate compared with SFR. In this section we evaluate the total traffic load, including the retransmitted traffic, taking into account the cell overflow at the buffers.

The evaluation was done using computer simulation with the following assumptions:

(1) Several terminals share the output buffer at the egress node of the network; cell loss occurs only at this point (Figure 11).

(2) The data flows in the same direction; there are no resource conflicts with the control information flowing in the reverse direction.

(3) The data frame length is fixed; frames are generated according to a Poisson process.

(4) Cells in a frame are sent to the network at a fixed rate-the maximum rate of the link between the terminal and the network.

(5) Processing time including propagation delay is assumed to be 5 ms in any direction.

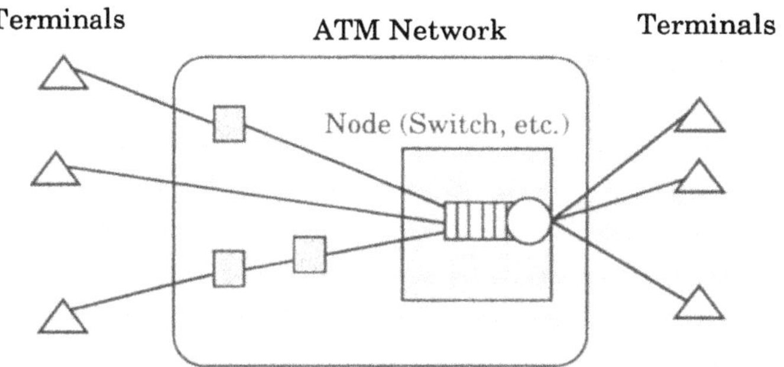

Figure 11. Simulation model for communicating terminals in B-ISDN.

Figure 12 shows the simulation results. The horizontal axis shows the initially generated total traffic load normalized by the output-link rate. The vertical axis shows the cell-loss rate at the output buffer. The frame length and the VC speeds are parameters. The shared-buffer space is assumed to be 127 cells.

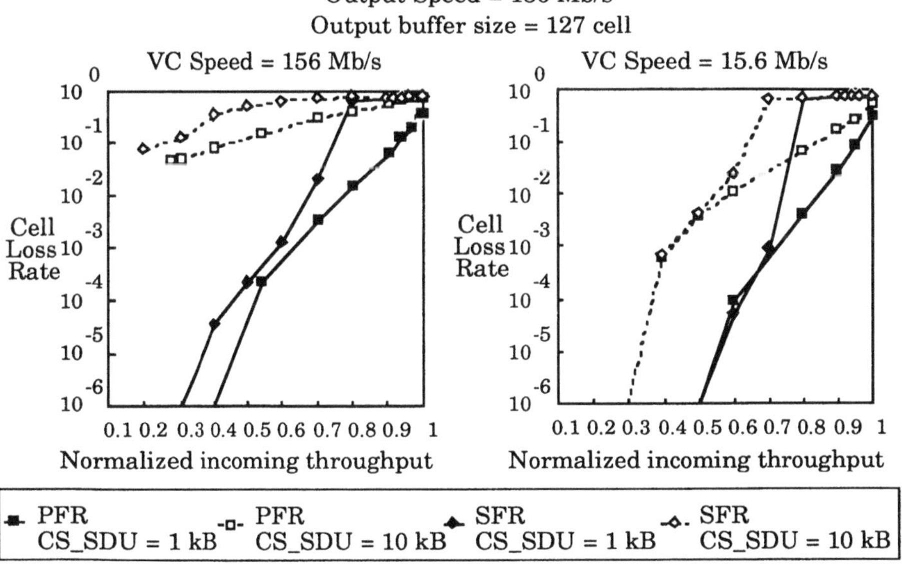

Figure 12. Simulated cell-loss rate depending on the incoming traffic.

The results show the following:

Dependency on the retransmission scheme

For all the variations in parameter values, PFR reduces the cell-loss rate by more than 90% when no congestion occurs. When the frame length is 1 kB, SFR generates congestion when the normalized incoming traffic load is 0.8 or greater. PFR keeps the cell-loss rate below 0.01.

Therefore, the cell-loss rate can be greatly reduced and congestion can be avoided for a wide variety of parameter values by using PFR.

Dependency on the frame length

Even if PFR is used, the network may become congested when the frame length is 10 kB, the VC rate is 156 Mbps, and the incoming traffic load is more than 0.2. This is also the case if SFR is used. To avoid congestion, the use of frames longer than 10 kB should be avoided when the VC rate is 156 Mbps or greater and the output buffer space is less than 127 cells. For frames that are 1 kB or less, a relatively low cell-loss rate can be provided for traffic load lighter than 0.6.

Dependency on the VC rate

The cell-loss rate can be reduced by reducing the VC rate. For a given cell-loss rate (for example 0.001), more traffic can enter the network with a lower VC rate. For SFR, the VC rate at which the network starts to become congested is independent of the incoming traffic load.

5. Conclusion

This paper described the functions needed for data communication in B-ISDN and classified the error recovery methods that can be used by the end-to-end protocol for data communication. The partial-frame retransmission (PFR) scheme was proposed; its performance was compared with that of selective-frame retransmission (SFR) from the viewpoints of implementation complexity, efficiency, total traffic load, and cell-loss probability. Compared to SFR, only a few additional function blocks are required for PFR.

The evaluation results showed that PFR can provide higher efficiency for a wide variety of parameter values compared with SFR. They also showed that PFR reduced the cell-loss rate to less than 0.1. In particular, when the incoming traffic load is relatively high, PFR can avoid the congestion that SFR allows.

More precise evaluations are needed that take into account the variations in incoming traffic load and such parameters as propagation delay. The proposed PFR with flow control based on cell-level transmission should be considered for use in the new SSCS protocol for data communication in B-ISDN.

Acknowledgments

We wish to thank Mr. Takeo Koinuma, Mr. Tatsuro Takahashi, and Mr. Hirokazu Ohnishi for giving us the chance to research this topic. We also wish to thank Dr.

Kou Miyake, Mr. Hiroyuki Ichikawa, Mr. Toshikazu Suzuki, and Mr. Michiharu Mito for their fruitful discussions.

References

[1] ITU-T Draft Recommendation M.3010, "Principles for a telecommunications management network."

[2] ITU-T Recommendation I.150, "Integrated services digital network (ISDN) general structure and service capabilities," 1991.

[3] ITU-T Recommendation I.361, "B-ISDN ATM layer specification," 1992.

[4] ITU-T Recommendation I.362, "B-ISDN ATM adaptation layer (AAL) functional description," 1992.

[5] ITU-T Recommendation I.363, "B-ISDN ATM adaptation layer (AAL) specification," 1992.

[6] ITU-T Recommendation Q.2110, "B-ISDN signalling ATM adaptation layer service specific connection oriented protocol (SSCOP)," 1994.

[7] J. Heinanen, "Multiprotocol Encapsulation over ATM Adaptation Layer 5," IETF RFC 1483, July 1993.

[8] J. Postel, "Transmission Control Protocol," IETF RFC 793, September 1981.

[9] J. Postel, "Internet Protocol," IETF RFC 791, September 1981.

[10] Protocol Engines; "XTP Protocol Definition," Rev. 3.6, January 1992.

[11] D. Cheriton, "VMTP: Versatile Message Transaction Protocol protocol specification," IETF RFC 1045, February 1988.

[12] ISO I IEC 4335; "Consolidation of Elements of Procedures (HDLC)."

[13] I. Inoue, S. Chaki, and N. Morita, "ATM Techniques for High Speed Data Communication," International Networking Conference 92 Proceedings, pp. 377-386, June 1992.

[14] ITU-T Recommendation Q.703, "Specifications of signalling system No. 7 signalling link," 1988.

[15] ITU-T Recommendation X.25, "Interface between data terminal equipment (DTE) and data circuit-terminating equipment (DCE) for terminals operating in the packet mode and connected to public data networks by dedicated circuit," 1992.

[16] ITU-T Recommendation Q.921, "ISDN user-network interface - data link layer specification," 1988.

[17] ITU-T Recommendation Q.922, "ISDN data link layer specification for frame mode bearer services," 1992.

[18] NTT, "Proposed SSCS Specification for AAL Type 3 Protocol," ITU-T contribution, COM XVIII, D.1848, December 1991.

[19] I. Inoue, N. Morita, and H. Ohnishi, "Proposal and Evaluation of an AAL Protocol with Error Recovery by Cell by Cell Retransmission," IEICE Autumn National Conference, B-296, September 1991 (in Japanese).

[20] I. Inoue, N. Morita, and H. Ohnishi, "Evaluation of an AAL Protocol Having an Error Recovery Function based on Retransmitting Lost Cells," IEICE Technical Report 91, SSE91-94, November 1992 (in Japanese).

21

Protocol Parallelization

Joseph D. Touch*

USC / Information Sciences Institute, 4676 Admiralty Way, Marina del Rey, CA, 90292-6695, U.S.A., (touch@isi.edu)

Abstract
There is increasing concern about the capability of existing protocols to keep pace with communication rates, as rates approach the gigabit range. This assumes a sort of "protocol bottleneck." Many similar bottlenecks are alleviated by the use of parallelism, so one hypothesis is to "parallelize" protocols. We examine the pros and cons of this hypothesis, and the dimensions to which parallelism might be applied. We distinguish the unique communication issues that result. Our conclusions indicate that conventional parallelism may not be applicable to protocols. New types of parallelism become more significant in this light. These include information parallelism (Parallel Communication) and packet-train parallelism.

Keyword Codes: C.2.2, C.2.4, C.5.0
Keywords: Network Protocols, Distributed Systems, Computer System Implementation

1. INTRODUCTION

High-speed networks have brought renewed emphasis on protocol performance optimization. Some optimizations focus on machine-specific implementation tuning [17], or protocol-specific implementation tuning [10]. Others have proposed "RISC" protocols, removing cumbersome functions from the protocol itself (XTP [4], VMTP [3], etc.). A number of projects have observed that bottlenecks in other disciplines are often alleviated by applying parallelism, either spatial or temporal (i.e., pipelining). Here we investigate the kinds of protocol parallelism possible, summarize the current attempts at parallelization, and make observations about their feasibility and limitations. We also observe some unconventional types of parallelism, such as information parallelism, which may be viable. We also consider whether the parallelization addresses protocol speed or latency.

First we define protocol parallelization, and what it requires. We present a framework in which to compare parallelization methods. Then we consider prior work regarding protocol optimizations in general, and with respect to parallelization in specific. Finally, we make some observations about gaps in the design space, and novel protocol methods which suggest types of parallelism not conven-

*. This research was partially sponsored by the Advanced Research Projects Agency through Ft. Huachuca Contract No. DABT63-91-C-0001. The views and conclusions contained in this document are those of the authors and should not be interpreted as representing the official policies, either expressed or implied, of the Department of the Army, the Advanced Research Projects Agency, or the U.S. Government.

tionally considered. We conclude that protocol processing itself may not benefit greatly from parallelization, and that parallelizing the control and feedback may prove beneficial in increasing channel utilization and reducing latency.

2. WHAT'S THE PROBLEM?

Even though the "communications bottleneck" is an observed phenomenon, its cause isn't well understood. There is a substantial difference between the bandwidth of a workstation backplane (typically near 1 Gbps), and the rate at which communication can occur to an external network (typically near 33 Mbps). The difference between these rates is a function of many factors, that include the costs of *asking* for state information, *thinking* about the question and forming a reply, and *answering* by sending the reply, i.e.,:

- *processing*
  (ability to think fast)
- *bandwidth within and between hosts*
  (ability to feed the questions to the thinker)
- *sourcing limits*
  (ability to have enough questions to think about)

Thinking is a processing bottleneck, exhibited in the processing speed of TCP/IP and the operating system (OS) interface involved in the transaction. The processing bottleneck for TCP has been addressed by parallelism [16], [1] (both discussed later), even though its performance has been shown not to be the predominant limitation to communication [17], [10]. Other components of the protocol stack have also been parallelized, e.g., via pipelining of the IP check-sum with the data transfer [5]. Processing in the OS has also been considered, although most of the focus has been on data transfer issues we denote as *answering* bottlenecks.

Answering is a data transport bottleneck, i.e., bandwidth limitations. This is currently being addressed by parallelizing the internal data paths of workstations. The basic idea is to replace the internal bus with a more general topology. The Cambridge Desk Area Network (DAN) [9], and ISI's NetStation [7] are examples.

One interesting question assumes nearly instantaneous *thinking* (protocol processing) and *answering* (bandwidth), and asks, "is there anything else?" Presume that TCP has infinite window sizes, and can run at 800 Mbps (it can on a CRAY [17]). Presume that we have a NetStation, in which each component of a workstation (disk, RAM, display, etc., [7]) can both source and sink at 800 Mbps. What then?

This is the question of *asking*. Latency is the final bottleneck [19]. Ultimately, answers can be given only as fast as questions arrive. If the next question depends on the current answer, round trip propagation latency is incurred between *asking* rounds. Assuming *thinking* and *answering* are not the bottlenecks implies *asking* is.

In terms of current protocols, even with nearly infinite bandwidth and processing, TCP can "fill the pipe" only so far as the source data (answer) exists. Once an entire data stream is sent, nothing more is communicated until the next query arrives. It is here that we believe parallelism is best applied, to the parallelization of possible next questions and thus answers. We call this Parallel Communication, and it is based on parallelization of the information stream [19], [21].

3. PARALLELIZATION ISSUES

Parallelization uses replicates to perform the work of a single entity, and involves considering replication dimension, mapping function, scale limitations, replicate interference, overhead, and expected gain. The type of entity replicated is the replication dimension. Common protocol dimensions include per-protocol, per-connection, per-packet, per-layer, and per-protocol function. This subsumes the difference between spatial and temporal (pipelining) parallelism, because temporal parallelism is spatial parallelism with head-to-tail interdependence between components, and usually implemented per function. From this point, we therefore do not consider pipelining as a distinct case.

Other dimensions not commonly considered include per-packet train, and per-information stream. Packet trains are sequences of packets that act as a unit in the protocol; a common example is a fragmentation group. Information streams are alternate packet sequences, such as would occur with breadth-first (BFS) source anticipation [21]. BFS source anticipation is parallelizing possible future questions and their answers, in order to keep the communications mechanism occupied between actual questions. Dimensions not considered are per-host address (issue for routers only), and per-application (equivalent to per-protocol).

The mapping function helps determine whether a replication dimension is feasible. Given a dimension, the map indicates how incoming data is switched to the appropriate replicate. Some dimensions are easy to map (per-protocol, per-layer, per-connection). Others require almost as much effort to map as would be required to emulate the protocol (per-protocol function).

Scale limitations indicate the bounds on the expected replication growth. If choose per layer, we are commonly limited to seven replicates (OSI model), or thereabouts. Per protocol function is often limited to five (TCP, TP-4), because the protocol definitions are often described as mostly serial processes.

Replicate interference describes the interaction required between the parallel components, and whether the work is increased as a result. Per-connection replicates are usually defined as not interacting, but redundant operations in protocol layers often implies that per-protocol and per-layer replicates interfere. Per-packet replicates interfere heavily, because they affect common connection and protocol parameters (i.e., state information).

Overhead includes the cost of replicate interference, as well as the general overhead of switching according to the mapping function, process creation and removal (if dynamic), and other costs of a parallel implementation.

Expected gain measures the overhead costs with the scale limitations, to determine whether the parallelization proposed is feasible or effective. Some types are feasible but not effective, because replication is easy and overhead is low (per-protocol), but the expected gain is minimal. Other types are neither effective nor feasible (per byte!), because the overhead outweighs any expected gain, or is at least as large as the protocol itself.

For each parameter, we consider whether a real bottleneck is being addressed Recall that TCP can run at 800 Mbps [17], however not across a real link. Is the limitation the processing, or the sourced information itself? Is speed a real issue, or is latency? Before we continue, we should reevaluate what the existing bottlenecks of a "heavyweight" transport protocol are.

Our conclusions are summarized in Table 2, near the conclusion of this paper.

3.1 Existing bottlenecks

Much work has been done to address existing protocol processing bottlenecks. This includes *thinking* optimizations, such as header processing optimizations, state optimizations, implementation optimizations, and augmenting the general processing power of the system. It also includes *answering* optimizations, including data transport path optimizations.

Header processing optimizations include "fast-path" optimizations and header prediction [10] are specific instances of general cross-product protocol optimizations such as Protocol Bypass [24]. The technique takes the cross product of all protocol functions and layers, factors out the statistically favored states and implements them as special (fast) cases. The remainder of the protocol is implemented as before.

State optimizations result in RISC-style 'lightweight' protocols. They implement the cross-product protocols (as above), and remove the non-favored states from the protocol. Examples include VMTP [3] and XTP [4].

Implementation optimizations include code tweaking such as has been done on the Cray TCP [17], and generic TCP [10], as well as *Integrated Layer Processing* (ILP) [5]. They also include optimizing the underlying OS interfaces, such as Jacobson's fast-sockets, and the x-Kernel optimizations [18].

Other optimizations address general processing issues, and would benefit conventional applications as well a protocol processing. These include fast context switching and hardware header processing. The latter is a flavor of support processor, other examples of which include FPUs, string processors, and graphic engines.

Protocols exhibits data transport limitations even after avoiding multiple data movement. This is evidence of conflict between the topology of the external network, and the internal backplane communication, an *answering* bottleneck. Proposed solutions involve "moving the network into the backplane", such as in the Cambridge Desktop Area Network (DAN) [9] or ISI's NetStation [7].

There are other limitations, *asking* bottlenecks, as have been recently observed [21]. In a high bandwidth-delay product network, the data source itself becomes a limit to the channel utilization (not enough questions). Assuming the window size limits of the current TCP specification are fixed, what will fill these windows? Measurements indicate that network bandwidth-delay products are increasing at a faster rate than that of the end system, so it's not clear that file sizes will increase in proportion to network rates, such that a larger window will be usable [22].

4. PRIOR WORK

Optimized TCP focused on increasing the window size to accommodate higher bandwidth-delay products, so-called "long delay" [11], "high-speed" [12], and "high-performance" [13] TCP. The large windows permitted flow control feedback to occur over byte blocks, rather than bytes themselves, permitting longer *answers*.

"High-performance" TCP was specifically designed for high bandwidth-delay environments, observing that the bandwidth-delay product was the issue, rather than the latency or the speed (as in prior proposals) [13]. A critical response indicated that some of these proposed optimizations were harmful, and suggested flow control occur over records, rather than bytes. This indicates a message structure beyond that normally assumed for TCP. A current proposal augments TCP to handle transactions [2], to accommodate some of the record-structure suggested before. This line of research is realizing that bandwidth-delay products are the primary issue (i.e., a question bottleneck), and that large windows don't solve the problem (because they solve answering). Additional structure in the information stream is required, beginning with a record structure

There are other proposals that observe the limitation of the exchange of state in a high bandwidth-delay product environment, and attempt to accommodate this.

Periodic exchange of state as a protocol mechanism is the basis of the Delta-t protocol [23], and has also been mentioned in TP++ [6] and the SNR "leaky-bucket" protocol [15].

A recent proposal indicated that high bandwidth-delay product environments change bandwidth-bound systems to be latency- or server-bound [14]. The sources become message limited (question-bound), and the channel utilization changes only if the message sizes increase. The proposal suggests that multiplexing will increase channel utilization, just as multiprocessing increases processor utilization in the presence of I/O communication latency.

A reply to this proposal observed that multiplexing would not so much solve the problem as define it away [20]. Deterministic multiplexing is equivalent to lower bandwidth-delay product links. The real issue is protocol state imprecision, induced by high bandwidth-delay products and variance in protocol function (nondeterministic protocols) or in latency (delay variance). Nondeterministic multiplexing just pushes the state imprecision problem to the multiplexer control level. The only known solution uses message stream structure (records, branches, and recursion) to permit source anticipation of remote state [19]. The message stream carries responses to "parallel futures" of the remote state. One application which supplies the requisite structure is distributed hypermedia, such as in the World-Wide Web.

5. CURRENT WORK

Recent work attempts to parallelize the processing of particular protocols. These include per-packet and per-function dimensions, both empirically and analytically derived. The results have been disappointing thus far, with a scale limit of 5 parallel processors for most experiments.

Early work tried per- layer parallelism, a sort of data flow architecture [25]. They decomposed TP-4 onto Transputers, where layers 3 and 4 were each assigned one processor for each of send and receive functions. The result was 4-way parallelism, with no further decomposition proposed.They concluded that the protocol definitions were not conducive to further parallel decomposition, especially because of the overlap of functions between layers.

Other work focused on per-function decomposition of OSI protocols [8]. They considered TP-4 and 802.2 LLC protocol decomposition. They decompose their protocols per-layer 4-ways, by allocating 2 processors each for send and receive. Further decomposition follows a dataflow diagram of the protocol, where functions are divided among the two processors. They observe that the interaction between the two processors on each side has significant overhead, and that further decomposition is not feasible because of increased communication costs. When they attempted to apply their technique to TCP/IP, they found a scale limit of 4 processors there as well. Some of their results may be an artifact of the communication topology of Transputers, which are 4-way connected.

More recent attempts factored TCP/IP as if its functional components were executing on a statically load-balanced compiler [16]. At first, they partitioned TCP into three functional components - send, receive, and timer. Their analysis concluded that TCP could be parallelized further by distinguishing between data and control, such that there were four components - *send_data*, *receive_data*, *control*, and *timer*. They found that the bulk of complexity lies in the control, which isn't surprising (this is the essence of a protocol). By making additional assumptions (not necessarily valid ones), they could reduce the work of the control protocol.

These simplifications include assuming a stable round-trip time (thus removing it from the protocol state), thus assuming a stable window size (also removing it from the state), and by performing control functions infrequently (not per-packet).

Unfortunately, this changes the operational semantics of TCP, and it's not clear whether their implementation will interoperate with standard TCPs. We observe that this is just the "big packet" solution to protocol optimization, such as in NETBLT (long answers). Low bandwidth protocols act like high bandwidth periodic protocols (Figure 1). High bandwidth eager-delivery (as soon as possible) protocols increase the rate of headers, thus the effort of protocol processing. Rate-based flow control reestablishes the header rate, slowing the protocol processing down (Table 1). The authors note that rate policing is required for their solutions to avoid the increase in header rate.

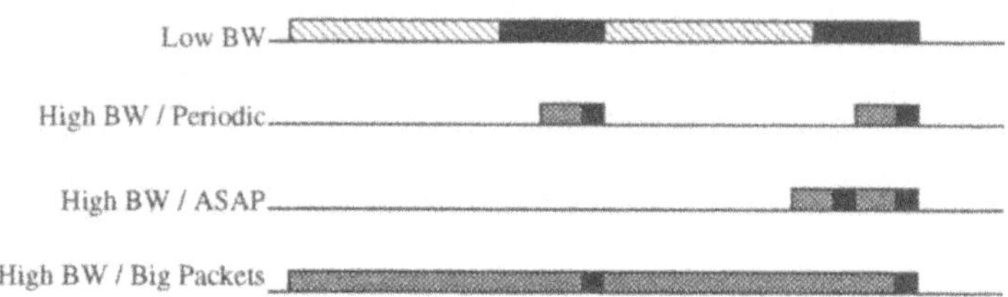

Figure 1.
Equivalence of types of protocols
- the top two are equivalent, the bottom two are not.

Increasing the packet size (as in NETBLT), or, equivalently, spacing the "real" headers out (ones that affect protocol state), results in large-packet protocols, which act like their more sluggish counterparts. A protocol is sensitive to the header processing rate. Allowing larger packets or fewer control headers is equivalent to assuming the stability of the slower protocol system. Large packets work only where state is a function of clock-time, not the bandwidth-delay product.

Table 1
Big packets are really slow/sluggish protocols

BW	rule	hdr/time	data/time	hdr/data
low bw	-	low	low	avg
high bw	periodic	low	low	avg
high bw	ASAP	high	high	avg
high bw	huge packets	low	high	low

They also observe that the parallelism is affected by state imprecision. Such imprecision is characterized by the number of possible vs. actual headers outstanding (i.e., effects of latency on imprecision), as well as the number of actual controls vs. data between the controls (effects of transmission and reception on imprecision).

Others considered per-packet parallelization [1]. They measured parallelism using simulations, and implementations of a multiprocessor x-Kernel implementation with spin-locks. They observed that the parallel processing contends for the shared protocol state (the Connection Control Block). They claim TCP saturates at a parallelism of 7 (measured as 5) with a speedup below 5x (measured as 3.5x), but that UDP didn't appear to saturate (through N=20). This is because TCP has significant shared state (the sliding window flow control), but UDP does not. We note that this per-packet experiment reaches a similar parallelism limit as per-function.

Other results were reported at the CNRI Gigabit Workshop in 1993. V. Jacobson reported that his fast TCP relies on state transition effects, rather than the full RFC-specification of the protocol. Inbound traffic processing was the bottleneck, especially where demultiplexing isn't stable; this corroborates our conclusion that Kleinrock's multiplexing solution isn't sufficient. The Nectar project tried a partitioned send/receive TCP, and encountered problems with shared CCB access as well. The IBM group reported on general parallelization using pipelining, and found that the component coupling was tight for TCP, and loose for IP. Other projects (HP/Witless, LANL CASA) examined using big packets to optimize throughput (as large as 4 Kbyte). They rely on the stability of the protocol and environment, as discussed before.

6. OBSERVATIONS

At this point we have several observations to make. First, parallelization per-packet or per-function does not help much. These results address some performance issues, but do not scale as the bandwidth, number of headers (i.e., packets), or available processors increases. As a result, they are of debatable utility. Other dimensions, including per protocol, per connection, per application, etc., are also of debatable utility, because we assume a required solution will provide high throughput for a single connection of a single protocol to a single application (1 Gbps to the user application).

These observations are important because they imply that *thinking* parallelism does not help protocol performance. *Answering* (bandwidth) bottlenecks are disappearing, as we remove simplifications (i.e. bus backplanes) from our workstation designs, and move towards a networked-backplane, thus parallelizing the *answers*. When these are solved, only one bottleneck remains - latency, i.e., a not enough *questions*.

We also made other observations. First, a protocol doesn't know what time it is, only how many bits are in transit. Consider slowing down existing hardware - from the CPU through the communications media interface. The protocol will work exactly as before. Even protocols with absolute clock timers will still function, because the clock time will be equivalently slowed.

So the rate issue is a red herring with respect to protocol processing. A protocol does, however, know how many bits are in transit, both in terms of the amount of state required, and the effort required to manage that state. Protocols are bandwidth-delay product sensitive [19]. Further, many existing parallelization techniques do not address bandwidth-delay product (the *question* bottleneck). One exception is [16], but their conclusion is that in order to avoid the problems that bandwidth-delay product causes, they exclude the high-frequency header domain.

In our work on an abstract model of communication latency (Mirage), and its protocol instantiation (Parallel Communication), we found a more direct correlation between header frequency, action, state, and stability. We conclude that one area of parallelism worth examining might be information parallelism (parallel *questioning*), where alternate streams of anticipatory data are present at the source.

Another area of consideration is the packet train. Packet trains are created when a large packet is fragmented into many smaller lower-layer packets. The control occurs at the head and tail of the train, and within the train. As a result, the difference between the intra-train state and inter-train state is well defined, and might be a useful dimension to parallelize. However, the benefits apply only to the segmentation and reassembly protocols.

6.1 LISN/MISN/WISN

We observed in our research that area isn't the issue, i.e., LAN, WAN, MAN. We find that protocols operate differently depending upon the information separation (IS), measured as bits in transit (bandwidth-delay product). That is, bandwidth-delay product is the formula, and information separation is what it describes.

So LAN/MAN/WAN become LISN/MISN/WISN. We define these distinctions based on the comparison between the messages sent and the IS. In LISN, the message is much larger than the IS, so existing sliding-windows protocols suffice. In MISN, the message is about the same size as the IS; this corresponds to RPC or remote evaluation (REV) domains. WISN requires Mirage / Parallel Communication to utilize the channel effectively, because the inter-message effects dominate, rather than intra-message. This is where structure in the information stream is used.

6.2 The protocol solution space has holes

When we considered the protocol solution space based on the IS, we found some holes in it (Figure 2). We could reduce the separation, or deal with it. We can remove the IS by moving the data to the code (RPC) or the code to the data (REV).

Alternately, we can deal with the IS directly. If the data is organized as a long linear stream, and the stream is longer than the IS (i.e., the LISN domain), we use sliding windows. Otherwise, we need to accommodate more complicated structures such as branching and recursion of the state space, to emulate the imprecision of state induced by the latency.

Examples of structured streams are hypermedia (interactive media), and reactive / proactive control systems. The resulting system has parallelism of action, and permits the state spaces to be partitioned while the protocol is running. This permits dynamic load balancing of the resulting processes among whatever set of processors area available.

As a result, Parallel Communication directly addresses the WISN domain, and scales to use additional processors as the bandwidth-delay product increases. The interaction between the partitions is well defined by the Mirage model. The source anticipation describes process factoring, and receiver feedback terminates source processes whose anticipation is not needed.

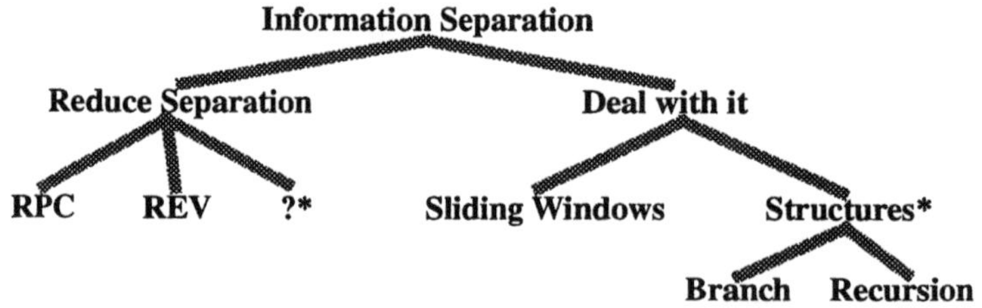

Figure 2.
Holes in the protocol state space (starred {*})

7. INFORMATION PARALLELISM

As we have discussed, the high information separation is the cause of inefficiency in the use of the communication stream. Parallelizing the protocol won't help solve this, because conventional protocols "run out of stuff to fill the pipe with". Information parallelism is a way to fill the pipe.

The protocol we use for information parallelism is called Parallel Communication. Simply put, it is sender-based preloading of a cache – in this case, the cache is the bandwidth-delay product (the information separation). The sender is sending preemptive replies to expected requests from the receiver. Each of these preemptive replies accommodates some possible state of the receiver; the set of replies represents a parallelism in the information between the sender and receiver.

A protocol for sender-based cache preloading isn't difficult, but when recursion of receiver state is considered, the mechanism becomes complex. We have developed such a mechanism for Parallel Communication.

We are currently proposing to use Parallel Communication to augment the WWW client/server (browser/server) interaction, as in Figure 3. The goal is to reduce the user-perceived response time, and permit the WWW browsers to act more like remote user-interfaces than the sluggish transaction-based mechanisms they currently are.

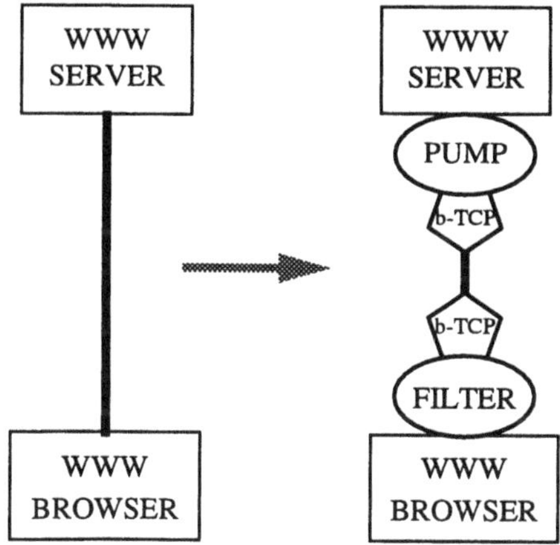

Figure 3.
WWW server and browser augmentation for Parallel Communication

For the WWW, the way in which information parallelism is determined is to filter the HTML (hypertext mark-up language) for links to other files. The other files are send as a set of information-parallel replies to the 'buttons' on that page.

8. CONCLUSIONS

We conclude that parallelizing protocols doesn't necessarily require parallel processing of existing state or packet data. It may be more useful to parallelize the possible states of the other side of the link, in order to scale as bandwidth-delay products increase. Our particular observations are summarized in Table 2.

Table 1
Summary of conclusions

DIM.	MAP	SCALE LIM.	REPL. INTFR.	OVRHD.	GAIN
protocol	easy	Y(1)	N	low	low
connection	easy	Y(1)	N	high	low
packet	easy	Y(5)	Y	high	low
application	easy	Y(1)	N	low	low
packet train	hard	Y(<10)	some	low	med.
function	none!	Y(5)	Y	high	low
layer	easy	Y(7)	N	low	low
information	moderate	N	some	some	high

We observe that other optimizations, such as big packets or periodic control, only require further assumptions about the stability of the state and state evolution. They do not address the protocol performance issue; rather, they attempt to make the protocol disappear.

Per-packet or per-function parallelization hits state precision limits, and does not scale beyond a factor of 5 or so. The better solution is to determine and support further structure in the communication stream, and the parallelism of possible future states that affords.

9. ACKNOWEDGEMENTS

This paper is the result of an ongoing discussion at ISI regarding the possibilities of parallelization of protocols. Herb Schorr proposed the question at ISI, and the members of the HPCC Division provided ample feedback. We especially appreciate the efforts of Jon Postel and Paul Mockapetris, both of ISI. Carolyn Nguyen of AT&T also provided feedback on the debate.

REFERENCES

[1] Bjorkman, M., and Gunningberg, P., "Locking Effects in Multiprocessor Implementation of Protocols," *Proc. ACM Sigcomm*, Oct. 1993, pp. 74-83.

[2] Braden, R., "Extending TCP for Transactions -- Concepts," Network Working Group RFC-1379, USC/ISI, Nov. 1992.

[3] Cheriton, D.R., "VMTP: A Transport Protocol for the Next Generation of Communication Systems," *Computer Communication Review*, Aug. 1986, pp. 406-415.

[4] Chesson, G., et. al., *XTP Protocol Definition,* Protocol Engines, Inc., Dec. 1988.

[5] Clark, D., and Tennenhouse, D., "Architectural Considerations for a New Generation of Protocols," *Proc. ACM Sigcomm*, Sept. 1990, pp. 200-208.

[6] Feldermeier, D., "Comparison of Error Control Protocols for High Bandwidth-Delay Product Networks," In participants proceedings, IFIP Workshop on Protocols for High Speed Networks, Nov. 1990.

[7] Finn, G., "An Integration of Network Communication with Workstation Architecture," *ACM Computer Communication Review*, V. 21, No. 5, Oct. 1991.

[8] Giarrizzo, D., Kaiserswerth, M., Wicki, T., and Williamson, R., "High-Speed Parallel Protocol Implementation," in *Protocols for High Speed Networks*, Rudin, H. and Williamson, R. Eds., Elsevier, 1989, pp. 165-180.

[9] Hayter, M., and McAuley, D., "The Desk Area Network," *ACM Transactions on Operating Systems*, Oct. 1991, pp. 14-21.

[10] Jacobson, V., "Congestion Avoidance and Control," *ACM Computer Communication Review*, Oct. 1988, pp. 314-329.

[11] Jacobson, V., and Braden, R., "TCP Extensions for Long-Delay Paths," Network Working Group RFC-1072, LBL and USC/Information Sciences Institute, Oct. 1988.

[12] Jacobson, V., Braden, R., and Zhang, L., "TCP Extensions for High-Speed Paths," Network Working Group RFC-1185, LBL and USC/ISI, Oct. 1990.

[13] Jacobson, V., Braden, R., and Borman, D., "TCP Extensions for High Performance," Network Working Group RFC-1323, LBL, USC/ISI, and Cray Research, May 1992.*IEEE Communications Magazine*, Vol. 30, No. 4, April 1992, pp. 36-40.

[14] Kleinrock, L, "The Latency / Bandwidth Tradeoff in Gigabit Networks," *IEEE Communications Magazine*, Vol. 30, No. 4, April 1992, pp. 36-40.

[15] Netravali, Arun N., Roome, W.D., and Sabnani, K., "Design and Implementation of a High-Speed Transport Protocol." *IEEE Transactions on Communications V. 38,* N.11 (Nov. 1990), pp. 2010-2024.

[16] Nguyen, C. and Schwartz, M., "Reducing the Complexities of TCP for a High Speed Networking Environment," *Proc. IEEE Infocom*, Mar. 1993, pp. 1162-1169.

[17] Nicholson, A., Golio, J., Borman, D.A., Young, J., and Roiger, W., "High Speed Networking at Cray Research," *ACM Computer Communication Review*, V. 21, N. 1, Jan. 1991, pp. 99-110.

[18] Peterson, L., Hutchinson, N., et. al., "The x-Kernel: A Platform for Accessing Internet Resources," *IEEE Computer*, V. 23, N. 5, May 1990, pp. 23-33.

[19] Touch, Joseph D., *Mirage: A Model for Latency in Communication*, Ph.D. dissertation, Dept. of Computer and Information Science, Univ. of Pennsylvania, 1992. Also available as Dept. of CIS Tech. Report MS-CIS-92-42 / DSL-11.

[20] Touch, J.D., and Farber, D.J., "Reducing Latency in Communication," letter to the editor in *IEEE Communications Magazine*, Feb. 1993, pp. 8-9.

[21] Touch, J.D., "Parallel Communication," *Proc. IEEE Infocom*, Mar. 1993, pp. 505-512.

[22] Touch, J.D., and Farber, D.J., "An Experiment in Latency Reduction," *Proc. IEEE Infocom*, June. 1994, pp. 175-181.

[23] Watson, R.W., "The Delta-t Transport Protocol: Features and Experience," *Protocols for High Speed Networks*, Elsevier, 1989, pp. 3-17.

[24] Woodside, C.M., Ravinadran, K., and Franks, R.G., "The Protocol Bypass Concept for High Speed OSI Data Transfer," In participant's proceedings, IFIP Workshop on Protocols for High Speed Networks, Nov. 1990.

[25] Zitterbart, M., "High-Speed Protocol Implementations Based on a Multiprocessor Architecture," *Protocols for High Speed Networks*, Rudin, H. and Williamson, R. Eds., Elsevier, 1989, pp. 151-163.

INDEX OF CONTRIBUTORS

KEYWORD INDEX

GPSR Compliance
The European Union's (EU) General Product Safety Regulation (GPSR) is a set of rules that requires consumer products to be safe and our obligations to ensure this.

If you have any concerns about our products, you can contact us on

ProductSafety@springernature.com

In case Publisher is established outside the EU, the EU authorized representative is:

Springer Nature Customer Service Center GmbH
Europaplatz 3
69115 Heidelberg, Germany

www.ingramcontent.com/pod-product-compliance
Ingram Content Group UK Ltd.
Pitfield, Milton Keynes, MK11 3LW, UK
UKHW012152240726
13966UKWH00002B/284
* 9 7 8 1 4 7 5 7 6 3 1 3 3 *